MW00676224

Sustaining Young Forest Communities

Managing Forest Ecosystems

Volume 21

Series Editors:

Klaus von Gadow

Georg-August-University,
Göttingen, Germany

Timo Pukkala

University of Joensuu,
Joensuu, Finland

and

Margarida Tomé

Instituto Superior de Agronomía,
Lisbon, Portugal

Aims & Scope:

Well-managed forests and woodlands are a renewable resource, producing essential raw material with minimum waste and energy use. Rich in habitat and species diversity, forests may contribute to increased ecosystem stability. They can absorb the effects of unwanted deposition and other disturbances and protect neighbouring ecosystems by maintaining stable nutrient and energy cycles and by preventing soil degradation and erosion. They provide much-needed recreation and their continued existence contributes to stabilizing rural communities.

Forests are managed for timber production and species, habitat and process conservation. A subtle shift from *multiple-use management* to *ecosystems management* is being observed and the new ecological perspective of *multi-functional forest management* is based on the principles of ecosystem diversity, stability and elasticity, and the dynamic equilibrium of primary and secondary production.

Making full use of new technology is one of the challenges facing forest management today. Resource information must be obtained with a limited budget. This requires better timing of resource assessment activities and improved use of multiple data sources. Sound ecosystems management, like any other management activity, relies on effective forecasting and operational control.

The aim of the book series ***Managing Forest Ecosystems*** is to present state-of-the-art research results relating to the practice of forest management. Contributions are solicited from prominent authors. Each reference book, monograph or proceedings volume will be focused to deal with a specific context. Typical issues of the series are: resource assessment techniques, evaluating sustainability for even-aged and uneven-aged forests, multi-objective management, predicting forest development, optimizing forest management, biodiversity management and monitoring, risk assessment and economic analysis.

For further volumes:
http://www.springer.com/series/6247

Cathryn H. Greenberg • Beverly S. Collins
Frank R. Thompson III
Editors

Sustaining Young Forest Communities

Ecology and Management of Early Successional Habitats in the Central Hardwood Region, USA

Editors
Cathryn H.Greenberg
USDA Forest Service
Southern Research Station
Bent Creek Experimental Forest
1577 Brevard Road
Asheville, NC 28806
USA
kgreenberg@fs.fed.us

Beverly S. Collins
Department of Biology
Western Carolina University
NS 132, Cullowhee, NC 28734
USA
collinsb@email.wcu.edu

Frank R. Thompson III
USDA Forest Service
Northern Research Station
202 Natural Resources Building
University of Missouri-Columbia
Columbia, MO 65211-7260
USA
frthompson@fs.fed.us

ISSN 1568-1319
ISBN 978-94-007-1619-3 e-ISBN 978-94-007-1620-9
DOI 10.1007/978-94-007-1620-9
Springer Dordrecht Heidelberg London New York

Library of Congress Control Number: 2011933842

Typical scene in the State of Durango where forests are managed by communities known as Ejidos: management is by selective tree removal, clear-felling is not allowed. Animals (ganado) are part of the multiple use system practiced there. (Photo by K. v. Gadow, autumn 2009)

Cover design: deblik

Printed on acid-free paper

Springer is part of Springer Science+Business Media (www.springer.com)

Preface

This edited volume addresses the rising concern among natural resource professionals that plants and animals associated with early successional habitats are declining in the Central Hardwood Region to undesirably low levels. The idea for this book was partially in response to a request by the USDA Forest Service's Southern Region, to the USDA Forest Service's Southern Research Station and partners, to identify top research synthesis needs, and to identify ecosystem restoration priorities for National Forests in the Southern Appalachians. Early successional habitat was identified as one of three top research synthesis needs. A full-day symposium, organized by the editors, at the 2010 Association of Southeastern Biologists conference in Asheville, North Carolina was the basis for this book. Our goal was to present original scientific research and knowledge syntheses covering multiple topics associated with early successional habitats. We strived for each chapter to include state-of-the-art, research-based knowledge and expert opinion, but also to identify research needs and discuss management implications for sustainable management in a landscape context.

Chapters were written by respected experts that include ecologists, conservationists, and land managers. The chapters provide current, organized, readily accessible information for scientists, the conservation community, land managers, students and educators, and others interested in the "why, what, where, and how" of early successional habitats and associated wildlife. Chapters cover concepts, management, plants and animals, ecosystem processes, and the future of early successional habitats. We provide a working definition of early successional habitats; examine where and why they occur over the landscape; and explore concepts related to their importance and sustainability. We examine the roles of natural disturbances, silviculture, and fire in creating and maintaining early succession habitats. We explore effects of these habitats on ecosystem processes and wildlife, and their role in producing forest food resources. We also explore management tools for early successional habitat, including use of novel places such as utility rights of way, and strategies for identifying priority species and implementing desired future conditions. The final chapter looks to the future, to project changes in forest age class diversity in relation to scenarios of land ownership, economics, demographics, and climate change. We attempted to provide a balanced view of past, current, and future scenarios on the extent and quality of early successional habitats within the Central Hardwood Region, and implications for ecosystem services and disturbance-dependant plants and animals.

We sincerely thank all those who encouraged and aided in the development of this book. Each chapter was peer reviewed by at least two outside experts and all co-editors, and we thank these colleagues for their useful suggestions: David Buehler, Josh Campbell, Dan Dey, Todd Fearer, Mark Ford, Jennifer Franklin, Charles Goebel, Margaret Griep, MaeLee Hafer, Chuck Hunter, Todd Hutchinson, Mike Jenkins, Jennifer Knoepp, Darren Miller, Mark Nelson, Chris Peterson, Jim Runkle, Ge Sun, Mike Ryan, Sarah Schweitzer, Ray Semlitsch, Bill Stiver, Bentley Wigley, and Mariko Yamasaki. We also thank the Association of Southeastern Biologists for allowing us to host a conference symposium on this important topic. We especially thank each author for contributing, and for timely chapter revisions, making this book possible.

Contents

Chapter 1
Introduction: What Are Early Successional Habitats, Why Are They Important, and How Can They Be Sustained?

Cathryn H. Greenberg, Beverly Collins, Frank R. Thompson III, and William Henry McNab

Abstract There is a rising concern among natural resource scientists and managers about decline of the many plant and animal species associated with early successional habitats. There is no concise definition of early successional habitats. However, all have a well developed ground cover or shrub and young tree component, lack a closed, mature tree canopy, and are created or maintained by intense or recurring disturbances. Most ecologists and environmentalists agree that disturbances and early successional habitats are important to maintain the diverse flora and fauna native to deciduous eastern forests. Indeed, many species, including several listed as endangered, threatened, sensitive, or of management concern, require the openness and thick cover that early successional habitats can provide. Management of early successional habitats can be based on the "historic natural range of variation", or can involve active forest management based on goals. In this book, expert scientists and experienced land managers synthesize knowledge and original scientific work to address critical questions on many topics related to early successional habitats in the Central Hardwood Region. Our aim is to collate information about early successional habitats, to aid researchers and resource management professionals in their quest to sustain wildlife and plant species that depend on or utilize these habitats.

C.H. Greenberg (✉) • W.H. McNab
USDA Forest Service, Southern Research Station, Bent Creek Experimental Forest, 1577 Brevard Rd., Asheville, NC 28806, USA
e-mail: kgreenberg@fs.fed.us; hmcnab@fs.fed.us

B. Collins
Department of Biology, Western Carolina University, NS 132, Cullowhee, NC 28734, USA
e-mail: collinsb@email.wcu.edu

F.R. Thompson III
USDA Forest Service, Northern Research Station, University of Missouri, 202 Natural Resources Bldg., Columbia, MO 65211–7260, USA
e-mail: frthompson@fs.fed.us

C.H. Greenberg et al. (eds.), *Sustaining Young Forest Communities*, Managing Forest Ecosystems 21, DOI 10.1007/978-94-007-1620-9_1,

1.1 Introduction

There is a rising concern among natural resource scientists and managers about decline of the many plant and animal species associated with early successional habitats, especially within the Central Hardwood Region (Litvaitis 1993, 2001; Thompson and DeGraaf 2001). Open sites with grass, herbaceous, shrub, or incomplete young forest cover are disappearing as abandoned farmland and pastures return to forest and recently harvested or disturbed forests re-grow (Trani et al. 2001). There are many questions about "why, what, where, and how" to manage for early successional habitats. Tradeoffs among ecological services such as carbon sequestration, hydrologic processes, forest products, and biotic diversity between young, early successional habitats and mature forest are not fully understood. Personal values and attitudes regarding forest management for conservation purposes versus preservation, or "letting nature take its course," complicate finding common ground regarding if and how to create or sustain early successional habitats.

In this book, expert scientists and experienced land managers synthesize knowledge and original scientific work to address critical questions sparked by the decline of early successional habitats. We focus primarily on habitats created by natural disturbances or management of upland hardwood forests of the Central Hardwood Region in order to provide in depth discussion on multiple topics related to early successional habitats, and how they can be sustainably created and managed in a landscape context.

1.2 Geographic Scope: The Central Hardwood Region

Broadleaved trees form the predominant forest cover type in parts of ten eastern states which Braun (1950) included in the Central Hardwood Region (Fig. 1.1). The boundaries of the region also are similar to ecoregions mapped by Bailey (1994) and bird conservation regions delineated by the US North American Bird Conservation Initiative (on the Breeding Bird Survey website (www.mbr-pwrc.usgs.gov/bbs/)). The canopy of mature upland forests is dominated by varying proportions of six broadleaf deciduous taxa. Oak (*Quercus*) and hickory (*Carya*), each represented by several species, are present in most stands. Yellow-poplar (*Liriodendron*) increases in importance east of the Mississippi River and usually dominates the canopy of moist sites in the Southern Appalachian Mountains, and maple (*Acer* spp.), beech (*Fagus grandifolia*), and birch (*Betula* spp.) occupy much of the canopy of forests in the northern and eastern parts of the region, particularly on the Allegheny Plateau. About 45% of the 130 million acres of forest land in this region is occupied by hardwood-dominated stands; mixtures of hardwoods and conifers account for an additional 5% (Smith et al. 2004). Conifers, primarily pine (*Pinus*), are minor components of many low-elevation stands on dry sites. The humid, continental climate of the region produces soil moisture regimes that are adequate for plant growth during much of the warm season, although minor water deficits can develop in late

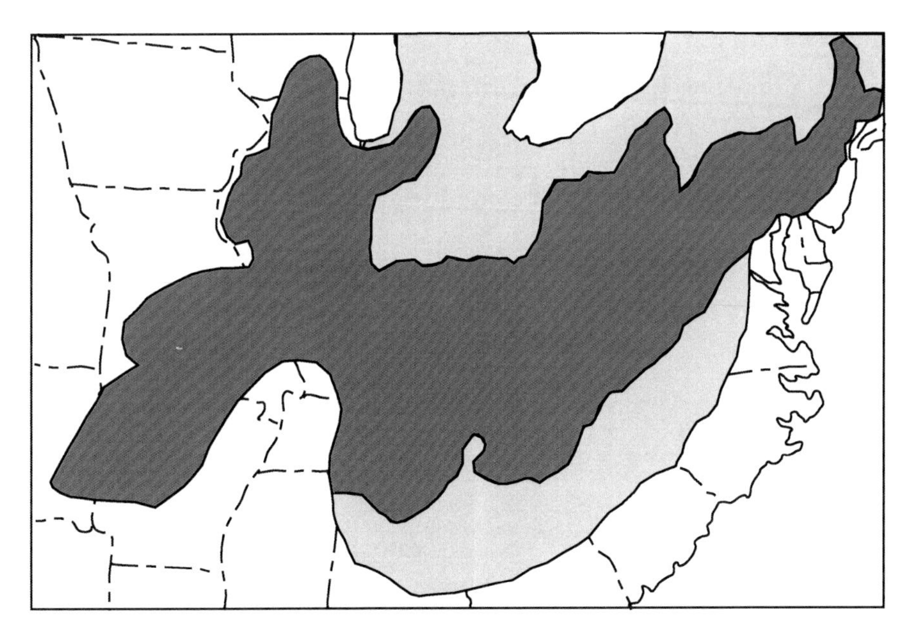

Fig. 1.1 Extent of the Central Hardwood Region in the eastern United States (dark shading). Transition to northern hardwoods occurs in the Lake States and to southern pines in the Appalachian Piedmont (light shading) (After Braun (1950))

summer. This characteristic climate (i.e., low soil moisture deficits and moderate levels of evapotranspiration) may be why forests of deciduous hardwoods dominate the Central Hardwood Region (Stephenson 1990). Detailed descriptions of forest composition and disturbance regimes characteristic of Central Hardwood Region subregions are provided in Chap. 2 (McNab).

1.3 What Are Early Successional Habitats?

Like most things ecological, there is no concise definition of early successional habitats. Early ecological studies and adoption of the term "succession" were based in part on secondary succession of abandoned farm fields (i.e., "oldfield succession"). In the southeastern USA, oldfields are first colonized by "pioneering" grasses and forbs, then gradually by pines or hardwoods, until closed forest develops (Clements 1916; Keever 1950, 1983; Odum 1960). Over time, the term "early successional" has taken on a broader meaning, to include recently disturbed forests with absent- or open-canopy and, often, transient, disturbance-adapted or pioneer species (many of them also found in old fields). Unlike oldfields, these recently disturbed forests generally do not undergo major shifts in woody species composition (Lorimer 2001). Similarly, we use the term "habitat" in this volume, as it is commonly used and understood in recent wildlife literature, to denote "a set of specific environmental features that, for a terrestrial

Plate 1.1 Examples of different types of early successional plant communities. From *left to right*: recently abandoned farmland, reclaimed surface mine, scrub-shrub, and recently harvested forest

animal, is often equated to a plant community, vegetative association, or cover type" (Garshelis 2000; but see Hall et al. 1997). We use 'early successional habitats' to refer to sets of plant communities, associations, or cover types for multiple wildlife species.

Vegetation structure in early successional habitats can range from scattered trees or snags to no canopy cover, or from an open, grass-forb understory to thickets of shrubs and vines (Plate 1.1). Abandoned farmlands, grassland, shrub-scrub, recently harvested forest, heavily wind-, fire-, or ice-damaged forests, and even ruderal habitats such as roadsides, utility rights-of-way, and restored coalfields are all early successional habitats from this functional perspective (e.g., Thompson and DeGraaf 2001). Plant composition and micro-physical structure differ considerably among these diverse early successional habitat types, and can be dominated by grasses, forbs, shrubs, seedlings, woody sprouts, or a patchy mix of herbaceous and developing woody cover. However, all have two structural attributes in common: they have a well developed ground cover or shrub and young tree component and they do not have a closed, mature tree canopy.

Recently disturbed, regenerating upland hardwood forests may not, strictly speaking, be "successional," in terms of species turnover, but they do change greatly in structure over time. Many hardwood tree species resprout after damage or harvest, such that there may be little change in woody species composition between the progenitor forest, the young regenerating forest, or the mature forest decades later. In these common cases, longer-term changes are due to change in physical structure and potential shifts in the relative abundance of species, rather than species loss and establishment over time (Lorimer 2001). In some cases, nonnative species colonize following disturbance, further altering the original forest composition (Busing et al. 2009). In this volume, Loftis et al. (Chap. 5) discuss dynamical changes in structure and woody species composition, and Elliot et al. (Chap. 7) discuss herbaceous layer response to different silvicultural or natural disturbances and across moisture or fertility gradients associated with topography and physiographic regions or subregions.

Another characteristic of early successional habitats is that they are created by intense or recurring disturbances and are transient if not maintained by disturbance. Different types and intensities of natural disturbances (such as wind- or ice storms,

Plate 1.2 Examples of variation in the structure of early successional habitats in the upland hardwood forest of the Central Hardwood Region. From *left to right*: an experimental gap in the first season following its creation; ice storm damage; hot prescribed burn

wildfire, or outbreaks of pathogens) or forest management practices (such as two-age harvests, clearcuts, group selections, or hot prescribed burns) can create early successional habitats ranging from homogeneous structure with no trees to highly heterogeneous structure with scattered standing trees, multiple windthrows, or standing boles with broken tops. The scale of early successional habitats can also range from canopy gaps to thousands of hectares (Plate 1.2).

Historical and current patterns of frequency, intensity, and scale of natural and anthropogenic disturbances that create early successional habitats vary across the Central Hardwood Region. For example, catastrophic hurricanes occur at 85–380 year intervals in upland hardwood forests of the mid-Atlantic and southern New England (Lorimer and White 2003). The proportion of the landscape in young forest in this region might have varied from 40% to 50% after a severe hurricane to <3% as the forest matured (Lorimer and White 2003). In contrast, further inland where the likelihood of catastrophic wind damage is small, the proportion of early successional habitats due to wind disturbance was likely low (1–3%) and maintained by canopy gaps from single-tree death (estimated at <1% annually) (Runkle 1990) and infrequent windstorms (Lorimer and White 2003). Widespread, frequent burning was used by Native Americans and (later) by European settlers to maintain an open understory and improve conditions for travel and game or livestock for about 14,000 years, and decades of fire suppression has contributed to today's decline of early successional habitats and a shifting forest composition (Lorimer 1993; Brose et al. 2001). This variation in disturbances over time and across the landscape certainly created "nonequilibrium" or irregularity in the availability of early successional habitats, and populations of plants and animals that utilize them also likely waxed and waned in response to their availability.

In this volume, White et al. (Chap. 3) discuss how types, intensities and frequencies of natural disturbance vary across the Central Hardwood Region, and implications of these disturbances for patterns and probabilities of early successional habitats being created or maintained. Spetich et al. (Chap. 4) discuss the historic role of fire in creating and maintaining early successional habitats, and how fire suppression policies of recent decades have reduced their extent in the Central Hardwood Region.

1.4 Why Are Early Successional Habitats Important?

Most ecologists and environmentalists agree that disturbances and early successional habitats are important to maintain the diverse flora and fauna native to deciduous forests of the Central Hardwood Region (Brawn et al. 2001). Patches of early successional habitat play a pivotal role in forest dynamics as foci for tree regeneration and maintaining disturbance-dependent plant species. Hunter et al. (2001) recognized 128 bird species associated with grasslands, shrub-scrub, savannah and open woodlands, or forest gaps in eastern North America. Indeed, many species, including several listed as endangered, threatened, sensitive, or of management concern, require the openness of reduced or absent overstory, tall grasses, or thick shrub cover that early successional habitats can provide (Hunter et al. 2001; Litvaitis 2001; Thompson and Degraaf 2001).

Disturbances across the landscape and through time create habitat heterogeneity and affect the spatial and temporal availability of food resources in a forest matrix (Thompson and Willson 1978). Different disturbance types and intensities shape the size, structure, and distribution of early successional habitat patches, which may be key factors for maintaining populations of wildlife species that depend on them. Canopy gaps or small patches of recently disturbed, young forest may be sufficient for some species, whereas others require larger areas (Thompson and DeGraaf 2001). Mobile species may be able to utilize a landscape of connected or recurring smaller patches, whereas species with limited dispersal ability may require larger or less ephemeral patches. Some disturbance-adapted bird species may require grass-dominated early successional habitats (e.g., Field Sparrows (*Spizella pusilla*) or Grasshopper Sparrows (*Ammodramus savannarum*)), whereas others require brushy areas (e.g., Eastern Towhees (*Pipilo erythrophthalmus*)); open areas with the presence of nesting cavities (e.g., Eastern Bluebird (*Sialia sialis*)); or high elevation early successional habitats (e.g., Chestnut-sided Warblers (*Dendroica pensylvanica*) and Golden-winged Warblers (*Vermivora chrysoptera*)). Thus, defining high- or low-quality early succession habitat must be tempered by the species or suite of species that require specific structural conditions.

Breeding bird density and richness generally are higher in disturbed habitats, including treefall gaps (Blake and Hoppes 1986; Greenberg and Lanham 2001), intensively burned forest (Greenberg et al. 2007a), and recently harvested young forests, particularly if some tree canopy is retained (e.g., Annand and Thompson 1997; Whitehead 2003). Many bat species use early successional habitats to forage for insects (e.g., Loeb and O'Keefe 2006). The density of many salamander species declines in recently disturbed early successional habitats (e.g., deMaynadier and Hunter 1995), but the abundance of some reptile species increases in response to the same conditions (e.g., Greenberg 2002). Indeed, many wildlife species forage opportunistically for insects and fruit in resource-rich young forest patches (Greenberg et al. 2007b).

In this volume Greenberg et al. (Chap. 8) discuss the ample availability of food resources, including native forest fruit, browse, and arthropod and small mammal prey for wildlife in recently disturbed upland hardwood forest. Franzreb et al. (Chap. 9) examine the relationship between availability of early successional (small-diameter) forest and population trends of 11 focal bird species associated with "scrub-shrub" forest structure. Loeb and O'Keefe (Chap. 10) discuss how young forest patch size, shape, distribution, and connectivity, as well as vegetation structure, influence use by different bat species in relation to roost sites, mature forest, and water sources. Moorman et al. (Chap. 11) synthesize information to provide an overview of amphibian and reptile response to forest disturbance and the creation of early successional habitats. Lanham et al. (Chap. 12) present a case for considering utility rights-of-way and other "novel" places as an option for managing bird and butterfly species associated with early successional habitats.

As noted earlier, all early successional habitats are ephemeral. For example, young upland hardwood forest reaches the stem exclusion stage within 10 or 15 years of harvest, when the density of young tree stems can exceed 20,000–25,000 stems/ha, and canopy closure reduces light availability at the forest floor (Dessecker and McAuley 2001). Habitat suitability for different wildlife species changes with changing forest structure; for example, there is rapid turnover of songbird species during this period (Thompson and DeGraaf 2001). Decline of Ruffed Grouse (*Bonasa umbellus*) also is attributed to paucity of the stem exclusion age class (6–15 years) in forests of the Central Hardwood Region (Dessecker and McAuley 2001); this age class declines with forest maturation and the absence of new disturbances. Disturbances are required to create early successional habitats and to maintain a forest with a mosaic of age classes and a structural heterogeneity that increases plant and animal diversity at local, landscape, and regional scales (Askins 2001, Shifley and Thompson, Chap. 6).

Ecosystem processes and services provided by forests, such as carbon storage and water resources, are altered by creating early successional habitats. In this book, Vose and Ford (Chap. 14) examine post-disturbance changes in water quality and quantity, and recovery over time in relation to forest management practices, woody species composition, and climate. Keyser (Chap. 15) examines how creating early successional habitats and forest regrowth affect carbon storage and sequestration at stand and landscape levels.

1.5 How Can Early Successional Habitats Be Sustained?

One approach to maintaining early successional habitats is to base forest management on the "historic natural range of variation" (Lorimer and White 2003). This requires us to determine a reference point or time period; understand both the natural range of variation and what is 'unnatural' (for example should pre-settlement clearing and burning by Native Americans be considered natural?); and be prepared to implement management actions toward the historical variation. For example, prescribed fire may be needed because wildfires are not allowed to burn the acreages they would have historically.

Alternative strategies for creating and maintaining early successional habitats include a proactive approach. We could 'look forward' by identifying desired future conditions or goals, such as amounts and characteristics of early successional habitats needed to maintain viable populations of dependant plants and animals, and create them accordingly. Chapters in this volume explore management tools for determining how much early successional habitat is needed, and how and where to create and sustain it on the landscape. Shifley and Thompson (Chap. 6), use long-term, landscape-level Forest Inventory and Analysis data to simulate management scenarios designed to create a "shifting mosaic" of age classes and sustain a target proportion of the landscape in a young forest condition. Warburton et al. (Chap. 13) focus on strategies being used to identify priority species and specific recovery goals, develop spatially explicit blueprints of desired future conditions, and implement them by creating early successional habitats to sustain target populations through regional initiatives, ventures, cooperatives, and State Wildlife Action Plans. This book concludes with a chapter using empirical forest forecasting models to project 50 year change in forest types and age distributions in relation to scenarios of land ownership, economics, demographics, and climate change (Wear and Huggett, Chap. 16).

1.6 Conclusions

Overall, our aim in this book is to collate information about early successional habitats, to aid researchers and resource management professionals in their quest to sustain wildlife and plant species that depend on or utilize these habitats. We focus primarily on early successional habitats generated by natural or anthropogenic disturbance in upland hardwood forests, which are the predominant ecosystem in the Central Hardwood Region. This focus is in part because of the rising concern over the decline of plant and animal species associated with early successional habitats in this region, and because large areas of upland hardwood forest are in public lands where, compared to privately owned lands, land management decisions can be influenced more easily by conservation concerns. Using information in this book, resource management professionals may elect to look to the past to guide management by the natural range of variation in disturbance types and frequencies, and the area and conditions of early successional habitats they created. Or, they might look forward to create conditions based primarily on an objective to sustain biodiversity and species associated with early successional habitats through future decades.

Literature Cited

Annand EM, Thompson FR III (1997) Forest birds response to regeneration practices in central hardwood forests. J Wildl Manage 61:159–171

Askins RA (2001) Sustaining biological diversity in early successional communities: the challenge of managing unpopular habitats. Wildl Soc Bull 29:407–412

Bailey RG (1994) Ecoregions of the United States [map]. USDA For Serv (scale 1:7,500,000, revised, colored), Washington, DC
Blake JG, Hoppes WG (1986) Resource abundance and microhabitat use by birds in an isolated east-central Illinois woodlot. Auk 103:328–340
Braun EL (1950) Deciduous forests of eastern North America. Blackburn Press, Caldwell, NJ. (Reprint of 1st ed)
Brawn JD, Robinson SK, Thompson FR III (2001) The role of disturbance in the ecology and conservation of birds. Annu Rev Ecol Syst 32:251–276
Brose P, Schuler T, Van Lear D, Berst J (2001) Bringing fire back: the changing regimes of the Appalachian mixed-oak forests. J For 99:30–35
Busing RT, White RD, Harmon ME, White PS (2009) Hurricane disturbance in a temperate deciduous forest: patch dynamics, tree mortality, and coarse woody detritus. Plant Ecol 201:351–363
Clements FE (1916) Plant succession: analysis of the development of vegetation. Carnagie Institute of Washington Pub 242. Washington, DC 17.3.3, 17.4.5
deMaynadier PG, Hunter ML Jr (1995) The relationship between forest management and amphibian ecology: a review of the North American literature. Environ Rev 3:230–261
Dessecker DR, McAuley DG (2001) Importance of early successional habitat to ruffed grouse and American woodcock. Wildl Soc Bull 29:456–465
Garshelis DL (2000) Delusions in habitat evaluation: measuring use, selection, and importance. In: Boitani L, Fuller TK (eds) Research techniques in animal ecology. Columbia University Press, New York, pp 111–164
Greenberg CH (2002) Fire, habitat structure and herpetofauna in the Southeast. In: Ford WM, Russell KR, Moorman CE (eds), The role of fire in nongame wildlife management and community restoration: traditional uses and new directions. Gen Tech Rep NE-288, USDA Forest Service Northeastern Research Station, Newtown Square, pp 73–90
Greenberg CH, Lanham DJ (2001) Breeding bird assemblages of hurricane-created gaps and adjacent closed canopy forest in the Southern Appalachians. For Ecol Manage 153:251–260
Greenberg CH, Livings-Tomcho A, Lanham AD, Waldrop TA, Tomcho J, Phillips RJ, Simon D (2007a) Short-term effects of fire and other fuel reduction treatments on breeding birds in a Southern Appalachian hardwood forest. J Wildl Manage 71:1906–1916
Greenberg CH, Levey DJ, Loftis DL (2007b) Fruit production in mature and recently regenerated forests of the Appalachians. J Wildl Manage 71:321–335
Hall LS, Krausman PR, Morrison ML (1997) The habitat concept and a plea for standard terminology. Wildl Soc Bull 25:173–182
Hunter WC, Buehler DA, Canterbury RA, Confer JL, Hamel PB (2001) Conservation of disturbance-dependent birds in eastern North America. Wildl Soc Bull 29:440–455
Keever C (1950) Causes of succession on old-fields of the Piedmont, North Carolina. Ecol Monogr 20:320–350
Keever C (1983) A retrospective view of old-field succession after 35 years. Am Midl Nat 110:397–404
Litvaitis JA (1993) Response of early-successional vertebrates to historic changes in land-use. Conserv Biol 7:866–873
Litvaitis JA (2001) Importance of early successional habitats to mammals in eastern forests. Wildl Soc Bull 29:466–473
Loeb SC, O'Keefe JM (2006) Habitat use by forest bats in South Carolina in relation to local, stand, and landscape characteristics. J Wildl Manage 70:1210–1218
Lorimer CG (1993) Causes of the oak regeneration problem. In: Loftis DL, McGee CE (eds) Oak regeneration: serious problems, practical recommendations. Gen Tech Rep SE-84, USDA Forest Service Southeastern Forest Experiment Station, Asheville, pp 14–39
Lorimer CG (2001) Historical and ecological roles of disturbance in eastern North American forests: 9000 years of change. Wildl Soc Bull 29:425–439
Lorimer CG, White AS (2003) Scale and frequency of natural disturbances in the northeastern US: implications for early successional forest habitats and regional age distributions. For Ecol Manage 185:41–64

Odum EP (1960) Organic production and turnover in old field succession. Ecology 41:34–49
Runkle JR (1990) Gap dynamics in an Ohio Acer-Fagus forest and speculations on the geography of disturbance. Can J For Res 20:632–641
Smith WB, Miles PD, Vissage JS, Pugh SA (2004) Forest resources of the United States, 2002. Gen Tech Rep NC-241, USDA Forest Service North Central Research Station, St. Paul, 137 p
Stephenson NL (1990) Climatic control of vegetation distribution: the role of the water balance. Am Nat 135:649–670
Thompson FR III, DeGraaf RM (2001) Conservation approaches for woody-early successional communities in the eastern USA. Wildl Soc Bull 29:483–494
Thompson JN, Willson MF (1978) Disturbance and dispersal of fleshy fruits. Science 200: 1161–1163
Trani MK, Brooks RT, Schmidt TL, Rudis VR, Gabbard CM (2001) Patterns and trends of early successional forest in the eastern United States. Wildl Soc Bull 29:413–424
Whitehead MA (2003) Seasonal variation in food resource availability and avian communities in four habitat types in the Southern Appalachian Mountains. Dissertation, Clemson University, Clemson

Chapter 2
Subregional Variation in Upland Hardwood Forest Composition and Disturbance Regimes of the Central Hardwood Region

William Henry McNab

Abstract Oaks and hickories characterize the Central Hardwood Region, with its temperate, humid climate and deep soils. Several xerophytic species characterize stands on xeric sites; mesic sites usually have greater diversity of oaks and hickories and include maple, ash, beech, and yellow-poplar. Ice and wind storms are common disturbances across the region; wildland fires ignited by lightning are uncommon and generally confined to small, stand-size areas. Variable environmental conditions, topography, and forest species compositions from the eastern Appalachians to the western Ozarks can require different silvicultural prescriptions to create early successional habitats, even in stands of similar appearance.

2.1 Introduction

Extensive temperate deciduous broadleaf forests are present in only three areas on earth: central Europe, eastern Asia, and the eastern USA (Rohrig and Ulrich 1991). These areas have a moderately humid, continental climate with ample summer rainfall and severe winters (Whittaker 1975). In the USA, deciduous forests dominated primarily by upland hardwoods (with minor amounts of coniferous species) occur mainly from latitudes 35° to 40°, between grasslands on the west and forests with a higher proportion of conifers on the north and south (USDA Forest Service 1967). Historically referred to as the Central Hardwood Region, these forests are among the most compositionally and structurally complex vegetative assemblages in eastern North America (Braun 1950; Barbour and Billings 2000). Embedded within this region are equally complex bottomland deciduous forests associated with some of the largest river systems in the

W.H. McNab (✉)
USDA Forest Service, Southern Research Station,
1577 Brevard Road, Asheville, NC 28806, USA
e-mail: hmcnab@fs.fed.us

C.H. Greenberg et al. (eds.), *Sustaining Young Forest Communities*,
Managing Forest Ecosystems 21, DOI 10.1007/978-94-007-1620-9_2,

Plate 2.1 Oaks and hickories typically dominate the canopy of mature forests on middle and upper slopes in the Central Hardwood Region, such as this stand on the Cumberland Plateau, in Scott County, Tennessee (US Forest Service photo by F.E. Olmsted, 1901. Source: US Forest Service Photograph Collection, #P9801, D.H. Ramsey Special Collections, UNC-Asheville)

USA, including the Ohio, Wabash, Cumberland, and Tennessee Rivers (Braun 1950). The Central Hardwood Region covers over 40 million ha of the central USA.

Two general associations of canopy species are apparent in upland forests of the Central Hardwood Region: xerophytic associations of dry sites and mesophytic associations of moist sites. Stands on dry sites (ridges, upper and middle slopes) are characterized by a high proportion of oaks (*Quercus*) and hickories (*Carya*) in the canopy (Plate 2.1). On moist sites (coves and lower slopes), however, oaks and

Plate 2.2 A 55-year old stand of almost pure yellow-poplar that regenerated on an old-field site in a mesic cove of the Southern Appalachian Mountains, in Union County of northeastern Georgia (US Forest Service photo by C.A. Abell, 1930. Source: US Forest Service Photograph Collection, #P9876, D.H. Ramsey Library Special Collections, UNC-Asheville)

hickories are less abundant and the canopy may be shared with red maple (*Acer rubrum)*, white ash (*Fraxinus pennsylvanicum*), beech (*Fagus grandifolia*), sugar maple (*A. saccharum*) or occasionally dominated by few mesophytic species after severe disturbance (Plate 2.2) (Schnur 1937; Braun 1950). Although the region is dominated by upland hardwoods several species of conifers, such as red cedar (*Juniperus virginiana*) and Virginia pine (*Pinus virginiana*), may be present as minor components in stands on dry sites, particularly after disturbance. Many stands have three or more vertical strata: the overstory canopy, usually 24 m or more in height; a midstory of shade tolerant species; and a low shrub layer that usually includes advance regeneration of overstory species (Braun 1950). Abundance of advance regeneration and new seedlings depends strongly on the severity and timing of natural disturbances, which vary on a gradient of intensity from relatively frequent mortality of single canopy trees to infrequent catastrophic stand replacing events (White et al., Chap. 3). Considerable knowledge is available on the ecology, regeneration, and response of upland hardwood stands to silvicultural manipulations (Barrett 1980; Hicks 1998; Loftis et al., Chap. 5).

In this chapter, I describe the extent of the Central Hardwood Region and composition of arborescent vegetation in relation to environmental gradients, and briefly review the major types of disturbance across the region. Although I use canopy tree composition and its association with soil moisture availability to subdivide the region, other authors have chosen to use equally appropriate methods of stratification, such as ecoregions, as a basis of analysis for their chapters. This chapter provides an overview of the Central Hardwood Region that will supplement and link material presented throughout the book.

2.1.1 Distribution of Upland Hardwood Forests

The Central Hardwood Region has been delineated with general agreement (Schnur 1937; Braun 1950; USDA Forest Service 1967; Hicks 1998; Barbour and Billings 2000; Fralish 2003) (Fig. 1.1). This region is bordered on the north by forests with fewer oaks and more northern hardwoods (e.g. beech, sugar maple) and conifers (e.g. eastern white pine [*P. strobus*], eastern hemlock [*Tsuga canadensis*]) that transition to the Northern Hardwood Region (Braun 1950). To the south, southern yellow pines such as loblolly (*P. taeda*) Virginia, and shortleaf (*P. echinata*) increase in importance and oaks decrease as forests transition to the Southern Pine Region in the Piedmont (Eyre 1980; USDA Forest Service 1967). Conifers or hardwood-conifer mixtures may occur because of local conditions. For example, stands of red spruce (*Picea rubens*) occupy small areas of high mountain tops in West Virginia (>1,000 m) North Carolina (>1,400 m), where altitude presents environmental conditions similar to boreal conditions of southern Canada. As with other major forest regions in the eastern USA, (e.g. Southern Pine, Hemlock-Northern Hardwood) the Central Hardwood Region may be defined by its canopy composition of deciduous upland hardwoods (Braun 1950; Eyre 1980), and particularly by the predominant oak-hickory forest cover type (Barrett 1980; Hicks 1998; Fralish 2003).

The Central Hardwood Region also can be described by its climate. Whittaker (1975) reported an association of temperate deciduous forests with average temperature range between 3°C and 20°C and annual precipitation from about 1,125 to 1,225 mm. Stephenson (1998) explained the distribution of deciduous hardwood forests by actual evapotranspiration and water supply: "(1) annual precipitation (water supply) must be greater than 600 mm, (2) annual potential evapotranspiration (energy supply) must be greater than 600 mm, and (3) the seasonal timing of available water and potential evapotranspiration must be such that at least 600 mm of both occur simultaneously." In general, the Central Hardwood Region is associated with a climate in which precipitation is about equivalent to potential evapotranspiration and deficiencies of precipitation are offset by stored soil water during the mid to late frost-free season of most years.

2.1.2 Environment of the Central Hardwood Region

The continental climate of the Central Hardwood Region is classified as humid-temperate (Bailey 1995), with long hot summers and cold winters. Mean annual temperature in most of the region ranges from 11°C to 16°C. The frost-free season ranges from about 120 days in the north to almost 200 days in the south.

Annual precipitation ranges from less than 750 mm near the Great Plains to more than 1,100 mm in mountainous areas, with more than half falling in the growing season. Throughout this region, potential evapotranspiration during the early to middle growing season is about equal to the precipitation

West of the Mississippi River, elevations in the Central Hardwood Region range from about 100 m in major river valleys to over 800 m in the Ouachita Mountains, which are part of the Ozark Plateau, an extensive area in Arkansas and Missouri underlain by limestone, sandstone, and shale bedrock. East of the Mississippi River, elevation ranges from 100 m in the northeastern part of the Central Hardwood Region to over 2,000 m in the Southern Appalachian Mountains of North Carolina. Topography on level-bedded limestone and sandstone formations in central Kentucky and the Highland Rim of Tennessee ranges from gently rolling to dissected and hilly. The Appalachian Plateau, from central Pennsylvania to northern Alabama, grades from a dissected plateau to high hills and subdued mountains underlain by shale, sandstone, coal, and some limestone. Eastward, the folded and faulted shale, sandstone, and limestone bedrock of the Ridge and Valley province form long sandstone capped ridges separated by valleys underlain by limestone. East of that province are the highly weathered, steep Blue Ridge Mountains and the hilly Piedmont, both of which are underlain by igneous and metamorphic rocks. Overall, the highly variable topography and geologic substrate of the Central Hardwood Region have a greater effect on the distribution of species at smaller, landscape scales than on limiting the broader regional extent of the oak-hickory cover type throughout the central USA.

Soils of the Central Hardwood Region are varied and include four orders with extensive distributions. Ultisols are present across about half the region, from Arkansas and southern Missouri, central Tennessee, and Kentucky southeastward to the Piedmont and north into Pennsylvania. These soils are acidic, generally low in fertility, have high clay content in the subsoil and low amount of moisture storage capacity, and often are eroded as a result of past land use. Inceptisols are young soils that are present on steep terrain of the Appalachian and Cumberland Mountains, West Virginia, Pennsylvania, and southwestern Indiana. Alfisols usually form over calcareous rock formations and are present in northern Missouri, northwestern and central Kentucky and much of Indiana, Ohio, and southern Michigan. With sufficient soil moisture, Alfisols can be highly productive. Mollisols are the principal soil order in northern Missouri and much of Illinois, where the "Prairie Peninsula" extends eastward from the Great Plains into the Central Hardwood Region. Although soil orders are variable across the region, properties such as solum thickness of the A and B horizons and texture are among the most important factors affecting species composition and productivity of stands on both dry and moist sites.

Much of the central part of the region north of the Ohio River, including most of Illinois, has been influenced by one or more periods of glaciation, the most recent being the Wisconsin, which reached its peak about 20,000 years before present. Forward movement of the ice sheet created a smoothed landscape while its retreat left a layer of till of varying thickness and local variation caused by end moraines. Soils are generally deep and fertile in the northern part of the glaciated area.

Soils on flat terrain in the southern parts of Ohio, Indiana, and Illinois generally have poor internal drainage, resulting in wet conditions in winter and dry conditions in summer. An indirect effect of glaciations in the northern part of the region was deposition of wind-blown loess, which ranges in thickness from less than 0.5 to over 5 m.

Major, landscape-scale, natural disturbances throughout the Central Hardwood Region are associated mostly with wind (Everham and Brokaw 1996) and ice storms (Lemon 1961), and somewhat less by fire. Wind storms, mainly tornados, are more likely to occur in the western parts of the region (Fig. 3.3). Ice storms tend to be more common in the Appalachian Mountains and in the north (Fig. 3.3). Drought is a subtle form of natural disturbance that affects species composition of stands, particularly on moist sites, in the low-elevation parts of the Central Hardwood Region. Although wildland fires ignited by lightning occur throughout the region, the humid climate and highly urban and agricultural landscape limit their numbers to fewer than 300 annually and total area burned of about 4,500 ha (www.nifc.gov/fire_info/lightning_human_fires, accessed 12 Jan 2011). However, anthropogenic fires, set by Native Americans and later European settlers, were common throughout much of the region until the past several decades (Spetich et al., Chap. 4). Two historic sources of natural disturbances to forests in the eastern part of the Central Hardwood Region include elimination of American chestnut (*Castanea dentata*) as a canopy species during the 1920s and the differential effects among species from defoliation by gypsy moth (*Lymantria dispar*). White et al. (Chap. 3) provide more detailed information on the type and extent of disturbances.

2.1.3 Subregions of the Central Hardwood Region

Several species of oaks are in the canopy of most upland stands in the Central Hardwood Region, particularly on sites that are drier than average; mesophytic species increase on upland sites that are wetter than average (Braun 1950). The east–west precipitation gradient, variable bedrock formations with differing geologies and associated soils, and variable topography combine to form four smaller subregions of more uniform vegetation composition within upland forests: (1) Western Dry Subregion, (2) Transition Dry-Mesic Subregion, (3) Central Mesic Subregion, and (4) Eastern Mesic Subregion (Fig. 2.1). These subregions are similar to geographical areas delineated by Braun (1950), as oak-hickory, western mesophytic, mixed mesophytic, and oak-chestnut, respectively. Although forests are generally dominated by oaks, species composition varies in relation to moisture gradients at the subregion and stand scales.

Much of the information in the following paragraphs was extracted with little change from descriptions of large ecosystems termed major land resource areas (Natural Resources Conservation Service 2006); arborescent species composition was condensed from Braun (1950). The climatic regime of each subregion is presented as a water balance diagram for a representative location that combines annual variation in temperature and precipitation (Stephenson 1998). Potential evapotranspiration and actual evapotranspiration were estimated using the function developed by the US Geological Survey. Thirty year normal temperature and

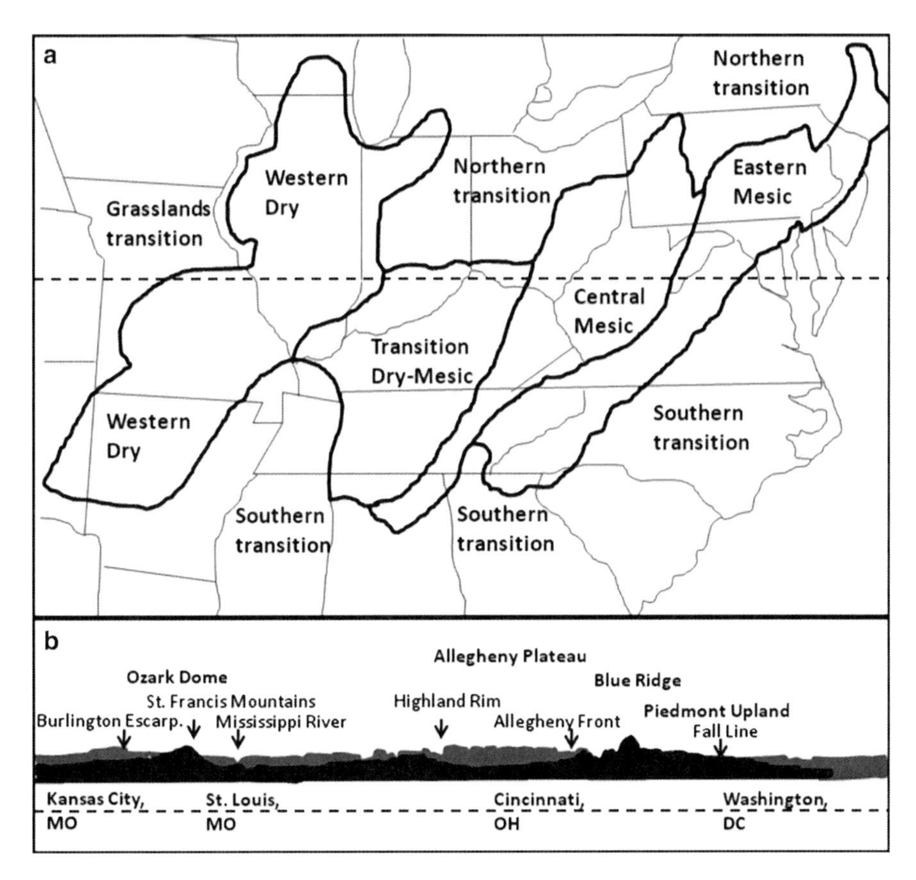

Fig. 2.1 (**a**) The Central Hardwood Region, its subregions, and transitions to adjoining subregions (From Braun 1950), and (**b**) profile of physiography along a transect (*dashed line*) from Kansas City, Missouri to Washington, DC and corresponding with the dashed transect line in (**a**) (from Lobeck 1957)

precipitation (1961–1990) were obtained from www.worldclimate.com. Soil field capacity of 200 mm was used for all locations.

2.1.3.1 Western Dry Subregion

This subregion extends in a broad diagonal band from northwestern Arkansas through south-central Missouri, and includes much of Illinois, which was about 60% prairie and oak savannah at the beginning of European settlement (Anderson 1970). The average annual temperature ranges from about 11.6°C to 15.5°C; the frost free season ranges from 175 to 245 days. Average annual precipitation ranges from about 1,000 to 1,150 mm, almost 60% falls during the growing season. Most of this subregion is a nearly smooth peneplain that was glaciated in the northern part (in Illinois) but not in the south, where it is slightly dissected by small streams.

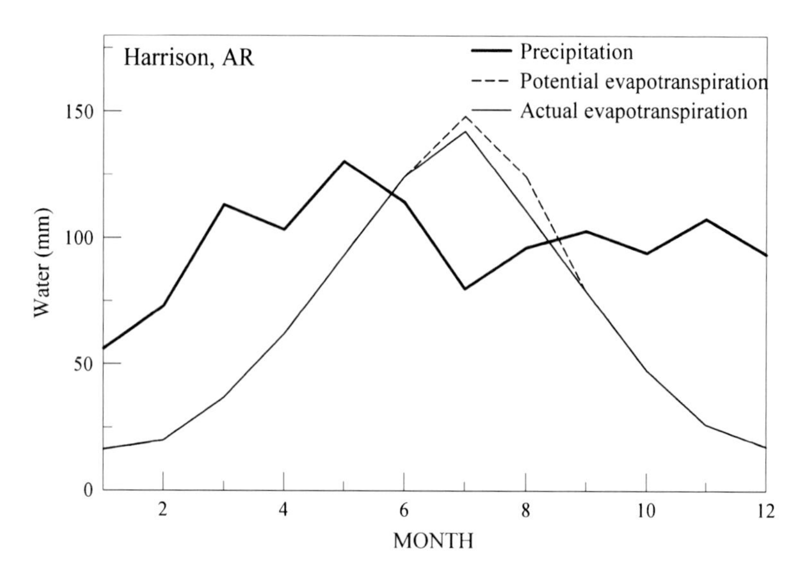

Fig. 2.2 A soil moisture deficit begins to develop in parts of the Western Dry Subregion during June in response to decreased summer precipitation as shown by a water balance diagram for Harrison, Arkansas

Elevation ranges from about 180 to 510 m. The Burlington Escarpment (Fig. 2.1) rises nearly 100 m above the peneplain and separates the Salem and Springfield Plateaus in southwestern Missouri. Bedrock geology is mostly sedimentary formations although intrusive granites are present in the St. Francis Mountains area of the Ozark Dome (Fig. 2.1), a large Precambrian uplifted and eroded region. Loess deposits range from several centimeters to almost a meter in thickness, with the greatest depth in the northern and eastern parts of the subregion. Soils are primarily Alfisols or Ultisols that have formed in material weathered from cherty limestone.

Tree species found in mesic ravines and gorges of this region include beech, northern red oak (*Q. rubra*), white oak (*Q. alba*), chinkapin oak (*Q. muehlenbergii*), sweetgum (*Liquidambar styraciflua*), winged elm (*Ulmus alata*), American elm (*U. Americana*), white ash, sugar maple, bitternut hickory (*C. cordiformis*), and basswood (*Tilia americana*). Dry site species on ridges and slopes include black oak (*Q. velutina*), white oak, shortleaf pine, scarlet oak (*Q. coccinea*), shagbark hickory (*C. ovata*), and in southern areas of the subregion, post oak (*Q. stellata*), blackjack oak (*Q. marilandica*), and southern red oak (*Q. falcata*). Before European settlement, much of this subregion in Illinois was a mosaic of open woodlands and tallgrass prairie maintained by frequent fire (Spetich et al., Chap. 4).

A water balance diagram for Harrison, Arkansas, in the northwest corner of the state near Missouri and the transition to the Great Plains, shows precipitation does not meet evapotranspiration requirements beginning in June and continuing during the growing season through October (Fig. 2.2). Water required for evapotranspiration, but not supplied by precipitation, is obtained from stored soil moisture. In a normal year, about 35% of available soil moisture remains at the end of the growing season. The steep decline in summer precipitation, particularly for July, is typical of climates associated with grasslands (Vankat 1979). Windstorms, particularly tornados, and drought are the main types of natural disturbance. Fire set by humans was also an important disturbance prior to the early 1900s (Spetich et al., Chap. 4).

2.1.3.2 Transition Dry-Mesic Subregion

This subregion includes much of the western and central parts of Kentucky and central Tennessee, and is a transition between the drier subregion to the west and mesic area to the east. The landscape is a plateau with low, rolling hills, upland flats, and narrow valleys. Steep slopes are present in the Nashville Basin and bordering the Coastal Plain on the west. Limestone is present in many areas, particularly the fertile Bluegrass Region of Kentucky (Fig. 2.1), which was an oak savannah before European settlement. The average annual temperature ranges from about 11.1°C to 15.5°C; the frost free season ranges from 185 to 235 days. Average annual precipitation ranges from about 1,100 to 1,600 mm. Precipitation is generally well distributed annually, but the monthly maximum occurs during late winter and early spring; the minimum occurs in fall. Bedrock geology is mostly Ordovician to Mississippian age limestone and dolomite. Thick clay covers much of the bedrock and areas of karst occur where clay is not present. Loess deposits of varying thickness cover much of the bedrock on uplands and ridges. Soils are mostly Alfisols, Inceptisols, and Ultisols that are deep to very deep, generally well drained, and loamy or clayey texture.

Some of the more important tree species of mesic sites in this subregion include beech, yellow-poplar (*Liriodendron tulipifera*), northern red oak, sugar maple, black walnut (*Juglans nigra*), and slippery elm (*U. rubra*); dry site species are white oak, northern red oak, shagbark hickory, black oak, pignut hickory (*C. glabra*), sassafras (*Sassafras albidum*) and, on basic soils, chinkapin oak. A water balance diagram for Columbia, Tennessee, shows a small deficit of precipitation developing in early summer that remains about constant until fall (Fig. 2.3). July precipitation increases sharply from the amount received in June, a trend which differs markedly from that of the adjoining Western Dry Subregion. Fire was less common here compared to neighboring subregions (Fig. 3.3). Drought is a recurring natural disturbance throughout this subregion; wind storms are more common in the north.

2.1.3.3 Central Mesic Subregion

This subregion extends from central Pennsylvania to northern Alabama (Fig. 2.1) and includes much of the Appalachian Plateau, which consists of the Allegheny Plateau and Mountains in the north and the Cumberland Plateau and Mountains in the south (Fenneman 1938; Bailey 1995). Its eastern and western boundaries are marked by abrupt changes in topography identified as the Allegheny Front and Highland Rim, respectively (Fig. 2.1). Elevations range from about 300 m along the western edge to over 1,200 m in the Allegheny Mountains. Topography in much of this subregion is

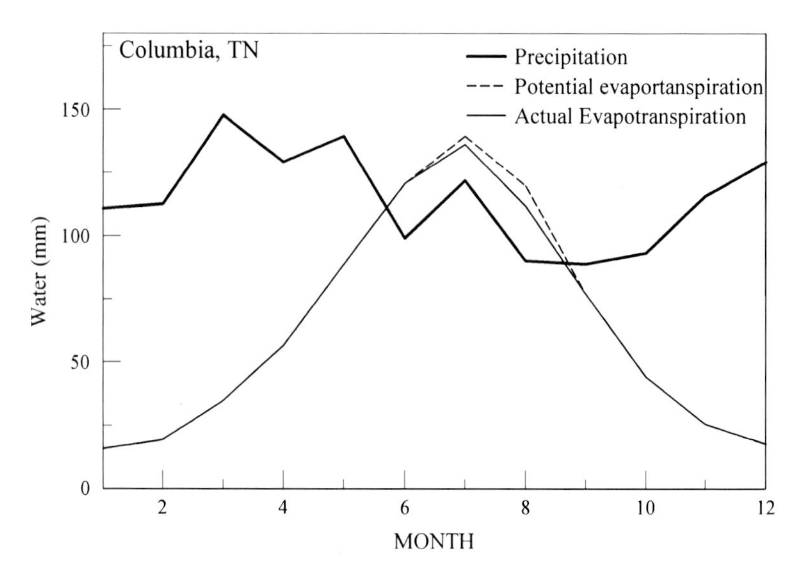

Fig. 2.3 As illustrated in a water balance diagram for Columbia, Tennessee, increased precipitation during July is the primary reason that soil moisture deficits are small in the Transition Dry-Mesic Subregion of the Central Hardwood Region

highly dissected plateau with steep side slopes between narrow ridge tops and mountains of moderate to high relief. The average annual temperature ranges from about 7.2°C to 15.6°C; the frost free season is variable depending on elevation and latitude, and ranges from 120 to 225 days. Average annual precipitation ranges from about 900 to 1,500 mm, and may exceed 1,900 mm in the mountains; about half occurs during the early growing season. Bedrock geology is mostly sedimentary formations of sandstone and shale, with small areas of limestone and coal in Virginia and Alabama. Ultisols form most of the soils in the undulating to rolling landscape of the Cumberland Plateau; shallow to deep and well drained to excessively drained Inceptisols are typical in areas of steep sandstone or shale residuum.

Canopy species composition is more varied here than in the other subregions and on mesic sites includes white oak, northern red oak, yellow-poplar, beech, sugar maple, buckeye (*Aesculus* spp.), black walnut, slippery elm, bitternut hickory and white ash; species on dry sites usually include white oak, chestnut oak (*Q. prinus*), black oak, shagbark hickory, pignut hickory, and redbud (*Cercis canadensis*). Stands of red spruce are present on the highest mountains in central West Virginia. A water balance diagram for Farmers, Kentucky, on the western edge of the subregion, shows a trend of actual evapotranspiration similar to the adjacent transition dry-mesic subregion to the west, but with very slight deficit of soil moisture as a result of the high July precipitation (Fig. 2.4). Ice storms are the most important type of natural disturbance, especially in the northern part of the subregion. Spetich et al. (Chap. 4) suggest that Native American use of fire was a common type of disturbance in this subregion that likely affected the regeneration dynamics of American chestnut, which was a component of many upland hardwood stands on non-calcareous soils,

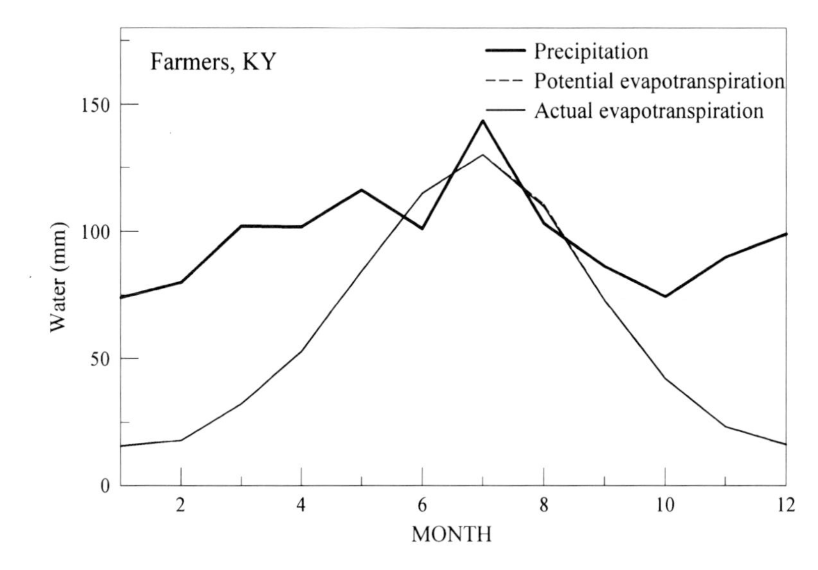

Fig. 2.4 Potential and actual evapotranspiration are almost equal in many parts of the Central Mesic Subregion, as shown by this water balance diagram for Farmers, Kentucky

before it was eliminated as a canopy species by the chestnut blight (*Cryphonectria parasitica*) in the early 1900s.

2.1.3.4 Eastern Mesic Subregion

Two large areas of differing physiology are included in this subregion: the low topography of the Ridge and Valley and higher peaks of the Blue Ridge Mountains (Fig. 2.1). Both of these areas extend from eastern Tennessee and western North Carolina through western Virginia into eastern Pennsylvania. Elevation ranges from about 30 m in the north to over 2,000 m in the southern mountains. Average annual temperature ranges from about 7.8°C to 15.6°C; the frost free season averages about 180 days, but is variable depending on elevation and latitude and ranges from 135 to 235 days. Average annual precipitation ranges from about 1,000 to 1,500 mm, and exceeds 2,500 mm along parts of the southern Blue Ridge escarpment in western North Carolina. Precipitation is generally evenly distributed annually with slightly reduced amounts in the fall. Ridge and valley bedrock geology consists of alternating beds of limestone, dolomite, shale, and sandstone; ridge tops are topped with resistant sandstone and the valleys have been eroded into less resistant shale and limestone. Southern Appalachian Mountain bedrocks consist mainly of Precambrian metamorphic formations of gneiss, schist, and small areas of amphibolites. The dominant soil orders are Inceptisols, Ultisols, and Alfisols, which are shallow to very deep, moderately well-drained to excessively drained and loamy or clayey.

Composition of the canopy in this subregion is almost as varied as in the central mesic subregion. Common mesophytic species of low to middle elevation stands are

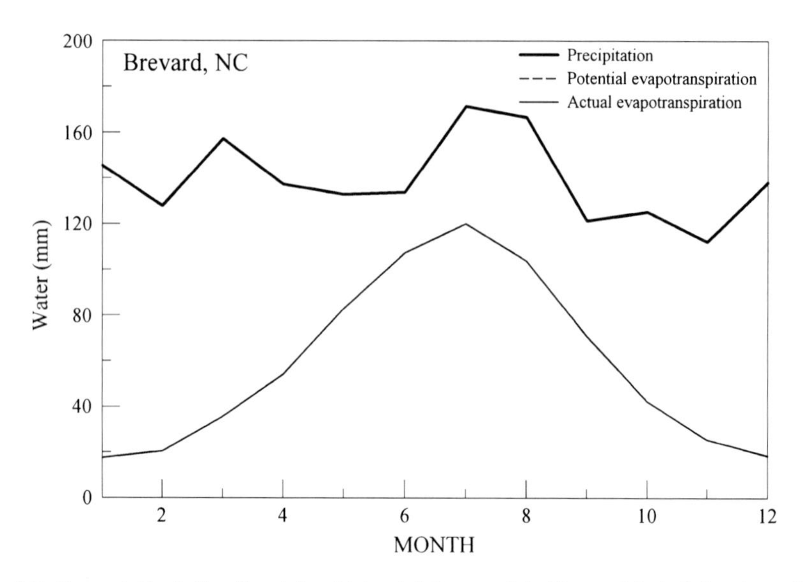

Fig. 2.5 Brevard, North Carolina, is in a high rainfall zone of the Eastern Mesic Subregion, where soil moisture deficits are rare

yellow-poplar, northern red oak, sweet birch (*Betula lenta*), white oak, red maple, white ash, silverbell (*Halesia carolina*), beech, and cucumber magnolia (*Magnolia acuminata*); xerophytic species include chestnut oak, scarlet oak, black oak, white oak, sourwood (*Oxydendrum arboreum*), pignut hickory, and mockernut hickory (*C. alba*). High elevation forests (above 1,300 m) are dominated by northern red oak on exposed slopes and ridges, and northern hardwoods (buckeye, sugar maple, yellow birch, and basswood) on protected slopes. Red spruce and Fraser fir (*Abies fraseri*) are extensive at still higher elevations (above 1,600 m). A water balance diagram for Brevard, North Carolina, indicates precipitation exceeds potential evapotranspiration throughout the growing season (Fig. 2.5). Ice storms are the major recurring type of natural disturbance although periodic drought (Hursh and Haasis 1931), the southern pine beetle (*Dendroctonus frontalis*) (Thatcher et al. 1981), and wind from remnants of hurricanes (Greenberg and McNab 1998) may also be important. Lightning caused fires are uncommon in the humid climate of this subregion, but human-caused fires were an important type of disturbance associated with American Indian and European settlement (Spetich et al., Chap. 4).

2.1.4 Subregional Comparisons and Subdividing the Region by Other Criteria

The four subregions of the Central Hardwood Region share a similar canopy composition dominated by upland oaks, and similar climates where rainfall is well distributed throughout the year and adequate to maintain soil moisture near field

Table 2.1 Composition by subregion of selected hardwood species on timberland of the Central Hardwood Region (Smith et al. 2004)[a]

Species or species group	Subregion			
	Western dry	Transition dry-mesic	Central mesic	Eastern mesic
Percent[b]				
All oaks	64	41	38	37
All hickory	12	11	7	4
White ash	3	4	2	3
Basswood	<1	1	3	1
Am. beech	<1	3	5	3
Yellow-poplar	<1	13	16	12
Sugar maple	1	5	6	4
Red maple	3	5	8	13
Others	16	17	15	23

[a]Timberland is forest land that is producing or is capable of producing crops of industrial wood in excess of 1.4 m^3/ha per year of industrial wood in natural stands
[b]Percent of total volume of hardwood growing stock >12.7 cm diameter breast height

capacity during much of the growing season. However, they differ in two respects: the relative proportions of mesophytic species and predominant types of natural disturbance. For example, yellow-poplar, an aggressive, shade intolerant mesophytic species that can hinder development of oak regeneration, is a minor component of stands in the Western Dry Subregion (Table 2.1). Ice storms are more likely to occur in the northern part of the Region (Fig. 3.3), and can cause almost continuous canopy disturbance across large areas. In contrast, early successional habitats resulting from wind-caused disturbances are more common in the western subregions and often result in small, discontinuous areas of tree blowdowns. Although the macro-climate may be similar throughout the Central Hardwood Region, differences in site quality, species composition, and natural disturbances among subregions should be considered by managers when planning silvicultural prescriptions to meet goals for creating and maintaining early successional habitats (Loftis et al., Chap. 5; Greenberg et al., Chap. 8). Potential climate change induced shifts in forest composition (Wear and Huggett, Chap. 16) and the frequency and intensity of natural disturbances (White et al., Chap. 3) will also require careful consideration in land management planning for early successional habitats at the local and regional level.

In addition, subregions delineated using Braun's (1950) major subdivisions as a basis may not be suitable for some purposes. Other methods and criteria for subdividing the Central Hardwood Region include physiographic provinces (Fenneman 1938), potential natural vegetation (Kuchler 1964), and large nested ecosystems of regional extent called ecoregions (Bailey 1983). Two widely used ecoregion maps have been developed by the USDA Forest Service (Cleland et al. 2007) and the US Environmental Protection Agency (2009) (Table 2.2). The versatility of these ecological maps is demonstrated by their use for purposes unintended in their development, such as delineation of bird conservation regions in the Central Hardwood Region used by some authors in this volume (Franzreb et al., Chap. 9; Warburton et al., Chap. 13) (US North American Bird Conservation Initiative 2000).

Table 2.2 Approximate correspondence among vegetation and ecological classifications of the Central Hardwood Region used by authors in this book

Subregions of Central Hardwood Region[a]	Divisions of the deciduous forest formation[b]	Ecoregions USDA-FS[c]	USEPA[d]	BCR[e]	SFFP[f]
Western Dry	3a, 3d	M231, 231G, 223A, 251D, M223, 223G	36, 37, 38 39, 72, 54	24 (west half) 22 (west half) 25 (north half)	Midsouth in Arkansas
Transition Dry-Mesic	2a, 2b, 2c, 2e, 2d	223D, 223E, 223F, 223B	55, 70, 71	24 (east half) 22 (east half)	Western part of App-Cumb.
Central Mesic	1a, 1b, 1c	221H. 221J, M221B, 221E M221C	68, 69, 70	28 (west half)	Central part of App-Cumb.
Eastern Mesic	4a, 4b, 4d, 4e	M221D, 231A, 221D, M221A 231I	66, 45	28 (east half) 29	Eastern part of App-Cumb. SA Piedmont

Note: Text in each column refers to units displayed on each map
Ecoregion assessment areas are abbreviated as: *App-Cumb* Appalachian-Cumberland, *SA Piedmont* Southern Appalachian Piedmont
[a]Chapter 2 of this book (Braun 1950)
[b]Braun (1950)
[c]Cleland et al. (2007)
[d]US Environmental Protection Agency (2009)
[e]US North American Bird Conservation Initiative (2000)
[f]Southern Forest Futures Project (Wear et al. 2009)

Authors of other chapters (Wear and Huggett, Chap. 16) have used a combination of ecoregions and administrative boundaries to define large areas of the Central Hardwood Region as a basis for forecasting of future forest conditions.

2.2 Conclusions

The Central Hardwood Region extends east from the Great Plains almost to the Atlantic Ocean and south from Lake Erie to the Piedmont Fall Line. The temperate, humid climate, well-distributed annual rainfall, and deep soils of the region form environmental conditions where moisture is usually available to vegetation throughout the frost-free season. At the regional scale, upland forests are characterized by oaks and hickories in the canopy. Species composition can be variable at the landscape scale, however, where topography and soil variation create moisture gradients occupied by a suite of vegetative species ranging from xerophytes (oaks and hickories) to mesophytes (maples and yellow-poplar). The relative abundance of some species, particularly yellow-poplar, varies considerable among subregions and should be taken into account when planning silvicultural activities to achieve resource management goals related to creating and maintaining suitable quantities and qualities of early successional habitats.

Literature Cited

Anderson RC (1970) Prairies in the prairie state. Trans Ill State Acad Sci 63:214–221

Bailey RG (1983) Delineation of ecosystem regions. Environ Manage 7:365–373

Bailey RG (1995) Description of the ecoregions of the United States. Misc Publ 1391, USDA Forest Service, Washington, DC, 108 p

Barbour MG, Billings WD (2000) North American terrestrial vegetation. Cambridge University Press, New York, p 708

Barrett JW (1980) Regional silviculture of the United States. Wiley, New York, 551 p

Braun EL (1950) Deciduous forests of eastern North America. Blackburn Press, Caldwell, 596 p

Cleland DT, Freeouf JA, Keys JE Jr, Nowacki GJ, Carpenter CA, McNab WH (2007) Ecological subregions: sections and subsections of the conterminous United States. [Map on CD-ROM] (AM Sloan, cartographer), USDA Forest Service, Washington, DC, Map scale 1:3,500,000

Everham EM III, Brokaw NVL (1996) Forest damage and recovery from catastrophic wind. Bot Rev 62:113–185

Eyre EH (ed) (1980) Forest cover types of the United States and Canada. Society of American Foresters, Bethesda, 148 p (with 1 map)

Fenneman NM (1938) Physiography of eastern United States. McGraw-Hill, New York, 714 p

Fralish JS (2003) The central hardwood forest: its boundaries and physiographic provinces. In: Van Sambeek JW, Dawson JO, Ponder F Jr, Loewenstein EF, Fralish JS (eds) Proceedings of the 13th central hardwood Forest conference, Gen Tech Rep NC-23, USDA Forest Service North Central Research Station, St. Paul, pp 1–20

Greenberg CH, McNab WH (1998) Forest disturbance in hurricane-related downbursts in the Appalachian mountains of North Carolina. For Ecol Manage 104:179–191

Hicks RR Jr (1998) Ecology and management of central hardwood forests. Wiley, New York, 412 p

Hursh CR, Haasis FW (1931) Effects of 1925 summer drought on southern Appalachian hardwoods. Ecology 12:380–386

Kuchler AW (1964) Potential natural vegetation of the conterminous United States. Special Publ 36, American Geographical Society. New York

Lemon PC (1961) Forest ecology of ice storms. Bull Torrey Bot Club 88:21–29

Lobeck AK (1957) Physiographic diagram of the United States. Geographical Press, Maplewood, 8 p

Natural Resources Conservation Service (2006) Land resource regions and major land resource areas of the United States, the Caribbean, and the Pacific Basin. Agri Handbook 296, USDA Natural Resources Conservation Service, 669 p

Rohrig E, Ulrich B (1991) Temperate deciduous forests – ecosystems of the world – 7. Elsevier, Amsterdam, p 635

Schnur GL (1937) Yield, stand, and volume tables for even-aged upland oak forests. Tech Bull 560. USDA Forest Service, Washington, DC, 87 p

Smith WB, Miles PD, Vissage JS, Pugh SA (2004) Forest resources of the United States, 2002. Gen Tech Rep NC-241, USDA Forest Service North Central Research Station, St. Paul, 137 p

Stephenson NL (1998) Actual evapotranspiration and deficit: biologically meaningful correlates of vegetation distribution across spatial scales. J Biogeogr 25:855–870

Thatcher RC, Searcy JL, Coster JE, Hertel GD (eds) (1981) The southern pine beetle. Tech Bull 1631, USDA Forest Service, Washington, DC, 267 p

US North American Bird Conservation Initiative (2000) Bird conservation region descriptions; a supplement to the North American bird conservation initiative. In: Bird conservation regions map. USDI Fish and Wildlife Service, Arlington, 38 p

USDA Forest Service (1967) Major forest types (map), the national atlas of the United States of America. Map scale 1:7,500,000. USDA Forest Service, Washington, DC

US Environmental Protection Agency (2009) Level III and IV ecoregions of the coterminous United States (map). Western Ecology Division. US Environmental Protection Agency, Corvalis. Map scale 1:7,500,000

Vankat JL (1979) The Natural Vegetation of North America: An Introduction. Wiley, New York, p 261

Wear DN, Greis JG, Walters N (2009) The southern forest futures project: using public input to define the issues. Gen Tech Rep SRS-115, USDA Forest Service Southeastern Research Station, Asheville, 17 p

Whittaker RH (1975) Communities and Ecosystems. Macmillan, New York, p 385

Chapter 3
Natural Disturbances and Early Successional Habitats

Peter S. White, Beverly Collins, and Gary R. Wein

Abstract Largely a legacy of stand-replacing human disturbances, today's central hardwood forests exhibit a narrower range of stand ages and structures than those in the presettlement landscape. Although natural disturbance types and frequencies vary within the region, large stand-replacing natural disturbances have always been infrequent; typical return intervals in excess of 100 years are longer than current forests have existed. Many present-day stands are dominated by early to mid-successional species in the overstory and late successional species in the understory; natural disturbances often serve to increase dominance of the understory late successional species, unless they are severe enough to disturb the canopy, forest floor, and soil. In any case, only the most severe natural disturbances or combinations of disturbances (including human disturbance) initiate large patches of early successional vegetation. Will the amount and spatial arrangement of early successional habitats created by natural disturbances be sufficient to meet management goals? We do not have the information to answer this question at present; the answer is further complicated by the potential effects of climate change on the rates and intensities of natural disturbances.

P.S. White (✉)
Department of Biology, University of North Carolina Chapel Hill,
Campus Box 3280, Chapel Hill, NC 27599–3280, USA
e-mail: Peter.White@unc.edu

B. Collins
Department of Biology, Western Carolina University,
NS 132, Cullowhee, NC 28734, USA
e-mail: collinsb@email.wcu.edu

G.R. Wein
Highlands-Cashiers Land Trust, P.O. Box 1703,
Highlands, NC 28741, USA
e-mail: hitrust@earthlink.net

C.H. Greenberg et al. (eds.), *Sustaining Young Forest Communities*,
Managing Forest Ecosystems 21, DOI 10.1007/978-94-007-1620-9_3,

3.1 Today's Forests – A Legacy of Human Disturbance

Today's central hardwood forests are largely a legacy of stand-replacing human disturbances that began in the 1700s and intensified in the 1800s and early 1900s (Lorimer 2001). Many of these forests owe their origin to large scale logging that took place between 1850 and 1940, while others date from farm abandonment that has occurred, at different times in different parts of our study area, from 1880 to the present (Fralish and McArdle 2009; Hart and Grissino-Mayer 2008). Peak agricultural clearing occurred between about 1880 and 1920, and post-farming stands from that period are similar in age to the post-logging forests.

Logging and agricultural disturbance were often accompanied by soil erosion, so the significance of these disturbances was more than a simple resetting of the successional clock; productivity and successional trajectories were affected on some sites. Burning and understory livestock grazing also were widespread during the 1800s and early 1900s, and occurred over landscapes variously cleared, farmed, or burned by Native Americans (Owen 2002).

Because of their roots in historical, widespread stand-initiating human disturbances, most of today's central hardwood forests are 70–100 years old, creating a landscape with reduced structural heterogeneity and age diversity compared to the presettlement landscape (Shifley and Thompson, Chap. 6). These forests are now reaching sawtimber size over large areas. Some stand characteristics, such as leaf area and basal area, have reached levels similar to presettlement forests, but composition, maximum tree sizes, and downed woody debris remain out of presettlement norms (Flinn and Marks 2007; Trani et al. 2001).

Present day stem densities generally are greater than densities in old-growth forests for three reasons: (1) Trees are mostly only about one-quarter to one-half their maximum sizes and forest understories were more open in the past due to (2) frequent fires, and (3) understory grazing. Shade-tolerant, fire sensitive, and mesic species often dominate in these denser forest understories and the forests are slowly converting from greater dominance by oaks (*Quercus* spp.) and hickories (*Carya* spp.) (with pines (*Pinus* spp.) in some areas) to maples (*Acer* spp.) and beech (*Fagus grandifolia*) as these species regenerate after the death of overstory trees (Cowell et al. 2010; Fralish and McArdle 2009; Hart and Grissino-Mayer 2008; Hart et al. 2008). Nowacki and Abrams (1997) refer to the widespread increase in mesic fire sensitive species across the deciduous forests of eastern North America as "mesophication." Although invasive pests and diseases (e.g., chestnut blight (*Chryphonectria parasitica*), gypsy moth (*Lymantria dispar*)) became important throughout the 1900s, they also served to increase canopy turnover rates and release advanced regeneration rather than initiate early succession composition and structure.

The maturation of central hardwood forests, the roughly synchronous nature of the large scale human disturbances that produced them, and the current smaller-scale disturbance regime, mean that early successional habitats within these forests are declining. This, in turn, raises concerns about the persistence of biodiversity supported by early successional habitats. In this chapter, we address the questions: What natural disturbances are important in these forests? Will these natural disturbances recreate the heterogeneity and patchiness of the past? Do natural disturbances initiate early successional habitats, which we consider to include new stands, young forest patches, or habitat within forests for open site, early successional plants, in the present landscape? Other chapters in this volume focus more specifically on vegetation response to disturbance; for example, Elliott et al. (Chap. 7) examine disturbance effects on herbaceous vegetation composition and diversity, and Loftis et al. (Chap. 5) examine effects of silvicultural disturbances on species composition of regenerating hardwoods.

In addition to natural disturbances within forests, there are other sources of open habitats and the biodiversity they support in the Central Hardwood Region. They include rock outcrops, glades, barrens, fire-dependent prairies that develop on certain bedrocks, and floodplains and stream channels affected by flood scour and beaver populations (Anderson et al. 1999). These habitats have slow rates of succession (rock outcrops, glades, and barrens), high rates of disturbance (floodplains, prairies) or both. For example, frequent fire can expand open grasslands and savannahs beyond the immediate boundaries of the bedrock islands that underlie some of these open communities. These open sites are also early successional habitats, but in this chapter we focus only on early successional habitats within upland forests, including new stands, patches of young forest, or open patches with early successional species. Anderson et al. (1999) have described the other kinds of open and successional communities in the North American forests.

3.2 Natural Disturbances and Early Successional Habitats

Large-scale or intense disturbances above a threshold of severity (Romme et al. 1998; Frelich and Reich 1999) initiate succession or maintain early successional forest habitats and allow the periodic regeneration of shade intolerant species. Frelich and Reich (1999) concluded severe or high cumulative disturbance maintain early successional species or initiate rapid conversion from late successional species to early successional species (a compositional catastrophe). Roberts (2004, 2007) linked disturbance severity to the percent of cover or biomass removed or disrupted through canopy, understory, and forest floor layers. We have adapted the Roberts model (Fig. 3.1, *left panel*) to link natural disturbance type and severity to early successional habitats. Disturbances are likely to have different impacts through forest strata (reviewed by Roberts 2004) and the threshold of severity to initiate succession is likely to differ both among strata and disturbances. For example, fire and flooding are ‘bottom up’ disturbances, with ground layer, understory, and canopy impacts at increasing severity. In contrast, wind disturbance, ice storms, and pathogens are often ‘top down’ disturbances. As windstorm severity increases, effects move from the canopy to soil and understory disturbance through tip-ups, thereby increasing the importance of seed dispersal relative to sprouting and seed bank in recruitment of understory stems (e.g., Busing et al. 2009; Clinton and Baker 2000). In general,

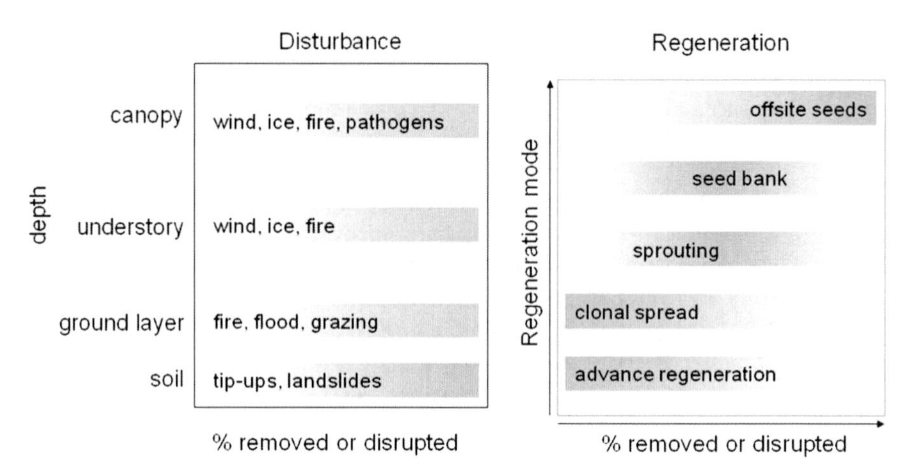

Fig. 3.1 On the *left*, a conceptual model (adapted from Roberts 2004, 2007) relates increasing severity of natural disturbance – as percent cover or biomass removed or disrupted through forest strata – to extent of early successional habitats, which is represented by the progressive shading and includes young forest and open patches with early successional plant species. Disturbance above some threshold of severity (Romme et al. 1998; Frelich and Reich 1999) may be required to initiate early successional habitats. On the *right*, the relative importance (*indicated by the shading*) of different regeneration modes changes with disturbance severity; regeneration from seed sources increases as disturbance severity increases

the establishment of shade intolerant species in the Central Hardwood Region depends on both canopy and ground layer disturbance.

Although severity of individual and multiple disturbances has been related qualitatively to forest conversion or maintenance of early successional species (e.g., Frelich and Reich 1999), few studies have quantified the severity of individual natural disturbance types needed to initiate succession or maintain early successional habitats in upland central hardwood forests. Most evidence is indirect. For example, hurricane damage that resulted, on average, in 25% basal area reduction in a mixed oak-hickory-pine forest did not shift composition toward shade-intolerant tree species (Busing et al. 2009). Natural disturbance alone also had little effect on habitat availability for early successional songbirds in a 60 year simulation study (Klaus et al. 2005). In a west-central Tennessee site that experienced moderate-severity windthrow and limited subsequent salvage logging, establishment of shade-intolerant tree species was more related to pre-disturbance forest composition than to disturbance severity (Peterson and Leach 2008). In contrast, however, Clinton and Baker (2000) found that gaps up to 4,043 m^2 could facilitate establishment of shade-intolerant species in Southern Appalachian forest. Vigorous sprouting (Clinton and Baker 2000) likely contributed to early successional forest structure, since these forests were young enough to have such species in the overstory. Elliott et al. (2002) also found that 84% reduction in basal area, through wind disturbance and subsequent salvage logging, allowed a heterogeneous mix of shade tolerant species, shade intolerant species, and opportunistic early successional understory species to establish in Southern Appalachian forests. Variation in forest composition, differences in disturbance severity over the landscape, and interaction of multiple disturbances (including interactions of natural disturbances and management) are most likely to create within-forest heterogeneity, with local patches of early successional habitats.

Differences in regeneration mechanisms among forest types and over disturbance severity gradients can contribute to the extent and, possibly, duration of early successional habitats. Figure 3.1 (*right panel*) is a conceptual model of the relationship between disturbance severity and predominant regeneration mechanism following disturbance. In general, contribution of seed sources increases with disturbance severity, although contribution from the seed bank will diminish if soil surface layers are removed (Aikens et al. 2007; Clinton and Baker 2000; Harrington and Bluhm 2001; Turner et al. 1998). Greater contribution from seed sources can increase abundance of early successional and shade-intolerant species, many of which regenerate from buried seeds or from seeds carried into the site by wind or animals. For example, regeneration after hurricane disturbance followed by salvage logging was characterized initially by many small-diameter stems and opportunistic species (*Rubus allegheniensis*) that regenerate from buried seeds (Elliott et al. 2002). Sites with a high abundance of species that resprout following disturbance are less likely to have new individuals establish, but may maintain young forest structure if early and mid-successional species dominate the canopy. Regeneration from seeds may also increase the time to canopy closure, when compared to sites with residual plants (those remaining after the disturbance) or a high abundance of species that resprout (Turner et al. 1998).

In general, only the most severe disturbances, such as catastrophic windstorms, fire, or landslides, create extensive early successional habitats. However, repeated natural disturbances, management following a disturbance event, or disturbance following management action could effectively increase disturbance severity or increase the duration of early successional species or structure (Elliott et al. 2002; Gandhi and Herms 2010; Kupfer and Runkle 1996). Frelich and Reich (1999) pointed out the importance of cumulative disturbance severity in maintaining early successional species or initiating catastrophic conversion of late successional to early successional species. Cumulative disturbances also are likely to maintain early successional habitats by preventing establishment of late successional species.

3.3 Disturbance Patterns Within the Central Hardwood Region

Some parts of the Central Hardwood Region are more susceptible than others to particular disturbance types. Understanding the variation in disturbance types and frequencies within the region can guide management actions to promote or sustain early successional habitats (see Shifley and Thompson, Chap. 6).

We used spatial information to examine the patterns of natural disturbances within the Central Hardwood Region. A Geographic Information System (GIS) coverage for ice storm potential (freezing rain) was derived by geo-referencing Fig. 3.1 (a map of the annual number of days with freezing rain as defined by 988 weather stations from 1948 to 2000) from Changnon and Bigley (2005). Line coverage of historical North Atlantic tropical cyclone tracks, 1851–2000 (NOAA 2009) was used to generate a density map of tropical storm occurrence within the region. Tornado density was calculated in ArcGIS (v. 9.3) using United States tornado touchdown points 1950–2004 (NWS 2005). A landslide coverage was based on a spatial index of landslide susceptibility and occurrence (Godt 1997). Raster digital data for mean fire return interval were obtained from LANDFIRE (US Forest Service 2006). The base maps for these disturbances are shown in Appendix I.

To evaluate the patterns of the combined disturbances, we first scaled each disturbance (0–100 scale) among 17 ecoregions (US Environmental Protection Agency 2009) contained within the larger Central Hardwood Region and calculated the mean value of each scaled disturbance weighted by the number of pixels that represented the disturbance within the ecoregion. We used principal components analysis (PCA) to identify linear combinations of the five disturbance types over the ecoregions. It is important to note here that base disturbance intensity differs among the disturbance types. For example, the landslide coverage includes both susceptibility and occurrence; ice storm potential is assessed through data on the days of freezing rain rather than ice storm damage; tropical storms vary in intensity; and mean fire return interval includes a range of severity from understory to stand-replacing fires.

The predominant disturbance type varies among ecoregions within the larger Central Hardwood Region (Figs. 3.2, 3.3). The first two principal components explained 77% of the variance in disturbances among the ecoregions. Axis 1 correlated positively with tornados (0.90) and negatively with landslides (–0.88) and tropical storms (–0.80). This axis represents an east–west gradient (Fig. 3.3) from tropical storms, the predominant disturbance in the east, to tornados in the west (Table 3.1, Fig. 3.2). The frequency of tropical storms decreases from the Piedmont (ecoregion 45, Table 3.1) and adjacent Blue Ridge (ecoregion 66) westward to the Ridge and Valley (67), Central Appalachians (69) and Western Allegheny Plateau (70), which are more susceptible to landslides (Figs. 3.2, 3.3; Table 3.1).

Principal component Axis 2 correlated positively with fire return interval (0.82) and negatively with freezing rain (ice storm potential) (–0.81). Not surprisingly, northern extensions of the region, including the Huron and Erie Lake Plains (57), Southern Michigan and Northern Indiana Drift Plains (56), and Eastern Corn Belt Plains (55) have the highest occurrence of freezing rain (Table 3.1; Figs. 3.2, 3.3). Western regions, from the Central Corn Belt Plains (56) south to the Ouachita Mountains (36), have the highest occurrence of tornados, but areas farther north (56) also experience freezing rain and more southern regions (36, 37, 38) experience frequent fire (5–15 year fire return intervals, Appendix I). The Appalachians and adjacent Plateau regions are an exception to the north – south gradient from freezing rain to high fire return intervals (Fig. 3.3); relatively high rainfall results in

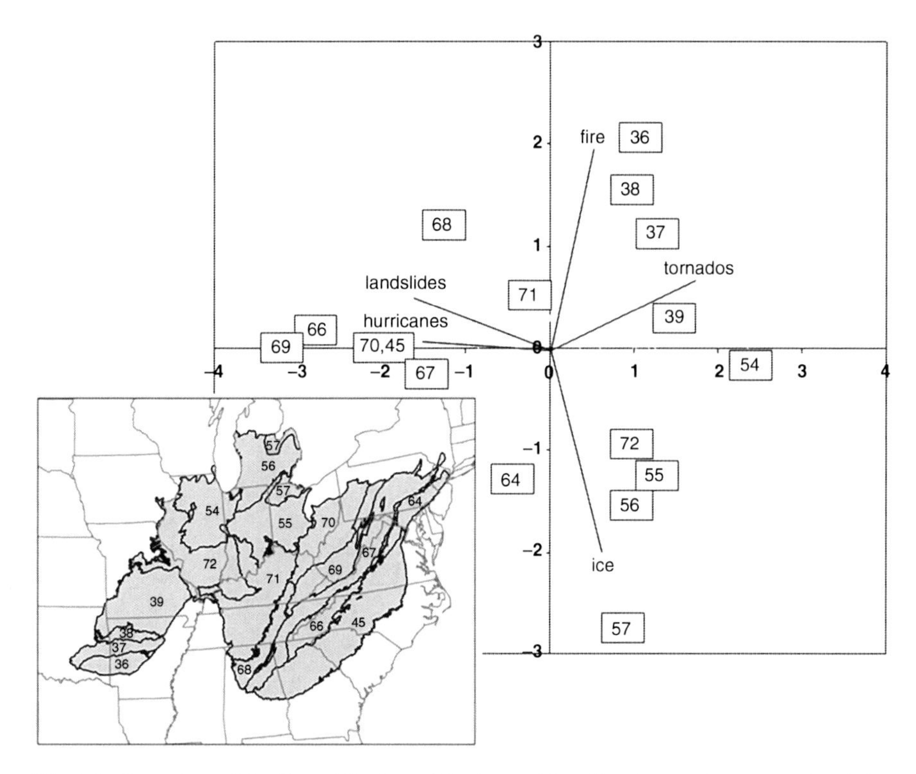

Fig. 3.2 Ecoregions of the Central Hardwood Region and five disturbance eigenvectors (*scaled to unit length*) plotted on the first and second principal component axes. Names of the numbered ecoregions are given in the text

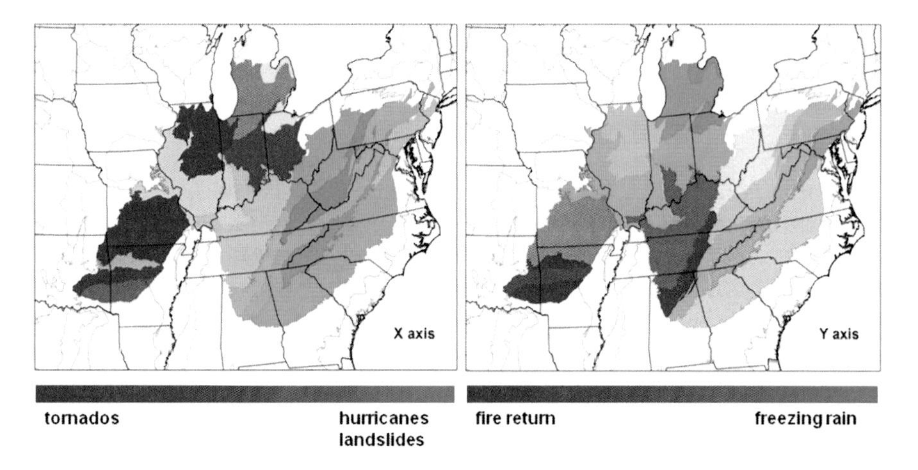

Fig. 3.3 Ecoregions of the Central Hardwood Region color-coded by their PCA scores (*first (X) and second (Y) axes*). First axis scores were positively correlated with tornados and negatively correlated with landslides and hurricanes. Second axis scores were positively correlated with fire return interval and negatively correlated with freezing rain

Table 3.1 The likelihood of experiencing disturbances within each ecoregion

Ecoregion	Freezing rain days/year	Tornados #/ km²/10 year (×10⁻³)	Trop. storms #/ km²/10 year (×10⁻⁵)	Fire return interval (years)
56	3.8	1.8	2.4	14.6
67	2.9	0.9	8.9	9.2
57	3.9	1.6	6.2	23.4
54	4.3	2.6	2.5	3.9
64	3.8	1.9	11.9	8.9
55	4.1	2.0	3.8	14.9
70	2.7	0.6	3.6	8.6
69	2.1	0.3	5.1	12.7
72	3.4	1.7	4.0	12.9
45	2.4	1.2	12.8	7.3
71	1.8	1.5	5.9	9.1
39	3.4	1.5	3.7	4.5
66	3.0	0.4	12.0	7.8
68	1.0	1.5	12.0	8.0
38	2.3	1.0	11.8	3.4
37	2.0	2.7	8.5	7.8
36	1.3	1.2	9.7	5.0

Information about the temporal scale and data sources for each disturbance is included in the text. Qualitative data for landslide incidence and susceptibility could not be averaged and thus were not included in the table. Averages for freezing rain (days/year) and fire return interval (years) were derived from area-based spatial data (Appendix 1) and were weighted by the proportion of area representing different values within the ecoregion. Tornados are the number of touchdowns points per km^2 per decade within the ecoregion. Tropical storm values were derived from storm tracks (line data, Appendix 1), and are reported as the number per km^2 per decade within the ecoregion

longer fire return intervals and higher elevations likely experience more frequent freezing rain or ice (Table 3.1).

Variation in natural disturbances over the Central Hardwood Region is likely to result in different patterns and probabilities of early successional habitats being created or maintained. Catastrophic windstorms, associated with tropical storms and hurricanes in the east and with tornados in the west, can create patchy and sporadic early successional habitats, although research suggests these storms generally are below the threshold needed for the initiation of extensive early successional stands unless followed by management (e.g., salvage logging) or a subsequent natural disturbance (Elliott et al. 2002; Gandhi and Herms 2010; Kupfer and Runkle 1996; Peterson and Leach 2008) that increases disturbance severity. In the Piedmont (eastern) and Ouachita (southwestern) ecoregions, fire is the most likely natural disturbance to act in concert with wind (Fig. 3.3). Historically, these fires were initiated by Native Americans, settlers, or lightning; today they are most likely to be initiated by land managers (see Spetich et al., Chap. 4).

In northern ecoregions, as well as on slopes and ridges of the Appalachians, ice storms are most likely to cause damage to the canopy. Susceptibility to ice storms may be greatest on steep slopes (Mou and Warrilou 2000) and damage can be more intense on edges (Millward and Kraft 2004). However, ice storms often do not lead to change in forest composition, although growth of understory species can slow recovery, especially in larger gaps (Mou and Warrilou 2000). Slopes of the Appalachians and adjacent Plateaus also are susceptible to, and have a high incidence of, landslides. These localized disturbances have high heterogeneity, with patches of unstable exposed soil, erosional and depositional zones, and an initial mix of surviving vegetation and early colonists (Myster and Fernandez 1995; Francescato et al. 2001; Walker et al. 2009). Rates and trajectories of succession can be highly variable on landslides (Francescato et al. 2001; Walker et al. 2009); early successional herbs and patches of shrubs can persist for decades or be replaced more rapidly by forest species (Francescato et al. 2001; Walker et al. 2009).

The presettlement forest landscape, except of course on sites of Native American cultivation, was largely forests whose dominant trees often survived to reach ages of 300–500 years. The mortality of canopy trees therefore occurred at low rates, probably varying from about 0.05% to 2% of canopy trees per year (Runkle 1982; Busing 2005). Large stand-replacing natural disturbances were always infrequent relative to tree lifespans, with return intervals in the 100s of years. Thus, return intervals are longer than the current forests have existed (Hart and Grissino-Mayer 2008; Lorimer 2001; Schulte and Mladenoff 2005). For example, Hart and Grissino-Meyer (2008) found evidence of only one stand release, in the 1980s, in an oak-hickory forest that established in the 1920s. Less severe disturbances, those that do not lead to stand replacement are, of course, more frequent.

Return intervals of particular disturbances at small scales are affected by local factors, such as topography, as well as regional factors such as climate. There are several challenges in predicting natural disturbance return intervals at a local scale. First, they are scale dependent. For instance, the return interval for tropical storms over the last 100 years in the state of North Carolina as a whole (139,396 km^2) is about 1.3 years (www.nc-climate.ncsu.edu). The return interval for Orange County, North Carolina, an inland county of 1,040 km^2 is about 50 years, while the return interval for a particular stand of trees within Orange County is in excess of 100 years (see also Busing et al. 2009). A second challenge is that disturbance rate and severity are contingent, proximately, on current structure and composition and, ultimately, on successional history. Thus, the disturbance rates in the homogeneous forests of the present, with their high densities and uniform canopy of trees that are smaller than old growth forests, are themselves a result of the synchronous origin of these stands some 70–100 years ago. A third challenge is that cumulative effects of repeat or multiple disturbances are more likely to produce early successional habitats than single events (Frelich and Reich 1999). A fourth is that invasive pest species are still spreading in this region. Finally, disturbance rates and severities are likely to change with changing climate and socioeconomic factors. Wear and Greis (Chap. 16) forecast how forest type and age class distribution might change over the next 50 years in response to biophysical and socioeconomic dynamics. Below, we discuss the linkage between natural disturbance and early successional habitats at the landscape scale.

3.4 Natural Disturbance and Early Successional Habitats on the Landscape

At landscape and regional scales, we can ask: how do natural disturbances affect the amount and distribution of early succesional habitats and is this pattern dynamically stable (i.e., in equilibrium and likely to be maintained) over time? A strict definition of equilibrium is "quantitative" equilibrium or "shifting mosaic steady state" in which disturbance rate is constant and the percentage of various patch types and stand ages, including early successional vegetation, is constant through time at large spatial scales. Given all the historic and present disturbances that impact forests of the Central Hardwoods Region, quantitative equilibrium is unlikely. A less stringent form of dynamic stability is "qualitative" or "persistence" equilibrium (see discussion in White et al. 1999) in which the rate of disturbance and size of disturbance patches vary, but within boundaries such that patch types, stand ages, and the species associated with these conditions fluctuate from year to year but do not disappear at large spatial scales. Qualitative equilibrium is more likely, and given that it suggests persistence of species dependent on all patch types, may be a reasonable standard for conservationists and managers.

Given (1) the narrow age range of current forests, (2) observations in the literature which suggest later successional species in understories increase after disturbance, and (3) the low probability of stand-replacing natural disturbances, large patches of early successional habitats may be declining on the landscape. However, disturbances do create edges. Light, nutrient, and seed dispersal gradients across edges allow open-site and early successional species to establish and persist in edge zones. For example, edges between forests and agricultural fields had a greater number of light-demanding species than forests interiors, and south-facing edges were as wide as 23 m (Honnay et al. 2002). In forest edges younger than 6 years, most edge-oriented species were close to the edge, with distributions related to light and light-related variables, but some species had peak density up to 40 m into the forest (Matlack 1994). Species composition and distribution patterns characteristic of edges persisted up to 55 years after edges were closed by succession (Matlack 1994).

Canopy gaps and similar disturbance patches also contain light, nutrient, and seed dispersal gradients that promote early successional forest composition and young forest structure. For example, canopy openness in 3-year-old experimental gaps greater than 20 m radius in bottomland hardwood forest declined linearly from the open center (>20% canopy openness) across the edge (>10% canopy openness) to more than 60 m (<5% canopy openness) into the surrounding forest (Collins and Battaglia 2002). Ten years after the gaps were created, the centers had a young forest canopy; species composition differed from gap centers into the surrounding forest, with wind-dispersed species more common in gap centers (Holladay et al. 2006). In a high-latitude Scots Pine (*P. sylvestris*) and Norway Spruce (*Picea abies*) forest, cumulative photosynthetically active radiation (PAR) was asymmetrically distributed around a canopy gap (deChantal et al. 2003). PAR decreased from 1,100 MJ m^{-2} in the gap to 300 MJ m^{-2} beneath surrounding forest over 20 m on the north side and over 36 m on the south side of the gap. After only two growing seasons, there was evidence that the asymmetric distribution of light and resources could contribute to Scots Pine and Norway Spruce becoming dominant in different parts of the gap.

Other mechanisms will also create refuges for early successional species at landscape and regional scales. Habitat fragmentation with urbanization and second home construction will increase edge habitat. Alien pests and pathogens that affect central hardwood forests, such as the emerald ash borer (*Agrilus planipennis*) and hemlock woolly adelgid (*Adelges tsugae*), will continue to create canopy openings. However, the relative homogeneity of stands ages in the Central Hardwood Region and current regeneration patterns in these forests suggest that early successional habitats will decline as these forests age. There are therefore concerns for particular management units, in terms of loss of heterogeneity and early successional habitats. Nonetheless, there are many processes that support the local regeneration of early successional species across this region. Unfortunately, data are not often collected at relevant scales to evaluate the net balance of these sets of processes.

3.5 Conclusion

The synchronous origin and narrow range of stand ages in the Central Hardwood Region will have implications for decades to come (see Shifley and Thompson, Chap. 6). Variation in the types and frequencies of natural disturbances creates a range of early succession and young forest species composition and structure; thus, scattered to connected patches of early successional habitats generated by natural disturbance are likely to be represented in the central hardwood forests of tomorrow. However, the narrow range of stand ages, reduced structural heterogeneity, current successional processes, and low frequency of disturbance at the local scale suggest loss of abundant early successional habitats, at least that generated by natural disturbance alone, at a scale relevant to conservation and management. We do not know if the frequency, patch size, and spatial distribution of natural disturbance-generated early successional habitats will be sufficient to sustain biological diversity (or for any other management goal). Additional research is needed on the scale-dependence of natural disturbance return intervals, the interactions among specific disturbance types, the impact of new invasive pests, and the potential influence of climate change on the frequency and intensity of natural disturbance events.

Appendix I: Base Maps of Natural Disturbances Within the Central Hardwood Region

The map of Hurricane Density within the Central Hardwood Region was derived from line coverage of historical North Atlantic tropical cyclone tracks, 1851–2000 (NOAA 2009). The Landslide map was based on a spatial index of landslide

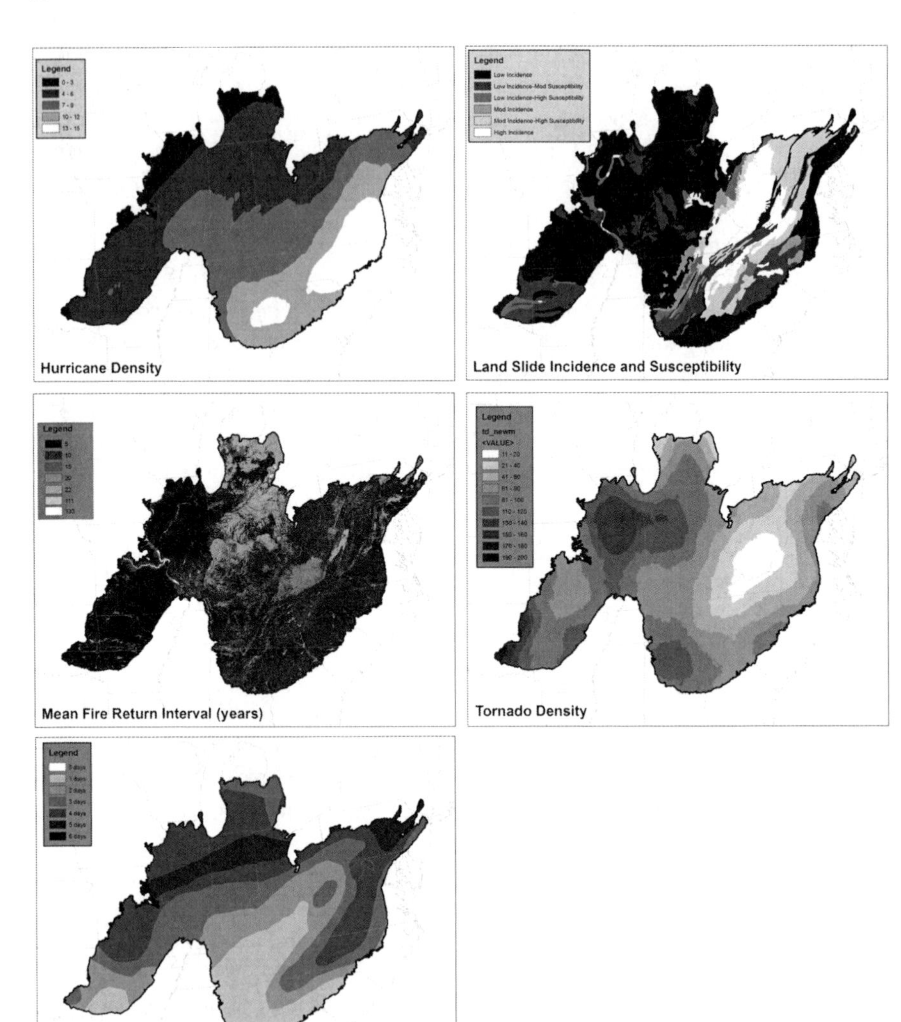

susceptibility and occurrence (Godt 1997). Raster digital data for Mean Fire Return Interval were obtained from LANDFIRE (US Forest Service 2006). Tornado density was calculated in ArcGIS using United States tornado touchdown points 1950–2004 (NWS 2005). The map of ice storm potential (Freezing Rain) was derived by geo-referencing Fig. 3.1 (a map of the annual number of days with freezing rain as defined by 988 weather stations from 1948 to 2000) from Changnon and Bigley (2005).

Literature Cited

Aikens ML, Ellum D, McKenna JJ, Kelty MJ, Ashton MS (2007) The effects of disturbance intensity on temporal and spatial patterns of herb colonization in a southern New England mixed-oak forest. For Ecol Manage 252:144–158

Anderson RC, Fralish JS, Baskin JM (eds) (1999) Savannas, barrens, and rock outcrop plant communities of North America. Cambridge University Press, Cambridge

Busing RT (2005) Tree mortality, canopy turnover, and woody detritus in old cove forests of the Southern Appalachians. Ecology 86:73–84

Busing RT, White RD, Harmon ME, White PS (2009) Hurricane disturbance in a temperate deciduous forest: patch dynamics, tree mortality, and coarse woody detritus. Plant Ecol 201(1): 351–363

Changnon D, Bigley R (2005) Fluctuations in US freezing rain days. Clim Change 69:229–244

Clinton BD, Baker CR (2000) Catastrophic windthrow in the southern Appalachians: characteristics of pits and mounds and initial vegetation responses. For Ecol Manage 126:51–60

Collins B, Battaglia LL (2002) Microenvironmental heterogeneity and *Quercus michauxii* regeneration in experimental gaps. For Ecol Manage 155:279–290

Cowell CM, Hoalst-Pullen N, Jackson MT (2010) The limited role of canopy gaps in the successional dynamics of a mature mixed Quercus forest remnant. J Veg Sci 21:201–212

de Chantal M, Leinonen K, Kuuluvainen T, Cescatti A (2003) Early response of *Pinus sylvestris* and *Picea abies* seedlings to an experimental canopy gap in a boreal spruce forest. For Ecol Manage 176:321–336

Elliott KJ, Hitchcock SL, Krueger L (2002) Vegetation response to large scale disturbance in a southern Appalachian forest: hurricane opal and salvage logging. J Torrey Bot Soc 129:48–59

Flinn KM, Marks PL (2007) Agricultural legacies in forest environments: tree communities, soil properties, and light availability. Ecol Appl 17:452–463

Fralish JS, McArdle TG (2009) Forest dynamics across three century-length disturbance regimes in the Illinois Ozark Hills. Am Midl Nat 162:418–449

Francescato V, Scotton M, Zarin DJ, Innes JC, Bryant DM (2001) Fifty years of natural revegetation on a landslide in Franconia Notch, New Hampshire, USA. Can J Bot 79:1477–1485

Frelich LE, Reich PB (1999) Neighborhood effects, disturbance severity, and community stability in forests. Ecosystems 2:151–166

Gandhi KJK, Herms DA (2010) Direct and indirect effects of alien insect herbivores on ecological processes and interactions in forests of eastern North America. Biol Invasions 12:389–405

Godt JW (1997) Digital representation of landslide overview map of the conterminous United States: US geological survey open-file report 97-289, scale 1:4,000,000. Available online at http://landslides.usgs.gov/html_files/landslides/nationalmap/national.html

Harrington TB, Bluhm AA (2001) Tree regeneration responses to microsite characteristics following a severe tornado in the Georgia Piedmont, USA. For Ecol Manage 140:265–275

Hart JL, Grissino-Mayer HD (2008) Vegetation patterns and dendroecology of a mixed hardwood forest on the Cumberland Plateau: implications for stand development. For Ecol Manage 255:1960–1975

Holladay C-A, Collins B, Kwit C (2006) Woody regeneration in and around aging southern bottomland hardwood forest gaps: effects of herbivory and gap size. For Ecol Manage 223:218–225

Honnay O, Verheyen K, Hermy M (2002) Permeability of ancient forest edges for weedy plant species invasion. For Ecol Manage 161:109–122

Klaus NA, Buehler DA, Saxton AM (2005) Forest management alternatives and songbird breeding habitat on the Cherokee National Forest, Tennessee. J Wildl Manage 69:222–234

Kupfer JA, Runkle JR (1996) Early gap successional pathways in a Fagus-Acer forest preserve: pattern and determinants. J Veg Sci 7:247–256

Lorimer CG (2001) Historical and ecological roles of disturbance in eastern North American forests: 9,000 years of change. Wildl Soc Bull 29:425–439

Matlack GR (1994) Vegetation dynamics of the forest edge – trends in space and successional time. J Ecol 82:113–123

Millward AA, Kraft CE (2004) Physical influences of landscape on a large-extent ecological disturbance: the northeastern North American ice storm of 1998. Landsc Ecol 19:99–111

Mou P, Warrillow MP (2000) Ice storm damage to a mixed hardwood forest and its impacts on forest regeneration in the ridge and valley region of southwestern Virginia. J Torrey Bot Soc 127:66–82

Myster RW, Fernandez DS (1995) Spatial gradients and patch structure on two Puerto Rican landslides. Biotropica 27:149–159

National Oceanic and Atmospheric Administration, National Hurricane Center (2009) Historical North Atlantic tropical cyclone tracks, 1851–2008. National Oceanic and Atmospheric Administration Coastal Services Center. Online_Linkage: http://maps.csc.noaa.gov/hurricanes/index.jsp

National Weather Service, Storm Prediction Center (2005) United States tornado touchdown points 1950–2004. National Atlas of the United States. Online_Linkage: http://nationalatlas.gov/atlasftp.html

Nowacki GJ, Abrams MD (1997) Radial-growth averaging criteria for reconstructing disturbance histories from presettlement-origin oaks. Ecol Monogr 67:225–249

Owen W (2002) The history of native plant communities in the south. In: Wear D, Greis J (eds) Southern forest resource assessment. Gen Tech Rep SRS-53, USDA Forest Service Southern Research Station, Asheville

Peterson CJ, Leach AD (2008) Limited salvage logging effects on forest regeneration after moderate-severity windthrow. Ecol Appl 18:407–420

Roberts MR (2004) Response of the herbaceous layer to natural disturbance in North American forests. Can J Bot 82:1273–1283

Roberts MR (2007) A conceptual model to characterize disturbance severity in forest harvests. For Ecol Manage 242:58–64

Romme WH, Everham EH, Frelich LE, Moritz MA, Sparks RE (1998) Are large, infrequent disturbances qualitatively different from small, frequent disturbances? Ecosystems 1:524–534

Runkle JR (1982) Patterns of disturbance in some old-growth mesic forests in eastern North America. Ecology 63:1533–1546

Schulte LA, Mladenoff DJ (2005) Severe wind and fire regimes in northern forests: historical variability at the regional scale. Ecology 86:431–445

Trani MK, Brooks RT, Schmidt TL, Victor RA, Gabbard CM (2001) Patterns and trends of early successional forest in the eastern United States. Wild Soc Bull 29:413–424

Turner MG, Baker WL, Peterson CJ, Peet RK (1998) Factors influencing succession: lessons from large, infrequent natural disturbances. Ecosystems 1:511–523

US Environmental Protection Agency (2009) Level III and IV ecoregions of the coterminous United States. Western Ecology Division, US EPA, Corvallis. http://www.epa.gov/wed/pages/ecoregions.htm

US Forest Service (2006) LANDFIRE. Mean fire return interval. USDA Forest Service, Missoula. http://landfire.cr.usgs.gov

Walker LR, Velazquez E, Shiels AB (2009) Applying lessons from ecological succession to the restoration of landslides. Plant Soil 324:157–168

White PS, Harrod J, Romme W, Betancourt J (1999) The role of disturbance and temporal dynamics. Chapter 2. In: Szaro RC, Johnson NC, Sexton WT, Malk AJ (eds) Ecological stewardship. Elsevier Science, Oxford, pp 281–312

Chapter 4
Fire in Eastern Hardwood Forests Through 14,000 Years

Martin A. Spetich, Roger W. Perry, Craig A. Harper, and Stacy L. Clark

Abstract Fire helped shape the structure and species composition of hardwood forests of the eastern United States over the past 14,000 years. Periodic fires were common in much of this area prior to European settlement, and fire-resilient species proliferated. Early European settlers commonly adopted Native American techniques of applying fire to the landscape. As the demand for wood products increased, large cutover areas were burned, sometimes leading to catastrophic fires and subsequent early successional habitats. By the early 1900s, these catastrophic fires resulted in political pressure leading to policies that severely restricted the use of fire. Fire suppression continued through the twentieth century due to an emphasis on commodity production and under-appreciation of the ecological role of fire. Without fire, fire-sensitive species were able to successfully outcompete fire-adapted species such as oak and pine while early successional habitats matured into older and more homogeneous forests. In the late twentieth century, land managers began reintroducing fire for ecosystem restoration, wildlife habitat improvement, hazardous fuel reduction, and forest regeneration. Responsible expanded use of prescribed

M.A. Spetich (✉) • R.W. Perry
USDA Forest Service, Southern Research Station, Arkansas Forestry Sciences Laboratory, P.O. Box 1270, Hot Springs, AR 71902, USA
e-mail: mspetich@fs.fed.us; rperry03@fs.fed.us

C.A. Harper
University of Tennessee, Ellington Plant Sciences, Room 280, 2431 Joe Johnson Drive, Knoxville, TN 37996–4563, USA
e-mail: charper@utk.edu

S.L. Clark
USDA Forest Service, Southern Research Station, University of Tennessee, Ellington Plant Science Bldg, Room 274, 2431 Joe Johnson Drive, Knoxville, TN 37996–4563, USA
e-mail: stacyclark@fs.fed.us

C.H. Greenberg et al. (eds.), *Sustaining Young Forest Communities*, Managing Forest Ecosystems 21, DOI 10.1007/978-94-007-1620-9_4,

fire and other management tools in the region could help mitigate past actions by increasing the amount and distribution of early successional habitats, plant and animal diversity, and landscape heterogeneity.

4.1 Introduction

We live in a changing environment. Climate, drought, fire, rainfall, ice storms, wind storms, insects, disease, and anthropogenic activities have caused changes in eastern hardwood forest species and structure for thousands of years. Of these, fire has been a major historical influence on forest composition, structure, and vegetation dynamics of the Central Hardwood Region (Lorimer 2001). Prior to European settlement, fire maintained the amount and distribution of early successional habitats such as young forest, savannah, and woodland (Ruffner 2006). At the time of Columbus' first contact, eastern forests were a shifting mosaic of woodlands, savannahs, forests, and prairies (Van Lear and Harlow 2002). The exclusion of fire over the past century has reduced the historic range, acreage, and distribution of these habitats as forest succession has taken place in the absence of fire (Dey and Guyette 2000).

Fire greatly influences biodiversity of landscapes by providing habitats for different plant and animal communities. Intense, stand-replacing fires with long intervals between burns can create ephemeral, early successional habitats dominated by shrubs, whereas more frequent burns (every 1–5 years) can maintain grasslands, woodlands, and shrubby habitats in an arrested state of succession. Different fire intensities and return intervals create different communities that each provide habitat for a diversity of species. For example, although 37% of forest birds in the Interior Highlands use forests as their primary breeding habitat, 8% use grasslands, 8% use savannahs or woodlands, and 21% use habitats consisting of shrubs and saplings (USDA Forest Service 1999).

In the last 50 years, the area of early successional habitats have remained relatively steady in the southeastern USA; however, these habitats have declined throughout other areas of the eastern USA, and are becoming scarce in areas such as the Great Lakes, Central Plains, and northeastern USA (Litvaitis 2001; Trani et al. 2001). Among communities that have declined by greater than 98% in the eastern USA, 55% are grassland, savannahs, or barrens and 24% are shrublands (Noss et al. 1995; Askins 2001; Thompson and DeGraaf 2001). Consequently, population numbers for many wildlife species associated with fire-dependent communities in the eastern USA have declined, including Ruffed Grouse (*Bonasa umbellus*), Grasshopper Sparrows (*Ammodramus savannarum*), Golden-winged Warblers (*Vermivora chrysoptera*), Prairie Warblers (*Dendroica discolor*), Brown Thrashers (*Toxostoma rufum*), and New England cottontails (*Sylvilagus transitionalis*) (Litvaitis 1993,2001; Thompson and Dessecker 1997; Hunter et al. 2001; Brennan and Kuvlesky 2005; Shifley and Thompson, Chap. 6). Shrub-nesting birds and species associated with shrubby, early successional habitats exhibit some of the steepest declines throughout portions of the eastern USA (Litvaitis 2001), and many eastern forest species dependent on fire-modified landscapes became endangered or extinct in the last 100 years (Hunter et al. 2001). Because of these declines, there is an increased interest in early successional forest habitats and the critical role fire plays in creating and maintaining these and other communities.

Thousands of years of fire resulted in the prevalence of fire-adapted tree species, such as oaks. Oaks are keystone species in the forests of the eastern USA (Fralish 2004), and acorns produced by oaks are an important food source for more than 100 wildlife species (Martin et al. 1951; Van Dersal 1940). After thousands of years of fire, oak-dominated forests represented 51% of eastern forests by 1993 (Spetich et al. 2002). Oaks have dominated these fire-mediated environments due to traits that make them resilient to fire, which gives them an advantage over many other species. They build large, belowground reserves of carbohydrates in their root systems which are protected from fire. When fire kills the aboveground shoot, belowground carbohydrate reserves are available for new shoots to resprout rapidly (Dey et al. 1996). Mature oaks also have relatively thick bark that helps protect them from damage from ground-level fires (Van Lear and Harlow 2002). Thus, oaks can outcompete less fire-adapted species.

Guyette et al. (2006b) identified five major factors controlling historic fire regimes – temporal, spatial, topographic, frequency of severe fires, and human activity. They found that drought and changes in human population density and culture were major influences on temporal differences in fire regimes of oak forest, and that human-caused fire is at least 200 times more frequent than lightning-caused fire in the eastern USA. Regional temperature, human population density, and topographic resistance to spread of fire (steep slopes, rivers, etc.) may be important to the spatial variability of fire regimes across the landscape (Guyette et al. 2006b). For example, as topographic roughness increases, the number of humans per km^2 necessary to reach the ignition saturation point increases. In the past, intense regional droughts that occurred in most of the eastern USA and Southern Ontario, Canada were strongly associated with severe fires (Guyette et. al. 2006a, b), and were highly influential during periods of numerous ignitions caused by humans. In fact, Guyette et al. (2006b) considered *Homo sapiens* a keystone species in many fire regimes, an idea particularly pertinent in eastern North America, where humans have been the primary cause of fires for thousands of years. Lightening fires account for relatively few fires annually (Shroeder and Buck 1970).

In this chapter, we divided fire history into three sections designated by the following time periods: (1) 14,000 years before present (BP) to 400 years BP, (2) 400 years BP to 1910 and (3) the past 100 years (1910–2010). These time periods are based on both the type of data available as evidence of fire and changes influencing fire regimes during each respective period. At the beginning of each section, we presented the types of data available as evidence of fire; we then synthesized evidence of fire for the respective time period. Our focus was on fire in hardwood forests throughout the eastern USA with emphasis on the Central Hardwood Region.

4.2 Fire 14,000 Years Before Present (BP) to 400 Years BP

Climate has a substantial effect on ecological communities, and the area occupied by eastern hardwood forests today has undergone dramatic changes over the past 14,000 years. Evidence of tree species and fire during this period is derived from pollen records, subfossil oaks, dendroecology and modeling, soil charcoal, and archaeological studies. Prior to 18,000 years BP, spruce (*Picea* spp.) and jack pine (*Pinus banksiana*) occupied most of what is now the eastern hardwoods region (Delcourt and Delcourt 1981). However, around 10,000 years BP, concurrent arrival of humans and a warmer, dryer climate allowed expansion of prairies, woodlands, and savannah into the eastern USA (Delcourt and Delcourt 1981). Most of the vegetation associations we see today were in place approximately 5,000–6,000 years ago, although structure of these forests may differ substantially (Van Lear and Harlow 2002). During sustained periods of drought, trees may succumb whereas prairie grasses, which are highly adapted to both fire and drought, may survive (e.g., Anderson 1990). In the last 5,000 years, changing climate throughout the region likely would have replaced many grasslands with forest if these areas had not been burned extensively by Native Americans (Anderson 1990).

Native Americans were an active and influential part of the landscape for more than 12,000 years (Delcourt et al. 1993). Delcourt et al. (1998) found evidence of a relationship between Native American use of fire and increases in fire adapted species such as pitch pine (*P. rigida*), oaks, black walnut (*Juglans nigra*), and American chestnut (*Castanea dentata*) after 3,000 BP. Others have noted Native American use of fire for agricultural clearing and driving game (DenUyl 1954; Campbell 1989; DeVivo 1990; Reich et al. 1990; Denevan 1992). Native Americans cleared land for agriculture in the southeast coastal plain by girdling and burning trees (Abrams 1992), a procedure that was likely practiced across much of the east. Fires set by Native Americans that were not contained by natural barriers burned surrounding habitats, and maintained fire-mediated early successional habitats.

Recent research has uncovered well-preserved trees termed subfossils. Subfossil oaks are ancient trees that were buried along rivers and creeks for thousands of years. These trees were probably buried during flood events that quickly covered the trees with sediment, thus preserving them. Recovery of these trees suggests oaks were present in the region 14,000 years BP based on ^{14}C dating (Guyette et al. 2004). Some of the earliest evidence of fire lies in soil charcoal.

A method which provides rough estimates of some past fire activity is the dating of soil charcoal. Soil charcoal in a mixed-hardwood forest of the Cumberland Plateau of Tennessee indicates that fire has occurred over the past 6,700 years or more, but only identified a minimum of five fire events during this time (Hart et al. 2008). In the Eastern Mesic Region (McNab, Chap. 2), Fesenmyer and Christensen (2010) examined soil charcoal in a North Carolina forest that included a gradient of mixed hardwood from yellow-poplar (*Liriodendron tupipifera*) and mesophytic oak species on mesic sites to chestnut oak (*Q. prinus*)-pitch pine forest on drier sites. Of 82 radiocarbon ages, only one (10,570 year BP) was older than 4,000 year BP. The rest of the samples dated from 0 to 4,000 year BP. They concluded that fire recurrently burned throughout the area over the past 4,000 years, and that fires regularly burned more frequently and over a broader landscape during the past 1,000 years than previously.

Recent models of mean fire interval based on temperature, human population, and precipitation for the past 14,000 years in the forest-prairie transition of northern Missouri indicate fire frequency intervals (mean number of years between fires) decreased from about once every 18 years during the Younger Dryas period to about 6 years today (Guyette, personal communication). The Younger Dryas period occurred 12,800–11,500 years ago when the climate was cold and wet, and there were relatively few humans in what is now the eastern USA. The main influences on this shortened fire return interval in recent times are increasing human population and increasing temperatures.

Native Americans in much of the eastern USA had a close relationship to forests and the wildlife within them. For example, 5,000–500 years ago in northern Arkansas, communities of at least 250 people were likely an important component of the ecosystem, especially where wood, water, and productive soils existed (Sabo et al. 2004). Oak was the most important species of fuel wood used by Native Americans during that period (Sabo et al. 2004). This was a time of Native American population growth in the area. During this period, Native Americans cultivated local plants like lambsquarters (*Chenopodium berlandieri*), little barley (*Hordeum pusillum*), and several types of squash; they also cultivated small quantities of corn (maize) (Sabo et al. 2004). Native Americans typically used fire to maintain agricultural areas (Abrams 1992; DeVivo 1990).

DeVivo (1990) hypothesized that prior to European contact, populations of Native Americans may have been much greater than previous estimates. He suggested that by the time Europeans began settling the interior, non-coastal areas of North America (nearly 200 years after Columbus landed on the continent), diseases transmitted over the previous 200 years dramatically reduced populations. Williams (1989) suggested that "in the late fifteenth century, the Western Hemisphere may have had a greater population than Western Europe."

Burning during this period provided habitats for a variety of wildlife species associated with open conditions maintained by fire. Frequent burning favors herbaceous vegetation such as grasses and forbs in the understory (Lewis and Harshbarger 1976, also see Greenberg et al., Chap. 8), and large herbivores such as bison (*Bos bison*) and elk (*Cervus canadensis*) were found throughout the Central Hardwood Region. Other species associated with fire-maintained habitats such as Heath Hens (*Tympanuchus cupido cupido*) and Greater Prairie Chickens (*Tympanuchus cupido*) occurred in the region but are now extirpated or extinct (e.g., Hunter et al. 2001). Species such as Bachman's Warbler (*Vermivora bachmanii*) (likely extinct) utilized other fire-maintained habitats such as canebreaks (Hunter et al. 2001).

4.3 Fire 400 Years BP to 1910

Knowledge of forests and fire during this period comes from dendrochronological methods, historic accounts, anecdotal journal entries, General Land Office (GLO) records, and census data. We examine the influence of Native American dynamics and the changes that occurred during and after Euro-American settlement.

By the 1600s, Native American populations likely had declined from diseases introduced by Europeans (Williams 1989), with depopulation occurring across most of North America (Dobyns 1983; Ramenofsky 1987). Native American populations continued to decline in states as far west as Arkansas, with large losses documented during the Native American Period (1680–1820). For example, an estimated 6,000 Quapaw lived near the convergence of the Arkansas, Mississippi, and White rivers prior to 1680, but were reduced to about 700 by 1763 (Baird 1980; Rollings 1995). Therefore, impacts of Native American fire recorded during this period may have been significantly less than those prior to the 1600s. Drought was also a significant factor in fire propagation during this time. For instance, in an Ozark fire history study, fire burned more than half of all sites in 1780, a major drought year in the Central Hardwood Region (Guyette and Spetich 2003). Other fire history studies have detected evidence of fire from the 1600s through the 1900s from Arkansas to Maryland (Shumway et al. 2001; Guyette et al. 2006a; Hutchinson et al. 2008).

Before Native Americans encountered Europeans, open forests were common. In the southern and central USA, oak and pine woodlands and savannahs existed that had developed through Native Americans using fires (Lorimer 2001). Forested communities were also more open than today, with tree densities about one-third of what they are in today's forests. Fires can decrease stand density by eliminating mid-story and understory woody species. If fires are severe enough, some overstory trees may also be eliminated, especially during drought years. General Land Office notes and land survey records for the Ozark Mountains of Arkansas from 1818 to the 1850s indicated tree densities ranged from 124 to 133 trees per ha, and oaks represented 70% of the survey record trees in the Boston Mountains (Foti 2004). Today, these forests are three times as dense, likely because of fire suppression that occurred after 1910 (Foti 2004). The historic low density of trees in the 1800s is further supported by Beilman and Brenner (1951) who described the forests as open woodlands with ground-level flora consisting of prairie species, also suggesting frequent fires. Before fire suppression occurred in oak forests of eastern North America, major fire years occurred approximately 3.6 times per century (Guyette et al. 2006b).

By the mid-1800s, Euro-Americans were well established across much of the eastern USA and had a significant impact on forest dynamics and conditions. European settlement included tree harvest for mine timbers, building materials, and charcoal, and land clearing for agriculture and grazing (Pearse 1876). Euro-Americans also adopted Native American practices of applying fire to the landscape. In the Boston Mountains of Arkansas, dendrochronological methods indicated return intervals of 11 years for the Native American Period (1680–1820), 2.7 years for the Euro-American Settlement period (1821–1880), 2.9 years for the Regional Development period (1881–1910), and more than 80 years for the period of Fire Suppression (1911–2000) (Guyette and Spetich 2003). Furthermore, historic fires were most frequent in the southern USA, decreasing in frequency as one moves northward (Guyette et. al 2010).

Vast areas of savannah and prairie were also maintained by fire. The tallgrass prairie region that extended from eastern North Dakota to central Texas to western Indiana once covered approximately 575,000 km^2 (Kuchler 1964). By the time settlers arrived on this landscape, oak savannahs had become a significant part of the tallgrass region, covering up to 13 million ha (Nuzzo 1986). However, these prairies and savannahs were only remnants of their former extent by the 1800s (Noss et al. 1995). After settlement, fire frequency and intensity decreased in the prairies and savannahs to a level favorable for oak establishment. Today, nearly all of what was once tallgrass prairie and savannah has been converted to farmland or forest (Abrams 1992), and only 0.02% of oak savannahs existing during European settlement remains (Dey and Guyette 2000).

Descriptions of frequent and widespread use of fire in the eastern USA are often found in historical accounts. For instance, Mudd's 1888 publication, "The History of Lincoln County, Missouri, from the earliest time to the present" states:

> Annually, after this rank growth of vegetation had become forested and dry, the Indians set fire to it and burned the entire surface of the country.

Other historical accounts in the region support the incidence of fire. For example, Brackenridge's Journal (1816) of a Voyage up the Missouri River in 1811 states that:

> …not withstanding the ravages of fire, the marks of which are everywhere to be seen, the woods, principally hickory, ash, and walnut formed a forest tolerably close

Most of those fires appeared to be of human origin. For instance, Guyette et al. (2006b) used a quantitative approach to estimate historic human-caused fires, determining that human-caused fires occurred at least 200 times as often as lightning-caused fires. This estimate was based on a conservative approximation for the number of humans per km^2 in the eastern USA at the time of first contact with Europeans, so actual frequency of fire at that time could have been higher. Lorimer (2001) also concluded most historic fires were caused by humans. However, only one dendrochronological study has quantitatively linked fluctuations of Native American populations with fire in North America. In northern Arkansas, the Cherokee population increased and decreased in unison with fire frequency (Fig. 4.1) (Guyette et al. 2006a). During drought years, fires burned large areas (26–58%) of Arkansas and Missouri Ozark sites at approximately 8.5 year intervals (Guyette et al. 2006a). In fact, extensive fire occurred more often during drought years in the 1700s, and fires likely achieved sizes unmatched during the century before.

A national census of fire in the USA for the year 1880 (Sargent 1884) further illustrates the impact of humans on the propagation of fire across the USA at that time. According to this census data, at least 98.5% of all fires in 1880 were human caused. Further, land clearing by farmers represented the greatest single source of

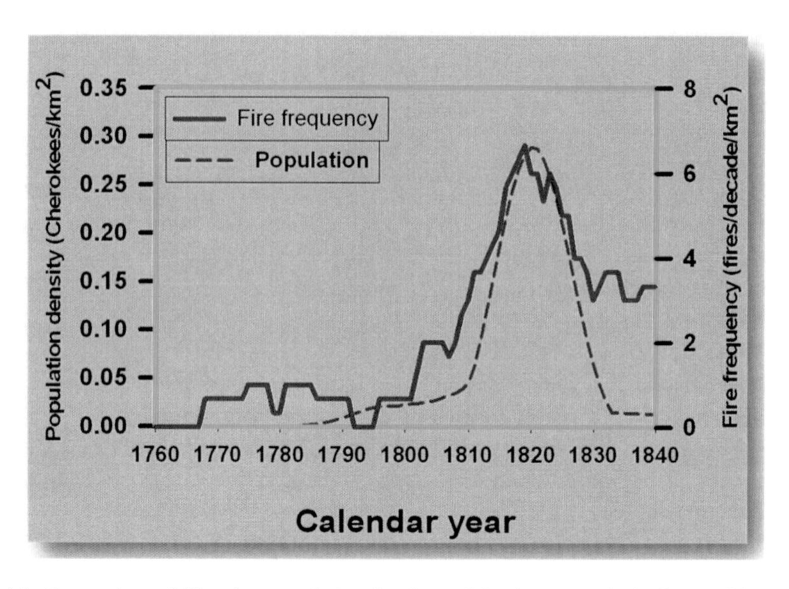

Fig. 4.1 Comparison of Cherokee population density and fire frequency in the Boston Mountains of Arkansas (figure adapted from data in Guyette et al. (2006a))

fire (38.6%) in 1880 because their brush fires often spread to surrounding forests. The second and third most important sources of fire were hunters (21%) and locomotives (17%). Fires caused by lightning represented only 1.1% of all fires, supporting the assertions by Guyette et al. (2006b) and Lorimer (2001) that lightning was of minor importance in fire propagation. Prairie fires, which could have been natural or human caused, represented 0.4% of all fires (Fig. 4.2). The second-most important damaging agent to forests in 1880 was browsing from domestic animals that were turned out into the forest to graze. Burning in these forests was done to improve forage in open-woods grazing (Sargent 1884). Grazing also likely further impacted forest structure and species dynamics.

Among states within the Central Hardwood Region, the states with the most acres burned in 1880 (Fig. 4.3) were Tennessee, Arkansas, and Missouri (Sargent 1884). At that time, western Tennessee was still largely forested, whereas the central area of Tennessee had been mostly cleared for farmland. In Arkansas, forests still covered the state with only isolated prairies north of the Arkansas River. However, Sargent (1884) noted the forests of Arkansas seemed to have relatively little damage from fires. The southeastern part of Missouri was still largely forested in 1880, but the southern and southwestern portions of the state, though once densely forested, were cut-over by 1880. By 1880, wooden barrel makers in Missouri were forced to obtain stock from Arkansas.

Lorimer (2001) reviewed disturbance in eastern North America and described differences in disturbance regimes among northern mesophytic hardwood forests and the oak-pine forests further south (also see White et al., Chap. 3). In northern hardwoods, catastrophic wind disturbance was the dominant disturbance regime; however, in central and southern USA oak-pine forests, fire was a dominant factor. Disturbances

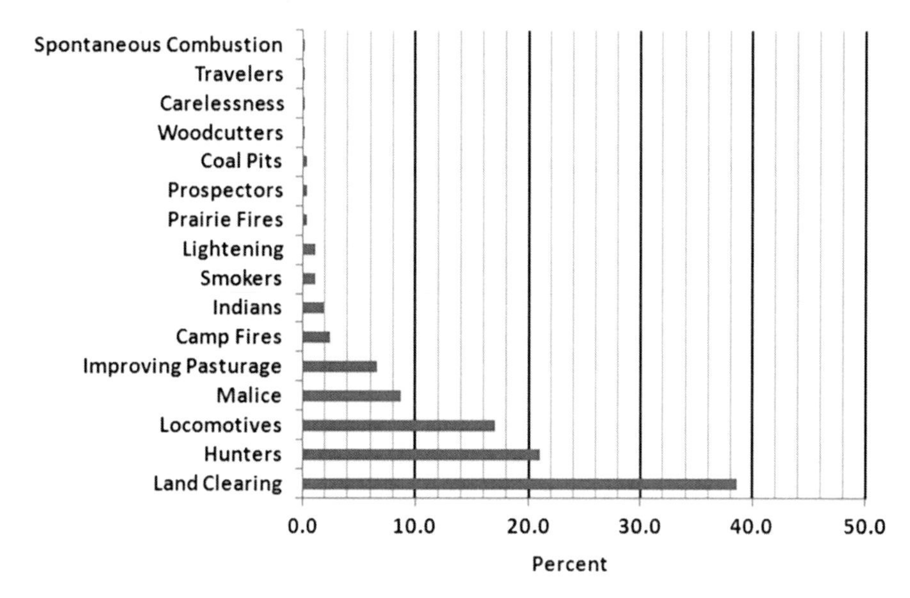

Fig. 4.2 The causes of fire in the USA in 1880 (data for this illustration from Sargent (1884))

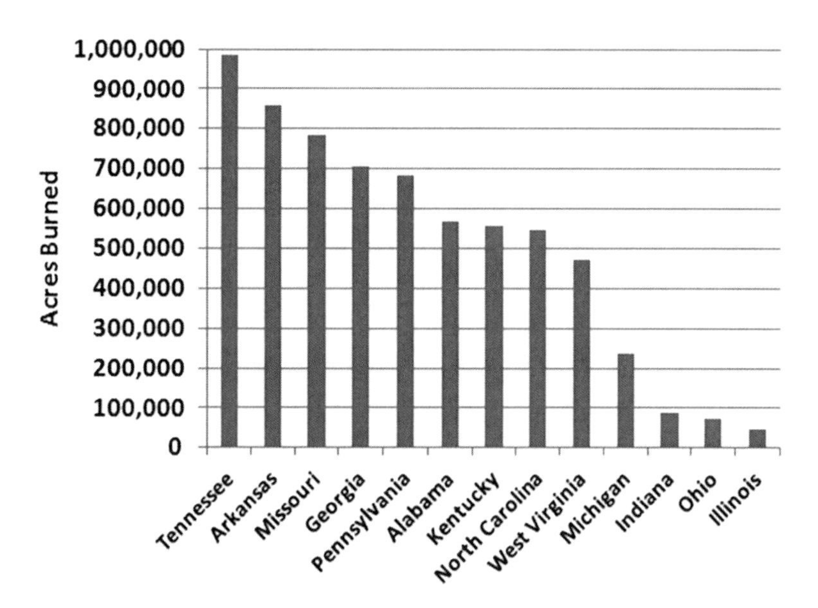

Fig. 4.3 Acres burned in 13 of the 14 states that fall within the Central Hardwood Region during 1880 (constructed from data in Sargent (1884). Note that Oklahoma did not return census data)

in northern forests resulted in relatively low amounts of early successional forest (less than 15 years old), which comprised less than 1–13% of the landscape. However, savannah and grassland occupied as much as 65% of the landscape in parts of the Midwest, where they were intermixed with oak woodlands and forest.

During this period wildlife was exploited by European settlers, and many species associated with fire-maintained communities were extirpated or became extinct. Timber harvest, land clearing, and wildfires created substantial areas of early successional habitats. From 1750 to 1880, forest cover in most of the eastern and Midwest USA declined to 10–30% of the landscape (Whitney 1994; Lorimer 2001). However, many of the remnant forests were young. For example, by 1885, over 75% of forest land in Massachusetts was less than 30 years old (Foster et al. 1998; Lorimer 2001). Overhunting of wildlife considered game in the 1800s reduced or eliminated many species, including Passenger Pigeons (*Ectopistes migratorius*) and Heath Hens. The last bison east of the Appalachian Mountains was killed in 1801 and native elk were extirpated from Pennsylvania by 1867 (Matthiessen 1987). Most bison and elk were gone from the Ozarks by the 1830s and 1840s (Schwartz and Schwartz 1981; Sealander and Heidt 1990). However, many nongame species associated with early successional habitats likely increased due to expanding habitat and reductions of predators such as bobcat (*Lynx rufus*), red wolf (*Canis rufus*), and mountain lion (*Puma concolor*). For example, naturalists in the late 1800s and early 1900s described dramatic population increases of several warblers associated with early successional habitats in response to increased habitat availability (Morse 1989; Litvaitis 1993).

As the country's demand for wood products increased, large areas were harvested, leaving miles of logging slash as potential fuel for severe fires to spread unchecked. Notable fires that occurred in the eastern USA during this time period included a fire in 1871 that burned 45,000 ha in Wisconsin and Michigan. This fire was actually a series of simultaneous forest fires fueled in part by large areas of logging slash. Ten years later, a similar fire consumed 40,000 ha in Thumb, Michigan. Thirteen years later, a fire consumed more than 1,000 km^2 around Hinckley, Minnesota. These fires, combined with fires that occurred in the early 1900s, helped prompt the development and implementation of policies on fire suppression.

4.4 The Past 100 Years (1910–2010)

Knowledge of fire, or lack of it, during this time period is derived from written records, research studies, and quantitative research information. Notable fires in the eastern USA just before and during this period included a fire in 1903 that burned more than 24,000 ha in the Adirondack Park of New York. In 1918, the Cloquet fire in northern Minnesota consumed 10,000 ha, killed 453 people, injured or displaced 52,000 people, and destroyed 38 communities. Forest fires also destroyed more than 7,000 ha of woodland during Maine's "Great Fires of 1947."

These eastern fires were important in influencing the suppression policies of the past century because they occurred in highly populated areas of the United States. However, fires in the western USA also had substantial influence on policy. For instance, the "Big Blowup of 1910," was a western fire that burned more than 2 million ha of forest in Washington, Idaho, and Montana. Eighty-seven people, including 78 fire fighters, died during this fire. The political impact of these and other major fires prompted the US Forest Service to make protecting the forest from such fires its mission (Egan 2009; Pyne 2001). This mission was later exemplified in the story and image of Smoky Bear.

As the forestry profession began to emerge in the late 19th and early 20th centuries, foresters educated and trained in Europe (particularly Germany) and the northeastern USA did not understand that certain local ecosystems depended on fire, and thus disdained the practice of woods burning (Donovan and Brown 2007). Although some foresters privately admitted that fire could be a useful management tool (Carle 2002), the US Forest Service worried that acknowledging the useful role of fire would be confusing to the public (Donovan and Brown 2007). By the 1920–1930s, an all-out campaign was waged by the federal government to curtail the use of fire across the landscape. State forestry agencies received funding through the Clarke-McNary Act of 1924, but these funds were withheld from states if they tolerated forest burning (Johnson and Hale 2002). Both the American Forestry Association and the US Forest Service distributed literature and provided movies discouraging the practice of woods burning (Bass 1981; Johnson and Hale 2002). However, some early professionals including foresters, botanists, and wildlife professionals had already established the importance of burning (e.g., Harper 1962; Schiff 1962). For example, Stoddard (1931) indicated that controlled burning was necessary to retain plentiful Bobwhite Quail (*Colinus virginianus*). Despite these policies, fire suppression wasn't effective or well accepted across many parts of the Midwest until the 1950s.

Due to these extensive fire-suppression efforts, between the 1920s and the 1940s, some areas of forest began to transform from open woodland conditions maintained by fire, harvesting, and grazing (Fig. 4.4) to more closed conditions (Fig. 4.5). One result of fire suppression was a gradual shift in composition of tree species, and oaks were replaced by other species (Abrams 1992). In what was once the tallgrass prairie region of Missouri, white oaks (*Q. alba*) are being replaced by sugar maples (*Acer saccharum*) (Pallardy et al. 1988), and in an Indiana old-growth forest, there has been a dramatic shift from oaks to sugar maples since 1926 (Spetich and Parker 1998). Spetich and Parker (1998) found biomass of regenerating oak was 14% and sugar maple was 12% of stand totals in 1926, but oaks represented only 1% of the regenerating biomass and sugar maple comprised 43% by 1992. Similar shifts in species composition of midwestern old-growth forests have been reported (e.g., Johnson et al. 1973; Cho and Boerner 1991; Shotola et al. 1992). Mortality in an Appalachian old-growth forest shifted species dominance from oak and hickory (*Carya* spp.) to forests dominated by sugar maple and yellow-poplar (McGee 1984).

During this period, forests of the eastern USA were also heavily impacted by timber harvest, and large areas of cutover forest existed by the 1920–1930s. Many of these areas, which were early successional forests at that time, are now at the complex or mixed stage (Johnson et al. 2002). Numerous other areas across the eastern USA are also currently undergoing shifts in species composition (Lorimer 2001).

Shifts in species composition related to fire exclusion occurred concurrently with the loss of another dominant tree species in the eastern USA, the American chestnut. The American chestnut was an important component of the eastern hardwood forest for thousands of years until a fungus (*Cryphonectria parasitica*) was introduced in

Fig. 4.4 Photograph illustrating the open conditions typical prior to fire suppression (photograph from John McGuire 1941 MS thesis on "The Sylamore Experimental Forest" Note: John McGuire later became Chief of the US Forest Service)

the early 1900s (Delcourt and Delcourt 1997; Dane et al. 2003). The increase of deadwood due to chestnut blight mortality was feared to create a fire hazard, spurring rapid removal of dead or dying trees, particularly in the southern states (Zeigler 1920). Change in fire regimes, coupled with other major land disturbances such as loss of American chestnut, land-clearing for agriculture, grazing by domesticated animals, and large-scale logging created an ideal habitat for further recruitment of disturbance dependent red oak (Abrams 2006).

Because of the heterogeneous nature of historic fires across landscapes, eastern forests were likely a mosaic of frequently burned and rarely burned areas, and species that were not fire-adapted, such as maples (*Acer* spp.), were likely retained throughout landscapes. How historic fire affected American chestnuts is unclear. American chestnuts may have benefited from fire, which reduced canopy cover and competition. One early account suggests that frequent fires may have been unfavorable for chestnuts (Russell 1987). However, more recent research suggests that fire may be required for optimal conditions for chestnut growth (McCament and McCarthy 2005).

The American chestnut shares similar regeneration and growth strategies with oaks, such as prolific sprouting capabilities (Matoon 1909; Paillet 1984, 1988), and chestnuts tended to sprout vigorously after cutting or burning (Russell 1987). Chestnut was frequently grown from coppice to provide timber because of its ability to resprout, and its dominance in the early 1900s in many areas may have been a result

Fig. 4.5 Photograph illustrating the typical closed vegetation conditions that occurred after fire suppression early in the 1900s (photograph from John McGuire 1941 MS thesis on "The Sylamore Experimental Forest")

of cutting in the eighteenth and nineteenth centuries (Russell 1987). Consequently, low-intensity fire likely contributed to its spread. However, chestnuts rarely reproduced from seed (Pinchot 1905). Reproduction and subsequent recruitment through seed regeneration was probably limited by frequent fire (Graves 1905). Once established, chestnut likely responded similarly to northern red oak (*Q. rubra*). It could grow quickly when exposed to canopy openings, but could also survive in a shaded understory for a number of years (Paillet 2002; McCament and McCarthy 2005).

Little information exists on how chestnuts responded to fire after planting, but historical and recent field and laboratory experiments indicate chestnut is a strong competitor in disturbed stands (Jacobs 2007). Additionally, a positive relationship was found between litter depth and presence of blight cankers on trees in Ontario (Tindall et al. 2004), suggesting that fire, which reduces litter depth, may help reduce infection from virulent strains of chestnut blight.

Changes in upland hardwood forests from fire suppression are likely altering species dynamics in eastern forests of the USA. On medium- to high-quality sites, successful regeneration and recruitment of oaks into forest overstories is lacking. Oak reproduction often fails to attain dominance under current disturbance regimes that lack fire; current regimes favor shade-tolerant species or species with faster juvenile growth. This has produced oak-dominated forests with natural oak regeneration and recruitment insufficient to sustain current levels of oak stocking. For instance, a review of historical data over the past century found that old-growth forests have a successional tendency toward shade tolerant or fire-sensitive species such as red (*A. rubrum*) and sugar maple (McCune and Mengis 1986).

Land managers, foresters, and wildlife biologists in the latter part of the twentieth century began to implement burning into management plans for multiple reasons, including reducing hazardous fuel loads, regenerating target tree species, and improving wildlife habitat. In the 1940s, controlled burning as a management tool was accepted in the southeastern USA, where fire-maintained pine woodlands were prevalent (Johnson and Hale 2002). As early as the 1950s, the US Forest Service was burning more than 100,000 ha, and in the 1960s they expanded controlled burning outside the longleaf pine (*P. palustris*) communities (a fire-dependent ecosystem) in the southeastern USA (Reibold 1971; Johnson and Hale 2002). Policies against burning by the National Park Service were reversed in 1967 to allow burning for ecosystem restoration and maintenance (e.g., Hendrickson 1972). The Endangered Species Act was passed in 1973, and one of the first species deemed endangered was the Red-cockaded Woodpecker (*Picoides borealis*), which was associated with fire-maintained pine woodlands of the southeastern USA. By the 1970s, managing habitats with fire for non-game wildlife became more widespread (Johnson and Hale 2002). The importance of fire to many ecosystems across the eastern USA is now recognized, and land managers are increasingly utilizing fire for ecosystem restoration and improved forest health throughout the region.

4.5 Conclusions

From 12,000 B.P. to 400 B.P., there is general agreement that Native Americans had a major influence on the vegetation of eastern North America through the use of fire. From 400 B.P. to 1910, fire frequency was relatively high throughout most of eastern North America. Catastrophic fires throughout the USA in the late 1800s and early 1900s led to fire suppression policies in USA forests. Fire suppression has contributed to shifts in species, structure, and density and the reduction of early successional habitats in many eastern forests over the past century. Science-based management of forest ecosystems in the latter part of the twentieth century began reintroducing fire to forest communities. It appears that prescribed fire will be needed to restore fire-dependent systems such as oak, oak/pine, woodlands, and savannahs; and to restore certain species such as the American chestnut.

Acknowledgement Thanks to Tod Hutchinson, Frank Thompson, and one anonymous reviewer for reviewing this manuscript and to Betsy L. Spetich for editorial guidance.

Literature Cited

Abrams MD (1992) Fire and the development of oak forests. Bioscience 42:346–353
Abrams MD (2006) Ecological and ecophysiological attributes and responses to fire in eastern oak forests. In: Dickinson MB (ed) Fire in eastern oak forests: delivering science to land managers. Gen Tech Rep NRS-P-1, USDA Forest Service Northern Research Station, Newtown Square
Anderson RC (1990) The historic role of fire in the North American grassland. In: Collins SL, Wallace LL (eds) Fire in North American tallgrass prairies. University of Oklahoma Press, Norman
Askins RA (2001) Sustaining biological diversity in early successional communities: the challenge of managing unpopular habitat. Wildl Soc Bull 29:407–412
Baird WD (1980) The Quapaw Indians: a history of the downstream people. University of Oklahoma Press, Norman
Bass SMW (1981) For the trees: an illustrated history of the Ozark-St. Francis national forests 1908–1978. USDA Forest Service, Southern Region, Atlanta
Beilman AP, Brenner LG (1951) The recent intrusion of forests in the Ozarks. Ann Missouri Bot Garden 38, St. Louis
Brackenridge HM (1816) Journal of a voyage up the river Missouri performed in eighteen hundred and eleven, 2nd edition. Reprinted in Thwaites RR (1904). Early western travels 1748–1846. vol VI. AH Clark, Cleveland
Brennan LA, Kulvesky WP (2005) North American grassland birds: an unfolding conservation crisis. J Wildl Manage 69:1–13
Campbell JJN (1989) Historical evidence of forest composition in the bluegrass region of Kentucky. In: Proceedings of the seventh central hardwood forest conference. Gen Tech Rep NC-132, USDA Forest Service North Central Forest Experiment Station, St. Paul
Carle D (2002) Burning questions: America's fight with nature's fire. Praeger, Westport
Cho DS, Boerner REJ (1991) Canopy disturbance patterns and regeneration of *Quercus* species in two Ohio old-growth forests. Vegetatio 93:9–18
Dane F, Lang P, Huang H, Fu Y (2003) Intercontinental genetic divergence of *Castanea* species in eastern Asia and eastern North America. Heredity 91:314–321
Delcourt PA, Delcourt HR (1981) Vegetation maps for eastern North America: 40,000 yr BP to the present. In: Romans RC (ed) Geobotany, vol II. Plenum, New York
Delcourt PA, Delcourt HR (1997) The influence of prehistoric human-set fires on oak-chestnut forests in the Southern Appalachians. Castanea 63:337–345
Delcourt PA, Delcourt HR, Morse DF, Morse PA (1993) History, evolution, and organization of vegetation and human culture. In: Martin WH, Boyce SG, Echternact AC (eds) Biodiversity of the southeastern United States: lowland terrestrial communities. Wiley, New York
Delcourt PA, Delcourt HR, Ison CR, Sharp WE, Gremillion KG (1998) Prehistoric human use of fire, the eastern agricultural complex, and Appalachian oak-chestnut forests: paleoecology of Cliff Palace pond, Kentucky. Am Antiquity 63:263–278
Denevan WM (1992) The pristine myth: the landscape of the Americas in 1492. Ann Assoc Am Geogr 82:369–385

DenUyl D (1954) Indiana's old-growth forests. Proc Ind Acad Sci 63:73–79
DeVivo MS (1990) Indian use of fire and land clearance in the southern Appalachians. In: Fire and the environment: ecological and cultural perspectives. Gen Tech Rep SE-69, USDA Forest Service Southeastern Forest Experiment Station, Asheville
Dey DC, Guyette RP (2000) Sustaining oak ecosystems in the Central Hardwood Region: lessons from the past–continuing the history of disturbance. In: McCabe RE, Loos SE (eds) Transactions of the 65th North American wildlife and natural resources conference, Wildlife Management Institute, Washington, DC
Dey DC, Johnson PS, Garrett HE (1996) Modelling the regeneration of oak stands in the Missouri Ozark Highlands. Can J For Res 26:573–583
Dobyns HF (1983) Their number become thinned: native American population dynamics in eastern North America. University of Tennessee Press, Knoxville
Donovan GH, Brown TC (2007) Be careful what you wish for: the legacy of smokey bear. Front Ecol Environ 5:73–79
Egan T (2009) The big burn: Teddy Roosevelt and the fire that saved America. Houghton Mifflin Harcourt, Boston
Fesenmyer KA, Christensen NL Jr (2010) Reconstructing holocene fire history in a southern Appalachian forest using soil charcoal. Ecology 91:662–670
Foster DR, Motzkin G, Slater B (1998) Land-use history as long-term broad-scale disturbance: regional forest dynamics in central New England. Ecosystems 1:96–119
Foti TL (2004) Upland hardwood forest and related communities of the Ozarks in the early 19th century. In: Spetich MA (ed) Upland oak ecology symposium: history, current conditions, and sustainability. Gen Tech Rep SRS-73, USDA Forest Service Southern Research Station, Asheville
Fralish JS (2004) The keystone role of oak and hickory in the central hardwood forest. In: Spetich MA (ed) Proceedings of the upland oak ecology symposium: history, current conditions, and sustainability. Gen Tech Rep SRS-73, USDA Forest Service Southern Research Station, Asheville
Graves HS (1905) Notes on the rate of growth of red cedar, red oak, and chestnut. For Q 3:349–353
Guyette RP, Spetich MA (2003) Fire history of oak-pine forests in the lower Boston mountains, Arkansas, USA. For Ecol Manage 180:463–474
Guyette RP, Stambaugh MC, Dey DC (2004) Ancient oak climate proxies from the agricultural heartland. Eos 85:483–483
Guyette RP, Spetich MA, Stambaugh MC (2006a) Historic fire regime dynamics and forcing factors in the Boston mountains, Arkansas, USA. For Ecol Manage 234:293–303
Guyette RP, Dey DC, Stambaugh MC, Muzika R (2006b) Fire scars reveal variability and dynamics of eastern fire regimes. In: Dickinson MB (ed) Proceedings of the fire in eastern oak forests: delivering science to land managers. Gen Tech Rep NRS-P-1, USDA Forest Service Northern Research Station, Newtown Square
Guyette RP, Stambaugh MC, Dey DC (2010) Developing and using fire scar histories in the Southern and Eastern United States. Project # 06-3-1-16. Joint Fire Science Program Final Report, web site: www.firescience.gov
Harper RM (1962) Historical notes on the relation of fires to forests. In: Komareb EV (ed) Proceedings of the first tall timbers fire ecology conference, Tall Timbers Research Station, Tallahassee, pp 43–52
Hart JL, Horn SP, Grissino-Mayer HD (2008) Fire history from soil charcoal in a mixed hardwood forest on the Cumberland Plateau, Tennessee, USA. J Torrey Bot Soc 135:401–410
Hendrickson WH (1972) Fire in the National Parks symposium. In: Komareb EV (ed) Proceedings of the tall timbers fire ecology conference. A quest for ecological understanding. Tall Timbers Research Station, Tallahassee, pp 339–343
Hunter WC, Buehler DA, Canterbury RA, Confer JL, Hamel PB (2001) Conservation of disturbance-dependent birds in eastern North America. Wildl Soc Bull 29:440–455
Hutchinson TF, Long RP, Ford RD, Sutherland EK (2008) Fire history and the establishment of oaks and maples in second-growth forests. Can J For Res 38:1184–1198
Jacobs DF (2007) Toward development of silvical strategies for forest restoration of American chestnut (*Castanea dentata*) using blight-resistant hybrids. Biol Conserv 137:497–506

Johnson AS, Hale PE (2002) The historical foundations of prescribed burning for wildlife: a Southern perspective. In: Ford WM, Russell KR, Moorman CE (eds) Proceedings of the role of fire in nongame wildlife management and community restoration: traditional uses and new directions. Gen Tech Rep NE-288, USDA Forest Service Northeastern Research Station, Newton Square
Johnson HS, Berkebile JS, Parker GR (1973) An ecological inventory of Bryan nature preserve. Proc Ind Acad Sci 83:167–172
Kuchler AW (1964) Potential natural vegetation of the conterminous United States. Am Geogr Soc Spec Pub 36, New York
Lewis CE, Harshbarger TJ (1976) Shrub and herbaceous vegetation after 20 years of prescribed burning on the South Carolina coastal plain. J Range Manage 29:13–18
Litvaitis JA (1993) Response of early successional vertebrates to historic changes in land use. Conserv Biol 7:866–873
Litvaitis JA (2001) Importance of early successional habitats to mammals in eastern forests. Wildl Soc Bull 29:466–473
Lorimer CG (2001) Historical and ecological roles of disturbance in eastern North American forests: 9,000 years of change. Wildl Soc Bull 29:425–493
Martin AC, Zim HC, Nelson AL (1951) American wildlife and plants. Dover Publishers, New York
Matoon WR (1909) The origin and early development of sprouts. For Q 7:34–47
Matthiessen P (1987) Wildlife in America. Viking, New York
McCament CL, McCarthy BC (2005) Two-year response of American chestnut (*Castanea dentata*) seedlings to shelterwood harvesting and fire in a mixed-oak forest ecosystem. Can J For Res 35:740–749
McCune BC, Menges ES (1986) Quality of historical data on midwestern old-growth forests. Am Midl Nat 116:163–172
McGee CE (1984) Heavy mortality & succession in a virgin mixed mesophytic forest. Res Paper SO-209, USDA Forest Service Southern Research Station, New Orleans
McGuire JR (1941) The Sylamore experimental forest. Thesis, Yale School of Forestry, New Haven
Morse DH (1989) American warblers. Harvard University Press, Cambridge
Mudd J (1888) History of Lincoln county, Missouri, from the earliest time to the present. Goodspeed Publishing Co, Chicago
Noss RF, LaRoe ET III, Scott JM (1995) Endangered ecosystems of the United States: a preliminary assessment of loss and degradation. Biol Rep 28, USDI National Biological Service, Washington, DC
Nuzzo VA (1986) Extent and status of the mid-west oak savanna: presettlement and 1985. Nat Areas J 6:6–36
Paillet FL (1984) Growth-form and ecology of American chestnut sprout clones in northeastern Massachusetts. Bull Torrey Bot Club 111:316–328
Paillet FL (1988) Character and distribution of American chestnut sprouts in southern New England woodlands. Bull Torrey Bot Club 115:32–44
Paillet FL (2002) Chestnut: history and ecology of a transformed species. J Biogeogr 29:1517–1530
Pallardy SG, Nigh TA, Garrett HE (1988) Changes in forest composition in central Missouri: 1968–1982. Am Midl Nat 120:380–390
Pearse JB (1876) A concise history of the iron manufacture of the American colonies up to the revolution, and Pennsylvania until the present time. Allen, Lane, and Scott, Philadelphia
Pinchot G (1905) A primer of forestry. Dep Bull 24, USDA Bureau of Forestry, Washington, DC
Pyne S (2001) Year of the fires: the story of the great fires of 1910. Penguin Putnam Inc, New York
Ramenofsky AF (1987) Vectors of death: the archaeology of European contact. University of New Mexico Press, Albuquerque
Reibold RJ (1971) The early history of wildfires and prescribed burning. In: Proceedings of the prescribed burning symposium. USDA Forest Service, Southeastern Forest Experiment Station, Asheville, 11–20
Reich PB, Abrams MD, Ellsworth DS, Kruger EL, Tabone TJ (1990) Fire affects ecophysiology and community dynamics of central Wisconsin oak forest regeneration. Ecology 71:2179–2190

Rollings WH (1995) Living in a graveyard: native Americans in colonial Arkansas. In: Whayne J (comp), Cultural encounters in the early south. University of Arkansas Press, Fayetteville
Ruffner CM (2006) Understanding of the evidence for historical fire across eastern forests. In: Dickinson MB (ed) Proceedings of the conference Fire in eastern oak forests: delivering science to land managers. Gen Tech Rep NRS-P-1, USDA Forest Service Northern Research Station, Newton Square
Russell EWB (1987) Pre-blight distribution of *Castanea dentate* (Marsh.) Borkh. Bull Torrey Bot Club 114:183–190
Sabo G III, Lockhart JJ, Hilliard JE (2004) The forest as a resource: from prehistory to history in the Arkansas Ozarks. Gen Tech Rep SRS-73, USDA Forest Service Southern Research Station, Asheville
Sargent CS (1884) Report on the forests of North America (exclusive of Mexico). GPO, Washington, DC
Schiff AL (1962) Fire and water: scientific heresy in the forest service. Harvard University Press, Cambridge
Schroeder MJ, Buck CJ (1970) Fire weather. Agri Handbook 360, USDA Forest Service, Washington, DC
Schwartz CW, Schwartz ER (1981) The wild mammals of Missouri, revised edition. University of Missouri Press, Columbia
Sealander JA, Heidt GA (1990) Arkansas mammals: their natural history, classification, and distribution. University of Arkansas Press, Fayetteville
Shotola SJ, Weaver GT, Robertson PA, Ashby WC (1992) Sugar maple invasion of an old-growth oak-hickory forest in southwestern Illinois. Am Midl Nat 127:125–138
Shumway DL, Abrams MD, Ruffner CM (2001) A 400 year history of fire and oak recruitment in an old-growth oak forest in western Maryland, USA. Can J For Res 31:1437
Spetich MA, Parker GR (1998) Distribution of biomass in an Indiana old-growth forest from 1926 to 1992. Am Midl Nat 139:90–107
Spetich MA, Dey DC, Johnson PS, Graney DL (2002) Competitive capacity of *Quercus rubra* L. planted in Arkansas' Boston mountains. For Sci 48:504–517
Stoddard HL (1931) The bobwhite quail; its habits, preservation and increase. Scribner's Sons, New York
Thompson FR III, Degraaf RM (2001) Conservation approaches for woody, early successional communities in the eastern United States. Wildl Soc Bull 29:483–494
Thompson FR III, Dessecker DR (1997) Management of early-successional communities in central hardwood forests with special emphasis on the ecology and management of oaks, ruffed grouse, and forest songbirds. Gen Tech Rep NC-195, USDA Forest Service North Central Research Station, St. Paul
Tindall JR, Gerrath JA, Melzer M, McKendry K, Husband BC, Boland GJ (2004) Ecological status of American chestnut (*Castanea dentata*) in its native range in Canada. Can J For Res 34: 2554–2563
Trani MK, Brooks RT, Schmidt TL, Rudis VA, Gabbard CM (2001) Patterns and trends of early successional forests in the eastern United States. Wildl Soc Bull 29:413–424
USDA Forest Service (1999) Ozark-Ouachita Highlands assessment: terrestrial vegetation and wildlife. Gen Tech Rep SRS-35, USDA Forest Service Southern Research Station, Asheville
Van Dersal WR (1940) Utilization of oaks by birds and mammals. J Wildl Manage 4:404–428
Van Lear DH, Harlow RF (2002) Fire in the eastern United States: influence on wildlife habitat. In: Ford WM, Russell KR, Moorman CE (eds) Proceedings of the role of fire in nongame wildlife management and community restoration: traditional uses and new directions. Gen Tech Rep NE-288, USDA Forest Service Northeastern Research Station, Newton Square
Whitney GG (1994) From coastal wilderness to fruited plain: a history of environmental change in temperate North America 1500 to present. Cambridge University Press, United Kingdom
Williams M (1989) Americans and their forests: a historical geography. Cambridge University Press, United Kingdom
Zeigler EA (1920) Problems arising from the loss of our chestnut. For Leaves 17:152–155

Chapter 5
Structure and Species Composition of Upland Hardwood Communities After Regeneration Treatments Across Environmental Gradients

David L. Loftis, Callie J. Schweitzer, and Tara L. Keyser

Abstract Early successional habitats can be created with a broad array of silvicultural techniques that remove all or most canopy trees in one to several cuttings and small to large patch sizes. Composition and early structural development of the resulting vegetation can be variable. Arborescent species composition is a function of regeneration sources already present and those that arrive during or after the cutting. The suite of species available for regeneration of a site, large or small, is a cumulative effect of disturbances and varies across multiple environmental gradients that include moisture, elevation (temperature), and soil chemistry.

5.1 Introduction

Regeneration activities in managed forests create forest stages with characteristics of early successional habitats such as low dense vegetation and openness overhead (Greenberg et al., Chap. 1). Where management is used to achieve multiple objectives (e.g., wood products and both young, open habitats and older, closed-canopy habitats for a range of wildlife species), regenerated areas provide transient early successional habitats during the early years of stand development. Oliver and Larson (1990) proposed a classification of stages of stand development that applies well to arborescent

D.L. Loftis (✉) • T.L. Keyser
USDA Forest Service, Southern Research Station, Bent Creek Experimental Forest, 1577 Brevard Road, Asheville, NC 28806, USA
e-mail: davidloftis@bellsouth.net; tkeyser@fs.fed.us

C.J. Schweitzer
USDA Forest Service, Southern Research Station, Bent Creek Experimental Forest, P.O. Box 1568, Normal, AL 35762, USA
e-mail: cschweitzer@fs.fed.us

C.H. Greenberg et al. (eds.), *Sustaining Young Forest Communities*, Managing Forest Ecosystems 21, DOI 10.1007/978-94-007-1620-9_5,

vegetation development following regeneration cuts that remove all or most of the overstory in one cut or several cuts within a decade or two. The stand initiation stage is the period of the early growth of individuals that existed in the understory prior to cutting (advance reproduction), stump and root sprouts from harvested trees, and seedlings that become established after cutting. Within a few years, growth of regenerating individuals and expansion of their crowns results in canopy closure, after which establishment of new seedlings ceases and many stems in a subordinate crown position die. This process marks the beginning of the stem exclusion stage, and typically occurs at about 10 years after cutting in hardwood forests of the Central Hardwood Region (Johnson et al. 2009; Nyland 1996). The stem exclusion stage can last many years and is followed by the understory reinitiation stage in which herbaceous, shrubs and tree regeneration become more prominent in the understory. The old-growth or complex stage (Johnson et al. 2009) follows, characterized by interspersion of the closed canopy environment, with canopy gaps created by overstory mortality. Where gaps are large enough to prevent crown closure by lateral crown growth of surrounding trees, the stand initiation stage may be triggered at the gap scale (Peet 1992). In hardwood forests of the Central Hardwood Region, early succesional habitats are provided in the stand initiation stage and the first few years of the stem exclusion stage, generally up to 15 years after the cutting that initiates regeneration.

5.2 Regeneration Methods

Tree harvests that replace an existing stand of trees with new age-classes (cohorts) are classified as regeneration methods. In the USA, these methods are usually classified by the number of cohorts created and maintained (even-aged, two-aged, and uneven-aged (3 or more)) and further distinguished by the timing and pattern of removal of the existing stand (SAF 1998). The following discussion deals with natural regeneration methods, or methods that rely on naturally occurring, rather than artificially-introduced regeneration sources.

5.2.1 The Clearcutting Method: Structure

Clearcutting is an even-aged method. The new cohort develops in the very open microenvironment created when all existing trees are cut. In the first 2–3 years after cutting, during the stand initiation stage, tens of thousands of stems per hectare become established from seedlings and saplings that existed on the site at the time of the cutting (i.e., advance reproduction), or from stump and root sprouts of harvested trees (Table 5.1). This response occurs over a wide geographic area, and, notably, over a broad range of site quality (as evaluated by the authors cited). As stands approach age 10, a distinct canopy begins to develop, and stands begin the transition to the stem exclusion stage. Stem density remains high as thousands of stems per

Table 5.1 Changes in stem density following clearcutting in upland hardwood forests of the Central Hardwood Region

Location	Years after harvest	Stems/ha	Size cut-off	Site index (m)[a]	Reference
Indiana	2	39,055	All	22.9–29.0	Sander and Clark (1971)
Kentucky	2	22,473	All	16.8–22.9	Sander and Clark (1971)
Ohio	2	55,217	All	16.8–22.9	Sander and Clark (1971)
Ohio	2	28,462	All	22.9–29.0	Merz and Boyce (1958)
Ohio	2	28,942	All	16.8–22.9	Merz and Boyce (1958)
North Carolina	2	45,851	All	27.4	Beck and Hooper (1986)
North Carolina	3	25,211	All	22.9	McGee (1967)
North Carolina	4	19,498	>1.4 m in height	27.4	Loftis (1978)
Kentucky	5	15,215	≥2 cm groundline diameter	Good	Arthur et al. (1997)
West Virginia	7	27,479	>0.3 m in height	24.4	Trimble (1973)
West Virginia	7	16,976	>0.3 m in height	21.3	Trimble (1973)
West Virginia	7	22,163	>0.3 m in height	18.3	Trimble (1973)
Alabama	8	24,245	>0.3 m in height	25.9	Schweitzer (*in prep.*)
Alabama	8	26,648	>0.3 m in height	18.3	Schweitzer (*in prep.*)
Ohio	10	5,519	>1.5 cm dbh	16.8–29.0	Sander and Clark (1971)
North Carolina	10	5,437	>1.4 m in height	27.4	Beck and Hooper (1986)
Kentucky	11	6,505	>2 cm dbh	Good	Arthur et al. (1997)
North Carolina	15	4,041	>1.4 m in height	27.4	Beck and Hooper (1986)

[a]Site index (SI) is the height (m) of a tree at 50 years of age and is used as an indicator of site quality

hectare compete for growing space and self-thinning begins. Again, at this age, there is no apparent relationship between stem density and site quality. Over the next few years, reduction in stand density through self-thinning and increasing canopy height (Beck and Hooper 1986) reduce the cover value for wildlife species such as Ruffed Grouse (*Bonasa umbellus*) (Dessecker and McAuley 2001).

5.2.2 The Shelterwood Method: Structure

In the shelterwood method, trees retained after an initial cut moderate the regeneration microenvironment. These overstory trees, or overwood, typically are removed in one or more removal cuts within a decade or two of the initial cut. In one variant of the shelterwood method, an overstory sparse enough to allow for the long term development of the new cohort is retained and results in a two-aged stand.

Shelterwood cuts can be separated into two structural and composition categories. In the first category, residual overstory density is low enough after the initial cut that regeneration density and species composition are similar to those after clearcutting (i.e., there is no differential species response to the environmental conditions created by the residual overstory density). In the second category environmental conditions created by a higher level of residual canopy density after the initial cut do result in differential species response of regeneration sources present at the time of the cut. Subsequent removal cuts in these shelterwoods will result in structural development similar to shelterwoods in the first category and to clearcuts, but with altered species composition due to the changes in the size structure and composition of the regeneration sources (see discussion of species composition below).

Regeneration development in the first category of shelterwood cuts and following the removal cuts in the second category of shelterwood cuts follows the same pattern as in clearcuts (Table 5.2). Very high initial stem densities diminish over time as the canopy of the new age-class closes. As a result, along with the early successional habitats provided by the substantial reduction in overstory trees, the shelterwood cut retains some remnant structure from the stand being harvested. It would be possible, for example, to harvest using a shelterwood method, providing both early successional habitats, and retention of some mature oak (*Quercus* spp.) trees to sustain some acorn production to benefit wildlife species that eat acorns (Greenberg et al., Chap. 8) that would not be possible with the clearcutting method. One might also retain trees that have cavities to benefit wildlife for either a few years in the case of conventional shelterwoods or for much longer periods of time in the case of two-aged stands.

5.2.3 Selection Methods: Structure

Uneven-aged methods create and maintain three or more cohorts using the single-tree and group selection methods. In the single-tree selection method, individual trees are removed periodically and more-or-less throughout the stand to provide

Table 5.2 Changes in stem density following the shelterwood method in upland hardwood forests of the Central Hardwood Region

Location	Years after cutting	Residual basal area (m^2/ha)	Stems ha^{-1}	Size cut-off	Oak site index (m)[a]	Citation
Indiana	2	10.1	54,276	All	22.9–29.0	Sander and Clark (1971)
Ohio	2	11.5	32,447	All	16.8–22.8	Sander and Clark (1971)
West Virginia	2	3.9	12,478[b]	0.3 m tall–2.3 cm dbh	21.3	Miller and Schuler (1995)
West Virginia	2	4.1	28,927[b]	0.3 m tall–2.3 cm dbh	21.3	Miller and Schuler (1995)
West Virginia	2	5.7	15,755[b]	0.3 m tall–2.3 cm dbh	24.4	Miller and Schuler (1995)
West Virginia	2	4.6	30,343[b]	0.3 m tall–2.3 cm dbh	24.4	Miller and Schuler (1995)
North Carolina	3	15.1	57,589	All	22.9	Loftis (1983a)
North Carolina	3	7.6	61,102	All	22.9	Loftis (1983a)
North Carolina	5	11.5	14,181	>1.4 m tall	24.4	Loftis (1983a)
West Virginia	2–5	<9.0	11,217	>0.3 m tall–2.3 cm dbh	24.4	Johnson et al. (1998)
West Virginia	2–5	<9.0	4,898	>0.3 m tall–2.3 cm dbh	21.3	Johnson et al. (1998)
Virginia	2–5	<9.0	1,516	>0.3 m tall–2.3 cm dbh	18.3	Johnson et al. (1998)
Alabama	3	5.7	12,372	>1.4 m tall	25.9	Schweitzer (*in prep*)
Alabama	3	11.5	12,236	>1.4 m tall	25.9	Schweitzer (*in prep*)
West Virginia	10	3.9	2,628[b]	>2.5 cm dbh	21.3	Miller and Schuler (1995)
West Virginia	10	4.1	2,309[b]	>2.5 cm dbh	21.3	Miller and Schuler (1995)
West Virginia	10	5.7	2,475[b]	>2.5 cm dbh	24.4	Miller and Schuler (1995)
West Virginia	10	4.6	2,405[b]	>2.5 cm dbh	24.4	Miller and Schuler (1995)
North Carolina	10	5.7	9,616	>1.4 m tall	25.9	Loftis (1983a)
North Carolina	17	15.1	1,441	Dominant/codominant	22.9	Loftis (1983a)
North Carolina	17	7.6	1,040	Dominant/codominant	22.9	Loftis (1983a)
West Virginia	17–22	5.3	794	Dominant/codominant	21–24	Miller et al. (2006)

[a]Site index (SI) is the height (m) of an oak tree at 50 years of age and is used as an indicator of site quality
[b]Commercial species only

Table 5.3 Stem density (stems/ha) following group selection in upland hardwood forests across the Central Hardwood Region

Location	Years after cutting	Opening size (ha)	Stems per ha	Size cut-off	Site index[a] (m)	Citation
Ohio	2	0.1	59,029	All	23–29	Sander and Clark (1971)
Ohio	2	0.2	44,253	All	23–29	Sander and Clark (1971)
Ohio	2	0.3	43,931	All	23–29	Sander and Clark (1971)
Ohio	2	0.1	29,626	All	17–22	Sander and Clark (1971)
Ohio	2	0.2	31,532	All	17–22	Sander and Clark (1971)
Ohio	2	0.3	31,655	All	17–22	Sander and Clark (1971)
Indiana[b]	6–10	0.07–0.23	10,300	≥2.5 cm dbh	N/A	Weigel and Parker (1997)
North Carolina	10	0.1–0.2	10,065	≥0.5 cm dbh	24	W.H. McNab (unpublished)
Indiana[a]	11–15	0.07–0.23	12,000	≥2.5 cm dbh	N/A	Weigel and Parker (1997)

[a]Site index (SI) is the height (m) of a tree at 50 years of age and is used as an indicator of site quality
[b]Estimated from graphical values

growing space for regeneration while still maintaining a relatively dense stand of trees. The relatively continuous canopy characteristic of single-tree selection and lack of canopy openness do not provide characteristic early successional habitats (Nyland 1996).

Group selection involves periodic removal of groups of trees, creating openings that might range between 0.1 and 0.4 ha, and results in a mosaic of relatively small patches of regeneration throughout the stand. As with the clearcutting and shelterwood methods, group selection provides the high stem density and overstory openness characteristic of early successional habitats (Table 5.3). In patch sizes from 0.1 to 0.4 ha (which meets the technical definition of group selection for eastern hardwoods (SAF 1998)), data generally do not suggest a strong relationship between stem density and patch size. In group selection, the condition of high density of small stems and overstory openness characteristic of early successional habitats is interspersed with more mature, closed-stand conditions. Further, within any given stand the aggregate amount of early successional habitats is less, by design, than in even-aged methods.

From a structural standpoint, clearcutting, shelterwood and group selection methods can all provide the low, dense vegetation and overstory openness characteristic of early successional habitats. In shelterwood methods with low residual basal areas after the initial cut, stem densities, although quite variable from stand to stand, show no relationship to the residual overstory basal area for the period of time the overwood remains intact. These stem densities are similar to those observed in clearcuts.

In group selection, stem densities show no strong relationship to patch size in the range of 0.1–0.4 ha and are similar those in the clearcutting and shelterwood methods. There is some evidence to suggest, however, that stem densities are related to site quality. Minkler's (1989) group selection study that compares north and south slopes, Sander and Clark's (1971) group selection data for good and fair sites, and Johnson et al.'s (1998) study of two-aged stands comparing stands on mesic sites and intermediate sites in West Virginia and more xeric sites in the Ridge and Valley Province of Virginia suggest higher stem densities on more productive sites. In contrast, graphical data provided by Weigel and Parker (1997) following group selection show no difference in stem density between openings created on northern and southern aspects with >10,000 stems per hectare >2.5 cm in diameter at breast height (dbh) observed regardless of aspect, and data from the Cumberland Plateau (Schweitzer, unpublished) show no difference in stem density between clearcuts on the mesic escarpment and the xeric top of the plateau.

Structural differences among the clearcutting, shelterwood and group selection methods are related in one way or another to retention, or lack of retention, of trees as stands are being regenerated. Many variations of the shelterwood method provide not only the same quality and scale of early successional habitats as the clearcutting method, but also the structural and compositional functions provided by residual overstory trees. The residual trees may be retained for a relatively short period of time – 5–15 years – or for an extended period of time at very low densities.

Group selection, over time, provides a mosaic of small patches in different stages of stand development. The relatively small patch size of group selection openings (0.1–0.4 ha) provides adequate early successional habitat for some animals but may not by sufficiently large for others (Dessecker and McAuley 2001; Shifley and Thompson, Chap. 6). The effect of differing spatial distributions of patches and road corridors that provide connectivity among them may be a fruitful area for research.

5.3 Species Composition of Regeneration

While some might argue that the structure of regeneration is the more important factor in early successional habitats, arborescent species composition of naturally regenerated stands has been a major focus of research in the Central Hardwood Region.

The past century's rich body of literature on secondary succession provides constructs for understanding species composition of regeneration following the application the various methods discussed above. One of the most interesting ideas related to timber harvest comes from Clements (1916):

> In all cases of burning or clearing the intensity or thoroughness of the process determines whether the result will be a change of vegetation or the initiation of a sere. The latter occurs only when the destruction of the vegetation is complete, or so nearly complete that the pioneers dominate the area. Lumbering consequently does not initiate succession except when it is followed by fire or other process which removes the undergrowth.
>
> Frederick E. Clements

Clements's observation raises the question of whether the use of successional terminology associated with seral stages (i.e., early succession, mid-succession, late succession) is appropriate for characterizing the compositional development of arborescent vegetation after timber harvest. His statements also highlight the role of propagules present in a stand prior to disturbance and that persist through disturbance in influencing the potential of the next stand. Other ecologists have also recognized the importance of propagules (or regeneration sources) present prior to disturbance. An extension of Egler's (1954) concept of Initial Floristics to forest management suggests that species composition after significant overstory removal is a function of regeneration sources that survive the disturbance plus new individuals that arrive during or after the disturbance (Kimmins 1997). In the vital attributes approach of Noble and Slatyer (1980) (Cattelino et al. 1979), the first vital attribute of a species is whether it persists through disturbance or arrives after disturbance. Other vital attributes of species deal with environmental conditions necessary for establishment and growth rates of regeneration sources. The silviculture literature has long recognized the multiple sources of hardwood regeneration. In hardwood silviculture application of natural regeneration methods typically attempts to utilize or enhance advance regeneration to influence stand composition.

An example of the importance of advance reproduction and influence of site quality (or position of a site along environmental gradients) in determining outcomes is oak regeneration, (see Johnson et al. 2009 for a complete discussion of regeneration and other topics in oak ecology and silviculture, as well as an extensive bibliography). It has long been known that oak regeneration depends on advance reproduction and sprouts from harvested trees in most upland hardwood forests (Leffelman and Hawley 1925; Liming and Johnson 1944; Sander and Clark 1971; Sander 1971; Loftis 1983b; Sander et al. 1984; Beck 1988; Johnson 1989). Further research established that success of oak advance reproduction is related to stem size at the time of canopy removal. In other words, the probability of an individual stem becoming dominant or codominant (dominance probability) in the new cohort that develops after substantial overstory removal increases with increasing groundline diameter or height of the advance reproduction (Sander 1971; Sander et al. 1984; Loftis 1990a). This research also established that the dominance probability for an advance oak stem of a given size decreased with increasing site quality, or more generally, from xeric to mesic sites. This inverse relationship is a function of increasing competition from other, faster-growing tree species with increasing site quality. The dominance probability for stump sprouts is also related to stem size (i.e., dbh). However, for stump sprouts, dominance probabilities decrease with increasing stem size, largely because the probability of sprouting decreases with increasing stem size (Johnson 1977; Weigel and Johnson 1998).

Johnson (1977) also notes that on more xeric sites with a more open canopy, oak seedlings that establish after good acorn (seed) crops tend to survive and accumulate over time beneath the canopy. On more mesic sites, disturbance, whether natural or silvicultural, is required for survival and growth of newly established oak seedlings (Loftis 1983b).

Based on these relationships, it is not surprising that successful oak regeneration is a common outcome of the regeneration methods listed above on xeric sites. Advance reproduction is relatively abundant after release, and competitive with regeneration sources of other species that occur on xeric sites. Stump sprouts (which are generally more abundant from stumps of smaller oak trees) provide an additional source of oak regeneration. The establishment of new oak stands after regeneration cuts is a likely outcome in the Western Dry Region of the Central Hardwood Region (see McNab, Chap. 2, Fig. 2.1) on all but the most mesic sites (on a relative scale)(Johnson et al. 2009). This outcome, however, is more restricted as one moves eastward to the Transition Dry Mesic and Central and Eastern Mesic subregions (Fig. 2.1), where a larger portion of the forested landscape would be classified as intermediate or mesic on the moisture gradient. Establishment of new oak stands or cohorts is still a common outcome after regeneration harvests, but is restricted to the more xeric end of the moisture gradient.

More species occur on intermediate and mesic sites than on xeric sites. In the Cumberlands and Southern Appalachians, stands on intermediate and submesic sites frequently have 20 or more arborescent species. A few of these, notably yellow-poplar (*Liriodendron tulipifera*), can regenerate and compete successfully from new seedlings established after harvest. Most species, however, depend on advance reproduction and stump sprouts as the source of future dominant and codominant stems (Johnson 1989; Beck 1988; Loftis 1989). Oaks often dominate the overstory of stands on intermediate sites and can be prominent on mesic sites. The application of regeneration harvests, however, frequently results in stands or even quite small openings (group selection) dominated by yellow-poplar (McGee 1975; McGee and Hooper 1975; Beck and Hooper 1986; Loftis 1983a, b, 1985, 1989; Beck 1988). Only where large oak advance reproduction is present prior to harvest does oak compete successfully after harvest with yellow-poplar (Loftis 1983a, b, 1990a). Dey (1991) found that pre-harvest size was also an important determinant of post-harvest growth for other advanced growth dependent species; like oak, these species must be present as large advance reproduction prior to disturbance to increase their likelihood of being represented in the new stand. It is likely that variation in species composition of regenerating stands is largely due to variation in the amount and size distribution of advance reproduction of all species present at the time of substantial overstory removal (Loftis 1989). For example, results from shelterwood cuts on the escarpment of the Cumberland Plateau illustrate the variability in regeneration outcomes on mesic sites (Schweitzer in prep; Table 5.4). In Table 5.4 yellow-poplar, which is commonly considered an "early successional species" is abundant in clearcuts, the 25% retention shelterwoods and 50% retention shelterwoods. Note, however, that white ash (*Fraxinus americana*) and sugar maple (*Acer saccharum*), both of which depend on advance reproduction and are usually associated with later stages of succession, are also abundant after application of all three methods (Table 5.4).

Table 5.4 Number of stems ha^{-1} >1.4 m in height and <15.2 cm diameter at breast height 8 years following a clearcut, a shelterwood with 50% basal area retention, and a shelterwood with 25% basal area retention on the Cumberland Plateau, Tennessee, USA (Schweitzer *in prep.*)

Species	Treatment clearcut	25% retention	50% retention
Fraxinus americana	2,140	764	1,082
Acer saccharum	1,282	1,046	1,470
Liriodendron tulipifera	3,175	964	1,529

As suggested in Sect. 5.1, there appears to be a level of basal area retention (category 1 shelterwoods) below which differences in species composition are due to differences in composition and size structure of advance reproduction prior to harvest rather than the basal area retained after harvest. This level appears to be about 50% of the pre-harvest basal area in the Southern Appalachians and the Cumberland Plateau. That is, species composition outcomes are essentially the same in cuts ranging from 0% basal area retained to 50% basal area retained, and is a function of composition and size structure of advance reproduction at the time of the initial harvest. This response assumes that overwood is removed in 5–15 years after the initial cut. Other areas within the Central Hardwood Region differ in the residual stocking level at which this response occurs.

If the amount and size distribution of advance reproduction of all species is important in determining species composition following a regeneration harvest, one must consider the factors that influence the variation. Variation in regeneration outcomes from stand to stand is a function not only of site quality, which determines the suite of species that grow on sites, but also the cumulative effect of disturbances and other events (e.g. seed production) that occur in the last few decades leading up to the regeneration harvest. We have shown experimentally that treatments that modestly alter the light environment on intermediate and mesic sites can substantially alter the size distribution of oaks and many other species that depend on advance reproduction (Loftis 1990b) (Table 5.5). Natural (and stochastic) disturbances such as ice (McNab et. al. 2006) and wind (Berg and Van Lear 2004) result in a range of light environments in stands subjected to these disturbances (see White et al., Chap. 3). Silvicultural treatments can be implemented in the decade (or more) prior to substantial overstory removal associated with a regeneration harvest to alter development of advance reproduction and, therefore, species composition after a regeneration harvest.

5.4 Summary

Early successional habitats—low dense vegetation and openness overhead—can be created using several regeneration methods. The clearcutting method, variations of the shelterwood method that leave mid- to low levels of residual trees and provide for timely removal cuts, and the group selection method all provide for early successional habitats. In all cases, structural changes associated with growth and self-thinning limit early successional habitats to about the first 15 years of stand or

Table 5.5 Density (trees ha^{-1}) of species by height class in control and treated plots (20–30% basal area reduction) in 1980 and 1995 (15 years post-treatment)

	Control		Treated	
	1980	1996	1980	1996
Liriodendron tulipifera				
<0.3 m	906	4,775	818	170
≥0.3 and <1.4 m	247	329	31	0
≥1.4 m	55	0	0	0
Betula lenta				
<0.3 m	55	165	0	124
≥0.3 and <1.4 m	192	137	0	463
≥1.4 m	0	55	15	494
Quercus-Carya[a]				
<0.3 m	6,010	4,967	895	942
≥0.3 and <1.4 m	1,125	1,290	509	1,204
≥1.4 m	55	165	124	679
Fraxinus americana/Tilia americana[a]				
<0.3 m	27	0	602	46
≥0.3 and <1.4 m	27	0	15	232
≥1.4 m	0	0	0	77
Acer rubrum[a]				
<0.3 m	19,431	29,997	20,161	3,041
≥ 0.3 and <1.4 m	274	274	602	1,328
≥1.4 m	55	0	15	340
Cornus florida[a]				
<0.3 m	631	82	247	0
≥0.3 and <1.4 m	439	110	324	31
≥1.4 m	110	82	15	93
Other[a, b]				
<0.3 m	1,811	2,113	1,853	679
≥0.3 and <1.4 m	659	796	1,050	710
≥1.4 m	27	384	31	247

[a]Heavily dependent on advance reproduction and stump sprouts
[b]Includes *Magnolia* spp., *Pinus strobus*, *Ostrya virginiana*, *Oxydendrum arboreum*, *Nyssa sylvatica*, *Halesia tetraptera*, *Carpinus caroliniana*, and *Sassafras albidum*

cohort development. Although there is some indication that density does vary with site quality (xeric to mesic), the vegetation, by any measure, is dense. The shelterwood and group selection methods, which retain mature trees in stands for varying periods of time, provide early successional habitats with the possibility of greater structural complexity.

Species composition resulting from the regeneration methods discussed is affected by the amount and size of advance reproduction present at the time of a regeneration harvest that removes a substantial part of the overstory. The amount and size of advance reproduction present at the time of a harvest is influenced by disturbance (including silviculture) and other events that occur prior to harvest.

Literature Cited

Arthur MA, Muller RN, Costello S (1997) Species composition in a Central Hardwood Forest in Kentucky 11 years after clear-cutting. Am Midl Nat 137:274–281

Beck DE (1988) Regenerating cove hardwood stands. In: Smith CH, Perkey AW, Kidd WE Jr (eds) Guidelines for regenerating Appalachian hardwood stands, vol 88-03. Soc Amer For Pub, Morgantown, pp 156–166

Beck DE, Hooper RM (1986) Development of a southern Appalachian hardwood stand after clearcutting. S J Appl For 10:168–172

Berg EC, Van Lear DH (2004) Yellow-poplar and oak seedlings density responses to wind-generated gaps. In: Connor KF (ed) Proceedings of the 12th biennial southern silvicultural research conference. Gen Tech Rep SRS-73, USDA Forest Service Southern Research Station, Asheville, pp 254–259

Cattelino PJ, Noble IR, Slatyer RO, Kessell SR (1979) Predicting the multiple pathways of succession. Environ Manage 3:41–50

Clements F (1916) Plant succession: an analysis of the development of vegetation, vol 242. Carnegie Inst Wash. Pub, Washington, DC

Dessecker DR, McAuley DG (2001) Importance of early successional habitat to ruffed grouse and American woodcock. Wildl Soc Bull 29:456–465

Dey DC (1991) A comprehensive Ozark regenerator. Dissertation, University of Missouri, Columbia

Egler FE (1954) Vegetation science concepts I. Initial floristic composition factor old field vegetation management. Vegetatio 4:412–417

Johnson PS (1977) Predicting oak sprouting and sprout development in the Missouri Ozarks. Res Pap NC-149, USDA Forest Service North Central Forest Experiment Station, St Paul

Johnson PS (1989) Principles of natural regeneration. In: Hutchinson JG (ed) Central hardwood notes. USDA Forest Service, North Central Forest Experiment Station, St Paul, pp 301–305

Johnson JE, Miller GW, Baumgras JE, West CD (1998) Assessment of residual stand quality and regeneration following shelterwood cutting in Central Appalachian hardwoods. N J Appl For 15:203–210

Johnson PS, Shifley SR, Roberts R (2009) The ecology and silviculture of oaks. CABI, London

Kimmins JP (1997) Forest ecology: a foundation for sustainable management. Prentice Hall, Upper Saddle River

Leffelman LJ, Hawley RC (1925) The treatment of advance growth arising as a result of thinnings and shelterwood cuttings. Yale University Sch For Bull 15, New Haven

Liming FG, Johnson JP (1944) Reproduction in oak-hickory forest stands of the Missouri Ozarks. J For 42:175–180

Loftis DL (1978) Preharvest herbicide control of undesirable vegetation in southern Appalachian hardwoods. S J Appl For 2:51–54

Loftis DL (1983a) Regenerating Southern Appalachian mixed hardwood stands with the shelterwood method. S J Appl For 7:212–217

Loftis DL (1983b) Regenerating red oak on productive sites in the Southern Appalachians: a research approach. Gen Tech Rep SE-24, USDA Forest Service, Southeastern Forest Experiment Station, Asheville, pp 144–150

Loftis DL (1985) Preharvest herbicide treatment improves regeneration in Southern Appalachian hardwoods. S J Appl For 9:177–180

Loftis DL (1989) Species composition of regeneration after clearcutting Southern Appalachian hardwoods. In Miller JH (ed) Proceedings of the fifth biennial southern silvicultural research conference. Gen Tech Rep SO-74, USDA Forest Service, Southern Forest Experiment Station, New Orleans, pp 253–257

Loftis DL (1990a) Predicting post-harvest performance of advance red oak reproduction in the Southern Appalachians. For Sci 36:908–916

Loftis DL (1990b) A shelterwood method for regenerating red oak in the southern Appalachians. For Sci 36:917–929
McGee CE (1967) Regeneration in southern Appalachian oak stands. Res Note, SE-72 USDA Forest Service, Southeastern Forest Experiment Station, Asheville
McGee CE (1975) Regeneration alternatives in mixed oak stands. Res Pap SE-125 USDA Forest Service, Southeastern Forest Experiment Station, Asheville
McGee CE, Hooper RM (1975) Regeneration trends 10 years after clearcutting of an Appalachian hardwood stand. Res Note SE-227, USDA Forest Service, Southeastern Forest Experiment Station, Asheville
McNab WH, Roof T, Lewis JF, Loftis DL (2006) Evaluation of landsat imagery for detecting ice storm damage to oak forests in eastern Kentucky. In: Buckley DS, Clatterbuck WK (eds) Proceedings of the 15th central hardwood forest conference. Gen Tech Rep SRS-101, USDA Forest Service Southern Research Station, Asheville, pp 128–138
Merz RW, Boyce SG (1958) Reproduction of upland hardwood in southeastern Ohio. Tech Pap 155, Central States Forest Experiment Station, Columbus
Miller GW, Schuler TM (1995) Development and quality of reproduction in two-aged central Appalachian hardwoods – 10 year results. In: Gottschalk KW, Fosbroke SLC (eds) Proceedings of the 10th central hardwood forest conference. Gen Tech Rep NE-197, USDA Forest Service, Northeast Forest Experiment Station, Radnor, pp 364–374
Miller GW, Kochenderfer JN, Fekedulegn DB (2006) Influence of individual reserve trees on nearby reproduction in two-aged Appalachian hardwood stands. For Ecol Manage 224: 241–251
Minkler LS (1989) Intensive group selection in central hardwoods. In: Rink G, Budelski CA (eds) Proceedings of the seventh central hardwood forest conference. Gen Tech Rep NC-132, USDA Forest Service, North Central Forest Experiment Station, St. Paul, pp 35–39
Noble IR, Slatyer RO (1980) The use of vital attributes to predict successional changes in plant communities subject to recurrent disturbances. Vegetatio 43:5–21
Nyland RD (1996) Silviculture: concepts and applications. McGraw-Hill, New York
Oliver CD, Larson BC (1990) Forest stand dynamics. McGraw-Hill, New York
Peet RK (1992) Community structure and ecosystem function. In: Glenn-Lewin DC, Peet RK, Veblen TT (eds) Plant succession: theory and prediction. Chapman-Hall, London
Sander IL (1971) Height growth of new oak sprouts depends on size of advance reproduction. J For 69:809–811
Sander IL, Clark FB (1971) Reproduction of upland hardwood forests in the Central States. In: Agri Handbook 405, USDA Forest Service, Washington, DC
Sander IL, Johnson PS, Rogers R (1984) Evaluating oak advance reproduction in the Missouri Ozarks. Res Pap NC-251, USDA Forest Service, North Central Forest Experiment Station, St. Paul
Society of American Foresters (1998) In: Helms JA (ed) The dictionary of forestry. Society of American Foresters, Bethesda
Trimble GR (1973) The regeneration of central Appalachian hardwoods with emphasis on the effects of site quality and harvesting practice. USDA Forest Service, Northeastern Forest Experiment Station, Upper Darby
Weigel DR, Johnson, PS (1998) Stump sprouting probabilities for Indiana oaks. Tech Brief NC-7, USDA Forest Service North Central Forest Experiment Station, St. Paul
Weigel DR, Parker GR (1997) Tree regeneration response to the group selection method in southern Indiana. N J Appl For 14:90–94

Chapter 6
Spatial and Temporal Patterns in the Amount of Young Forests and Implications for Biodiversity

Stephen R. Shifley and Frank R. Thompson III

Abstract Forest inventory data provide simple indicators of forest structural diversity in the form of forest age distributions and their change over time. A result of past land use and disturbance, more than half of the 51 million ha of forest in the Central Hardwood Region is between 40 and 80 years old and young forest up to 10 years old constitutes only 5.5% of the area. Simulations of a sustained level of management over time produce more uniform (flatter) age-class distributions. A management scenario designed to maintain about 7% of total forest area as young habitat results in a region-wide young forest deficit of one million ha relative to current conditions. However, management activities that create an average of 200 ha of additional young forest per county per year would be sufficient to erase that deficit.

6.1 Introduction

Concern over the quantity, spatial distribution, and temporal distribution of early successional forest habitats is a direct result of conservation concerns for wildlife species dependent on these forest communities and for biodiversity in general. Sustaining diverse, early successional forest habitats through time and across forest landscapes is expected to improve—or at the very least to not diminish—opportunities for sustaining wildlife communities that depend upon access to them.

Early successional forests can consist of pioneer tree and shrub species in association with annual and perennial herbaceous plants colonizing former agricultural or other nonforest land. They can also be early stages of forest regeneration following harvests or other major disturbances in established forests (see Greenberg et al., Chap. 1).

S.R. Shifley (✉) • F.R. Thompson III
USDA Forest Service Northern Research Station, University of Missouri,
202 Natural Resources, Columbia, MO 65211–7260, USA
e-mail: sshifley@fs.fed.us; frthompson@fs.fed.us

C.H. Greenberg et al. (eds.), *Sustaining Young Forest Communities*,
Managing Forest Ecosystems 21, DOI 10.1007/978-94-007-1620-9_6,

Early successional forests include young forest plantations as well as natural forest regeneration, although plantations comprise only 5% of the forest area in the Central Hardwood Region. Managers often refer to these young forests as seedling- or sapling-sized forests. In the context of forest stand development, these forests are in the stand initiation stage of development (Loftis et al., Chap. 5). All these young forest communities or vegetation stages can, to varying degrees, provide habitat for early successional wildlife. If undisturbed, they will eventually progress to the stem exclusion, understory reinitiation, and old-growth stages of stand development (Oliver and Larson 1990; Johnson et al. 2009; Loftis et al., Chap. 5).

Our focus is on young or regenerating forests that are in the stand initiation stage of development. For simplicity we refer to them as young forests. In the Central Hardwood Region the dominant tree cover in these young forest habitats typically is not more than 10 years old and never more than 20 years old. Young forests are an important component of all early successional habitats in the Central Hardwood Region. Young forests can be readily established and maintained through forest management, and data on habitat abundance over time are available from the US Forest Service (Forest Inventory and Analysis 2010b). Other vegetation types, such as shrublands, provide additional habitat for early successional species, however these types are poorly inventoried (Greenberg et al., Chap. 1; Warburton et al., Chap. 13).

This chapter is organized around three topics. First, forest management practices greatly impact the amount of young forest in the landscape and how it changes over time with stand development. Whether forests are managed for wood products, biological diversity, or other ecological services or commodities, there is growing interest in sustainable forestry. We examine the availability of young forests in the broader perspective of sustainable management. We discuss how age class distributions are an indicator of sustainable management and a coarse filter for habitat diversity. Second, we consider the age distribution of forest area to be an indicator of the past, present, and future quantity of young forest habitats. We summarize forest age class distributions by state and forest type for the Central Hardwood Region. We determine the temporal trends in forest age class distributions that are likely to result from continued current patterns of disturbance, and explore management scenarios that would sustain a constant availability of young forests over time. Third, we consider how spatial and temporal distributions of young forests affect wildlife habitat quality. Using simulation modeling, we demonstrate how forest management practices and natural disturbances affect the amount and spatial pattern of forest age classes at the landscape level. For the second and third topics, we provide examples of how these patterns are related to early successional wildlife species.

6.2 Sustainable Forest Management and Age Class Diversity

Internationally accepted criteria and indicators for sustainable forest management developed under the Montréal Process (http://www.rinya.maff.go.jp/mpci/evolution_e.html) include 64 indicators (informative metrics) organized into seven Criteria (broad classes). The first criterion is Conservation of Biological Diversity. Its nine associated indicators deal with landscape-scale or ecosystem diversity, species diversity, and genetic diversity:

Ecosystem diversity

1. Area and percent of forests by forest ecosystem type, successional stage, age class, and forest ownership or tenure.
2. Area and percent of forests in protected areas by forest ecosystem type, and by age class or successional stage.
3. Fragmentation of forests.

Species diversity

4. Number of native forest associated species.
5. Number and status of native forest associated species at risk, as determined by legislation or scientific assessment.
6. Status of on-site and off-site efforts focused on conservation of species diversity.

Genetic diversity

7. Number and geographic distribution of forest associated species at risk of losing genetic variation and locally adapted genotypes.
8. Population levels of selected representative forest associated species to describe genetic diversity.
9. Status of on-site and off-site efforts focused on conservation of genetic diversity.

Although these nine indicators provide information important for characterizing forest biodiversity, there is no general agreement on which values or thresholds for a given indicator characterize forest sustainability (or a lack thereof). This situation is not particular to the Montréal Process Criteria and Indicators. Most definitions of sustainable forestry lack quantitative specificity (e.g., World Commission on Environment and Development 1987; Helms 1998, National Archives and Records Administration 2007).

Management to sustain diverse forest habitats can be a coarse filter approach for managing biological diversity. Rather than attempting to monitor individual populations of all forest associated species, the coarse filter approach provides a diversity of forest habitats. This diversity is expected to provide suitable habitats for forest-associated species, including those for which inventory data are limited (e.g., invertebrates and fungi). Coarse filter screening can be adapted to take into account the distribution of habitats across ecoregions or other relevant strata in an attempt to ensure greater dispersion of habitats (e.g., to ensure that focal habitats such as woodlands or old-growth forests are spatially distributed across multiple ecoregions).

The coarse filter approach has been used to guide establishment of protected forest areas, but it is not limited to allocation of protected forests. Coarse filter approaches are equally applicable to actively managed forests where they can be used to quantify and maintain habitat diversity. Because forest habitats change over time due to endogenous and exogenous processes and disturbances, any coarse filter approach to managing biological diversity must take into account the spatial and temporal patterns of disturbances necessary to maintain future habitat diversity (Haufler et al. 1996). The historic range of variation in habitat diversity and/or ecosystem processes (e.g., historic forest fire frequency) is sometimes used as a benchmark for judging adequacy of current habitat diversity, but that presumes a rationale for selecting a particular time in history as a desirable benchmark (see Warburton et al., Chap. 13). Benchmarks for the range of historic variation in forest ecosystems are often linked to years for which historic data happen to have been collected (e.g., during initial land surveys) rather than by other objective criteria.

Although coarse filter screening is a suitable starting point for evaluating present and future forest habitat diversity, characterization and maintenance of habitat diversity requires assessment at finer spatial scales. Thus, a hierarchical approach also includes meso-scale evaluations of habitat components (e.g., down logs, snags, seeps, riparian forests), and fine-scale evaluation of individual species, with particular emphasis on species that are at risk of extinction (Schulte et al. 2006).

A coarse filter approach to evaluating and conserving forest diversity could be applied using variables that describe forest vegetation types or stages such as age structure, overstory species composition, understory species composition, size structure, tree density, stocking percent, past disturbance regime, patch size, geographic location, ownership, or spatial juxtaposition of habitat components. However, there are some practical constraints on the process. First, variables used in a coarse filter analysis must have been inventoried across the landscape of interest. Second, the number of variables analyzed simultaneously when applying a coarse filter approach to a specific landscape must be commensurate with the capacity to manage the landscape for habitat diversity. For example, in a given forest landscape it might be desirable to maintain 20 forest age classes across each of five forest cover types with multiple spatially dispersed forest patch sizes. However, the logistics of creating and maintaining that range of structural and compositional diversity over time would likely be daunting, especially in forest landscapes with a large proportion of private ownership.

Data are readily available to support a coarse filter examination of forest age class diversity at regional and national scales (millions to hundreds of millions of hectares). Age classes are indicative of forest size class, structural characteristics, and successional stages, and all these age-associated characteristics are in turn indicative of habitat quality for key wildlife species. In the past decade, publications at local, state, regional, and national scales have reported forest age class diversity as an indicator of forest biodiversity using the Montréal Process Criteria and Indicator framework (e.g., Baltimore County 2005; Carpenter 2007; USDA Forest Service 2004).

The distribution of forest area by age class is among the simplest large-scale indicators of forest structural diversity. Such information has been applied routinely for more than a century to develop sustained yield timber management plans, and in some conservation approaches forest structural diversity is assumed to be a surrogate for wildlife diversity. Moreover, forest age or size class distributions also can be used to directly model wildlife habitat or abundance. For example, Rittenhouse et al. (2007) and Tirpak et al. (2009) used forest age class or size class for individual sites (e.g., 0.01 ha) and for collections of sites describing the surrounding landscapes as key variables in habitat suitability models for wildlife. Twedt et al. (2010) used county-level estimates of forest area by size-class from the Forest Inventory and Analysis program as covariates to predict bird abundance as measured by the North American Breeding Bird survey. In this volume, Franzreb et al. (Chap. 9) relate trends in bird abundance to trends in forest age classes.

6.3 State and Regional Forest Age Class Distributions

We assessed forest age class distributions for ten states that fall mostly in the Central Hardwood Region: Arkansas, Illinois, Indiana, Kentucky, Michigan, Missouri, Ohio, Pennsylvania, Tennessee, and West Virginia (see McNab, Chap. 2; Fig. 2.1). We utilized state-wide inventory data to examine forest age-class structure by state, for multiple states, and by forest type. The Forest Inventory and Analysis (FIA) division of the US Forest Service inventories the nation's forests in detail and publishes state, regional, and national summaries (e.g., Woodall et al. 2006; Smith et al. 2009, Forest Inventory and Analysis 2010a). Moreover the data and associated summary, mapping, and analysis tools are publicly available online (Forest Inventory and Analysis 2010b). In addition to measures of current forest conditions, there are temporal records for some forest characteristics that extend back as far as 50 years for some states. To the extent possible we utilized the FIA EVALIDator tool (Miles 2010) for data summaries; it automatically applies the appropriate weights to expand plot-level data to area wide estimates while taking into account the three-phase, stratified, FIA sampling design.

Estimation of forest characteristics using FIA data is not restricted to individual state totals. Forest area, volume, and number of trees can be readily summarized by county, groups of counties within or among states, groups of states, congressional districts, ecoregions, or other user-specified geographic regions. Summaries for any geographic region can be further subdivided by species group, age class, tree size, site quality, or other variables of interest. However, as the sampled area and number of inventory plots included in any subcategory decreases, the standard error of the estimate for that subcategory increases. We concentrated primarily on state-level summaries for two reasons. First, some historical forest resource data are available only as state-wide summaries, so some historical trends can only be summarized at the state scale. Second, state forestry or natural resource agencies will be key to designing and implementing any future policies or management practices intended to create or maintain early successional forest habitats within their jurisdiction.

We summarized the current forest area by age class for each state and by forest cover type across all 10 states. For the various age class distributions we computed rates at which forest regeneration would need to occur (i.e. hectares regenerated per decade) to periodically replenish and maintain alternative amounts of young forest habitat. Specifically, for scenarios that set aside differing levels of undisturbed forest (e.g., 10% or 20% of all hectares) we computed the amount of additional young forest that must be created (or avoided) to sustain a constant area of young (20 years or less) forest over time. Finally, for one selected state within the study region—Indiana—we examined changes in forest age distributions over time by compiling time series information from previously published inventory reports.

6.3.1 Forest Age Classes by State

The ten states examined in this analysis have, combined, 51 million ha of forest land, or 17% of all US forest land (Table 6.1). Forests cover 43% of the land area in the 10-state region, and oak-hickory (*Quercus-Carya*) and maple-beech-birch (*Acer-Fagus-Betula*) forest types cover more than two-thirds of the forest area. The distribution of forest area by age class across the study region is distinctly unimodal (Fig. 6.1). Across the ten states, 60% of the total forest area is between 40 and 80 years of age. State entries in Fig. 6.1 are ordered from those with the least forest area in the 0–10-year old age class (Illinois) to the greatest (Arkansas).

Individual states vary from two to eight million ha in total forest area. Viewing the distribution of forest area by age class on a percentage basis normalizes differences in total forest area (Fig. 6.2). The shapes of the forest age distributions for the ten states are remarkably similar. Maximum forest area by 10 year age classes ranges from 17% to 21% and consistently occurs in ages classes from 50 to 80 years.

We grouped states with similarly shaped age class distributions for forests less than 30 years of age (i.e., similar left tails of the age class distribution) (Fig. 6.2). Arkansas, Tennessee, and Ohio have greater area in the 0–10 year age class than in the 11–20-year age class (Fig. 6.2a). That pattern is less pronounced for Kentucky, Indiana, Illinois, and West Virginia (Fig. 6.2c). For Michigan and Missouri the proportion of forest area increases with increasing age through at least age 50 (Fig. 6.2c), and Pennsylvania forests have a similar trend.

These age class distributions are the cumulative effect of past forest disturbances. Stand-initiating disturbances (Oliver and Larson 1990; Johnson et al. 2009) are those that remove enough of the existing forest overstory to regenerate a new stand and reset the age of the dominant forest cover to zero. Based on the age-class distributions (Fig. 6.2), stand-initiating disturbances in this region were common 50–80 years ago, affecting at least 15–20% of the area per decade. In the past two decades, stand-initiating disturbances affected 3–10% of the forest area per decade with considerable variation among states.

Table 6.1 Forest area by state, ownership and forest type group in the Central Hardwood Region, 2011. States are ordered from greatest to least proportion of forest land

State	Land area (hectares)	Forest land area (hectares)	Forest land %	Private forest land ownership %	Oak-hickory	Maple-beech-birch	Elm-ash-cottonwood	Loblolly-shortleaf pine	Oak-pine	Aspen-birch	Spruce-fir	White-red-jack pine	Other forest type groups
West Virginia	6,240,891	4,847,917	78	87	75	18	1	1	2	0	0	1	2
Pennsylvania	11,612,551	6,741,788	58	71	54	33	2	1	1	2	0	2	5
Arkansas	13,491,498	7,497,856	56	80	42	0	5	29	11	0	0	0	13
Michigan	14,686,235	8,024,840	55	62	16	29	10	0	3	17	13	10	2
Tennessee	10,684,211	5,619,919	53	85	72	2	5	6	8	0	0	1	6
Kentucky	10,293,927	5,007,784	49	89	75	8	6	2	5	0	0	0	4
Missouri	17,851,417	6,233,550	35	83	82	1	7	1	6	0	0	0	3
Ohio	10,610,121	3,256,615	31	87	63	21	9	1	1	1	0	1	3
Indiana	9,303,644	1,920,712	21	84	73	8	12	1	2	0	0	1	3
Illinois	14,416,194	1,942,376	13	83	71	1	23	1	1	0	0	1	2
Total All States	119,190,688	51,093,358	43	79	58	13	6	6	5	3	2	2	5

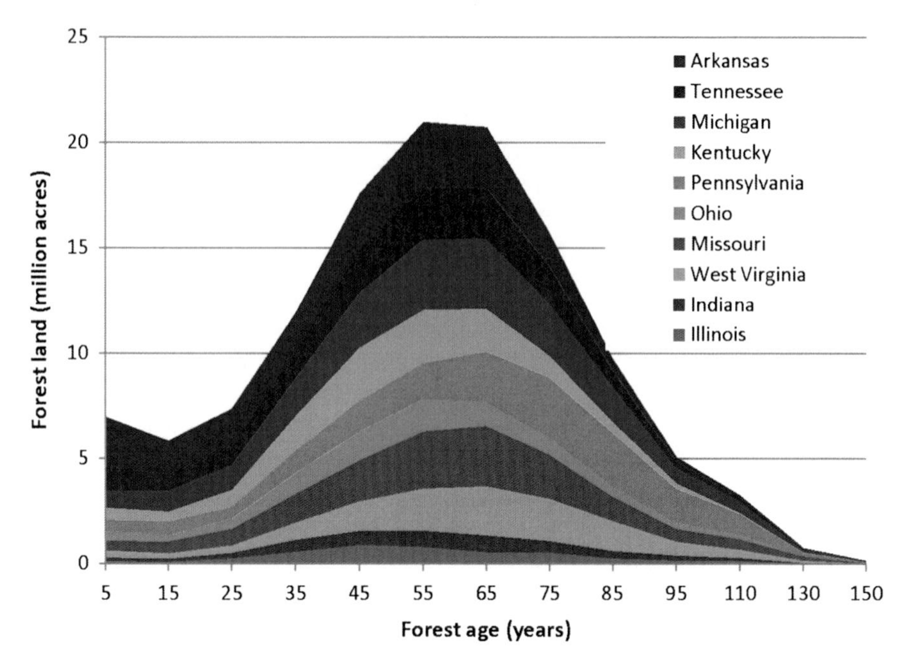

Fig. 6.1 Forest area by 10-year age classes for ten states in the Central Hardwood Region. Individual states are shown as stacked colored areas and the total area is represented by the top of the distribution States are ordered from the least (Illinois) to the most (Arkansas) total area in the youngest age class (0–10-years old)

6.3.2 Forest Age Classes by Forest Cover Type

Forest cover types are based on species composition of the dominant tree cover (Helms 1998; Eyre 1980). The forest cover type—sometimes referred to simply as the cover type or the forest type—is determined by the group of associated species with a plurality of the stand stocking or basal area. We used FIA naming conventions to summarize forest area by eight forest type groups (Table 6.1) that consolidate rare forest types with more common ones. These groups are the focus of the region-wide analyses reported in this section. We omitted nine additional groups that occur within the study region but make up 1% or less of the total forest area.

The age distributions for most forest type groups in the Central Hardwood Region (Fig. 6.3) have unimodal shapes similar to those observed for state-wide summaries across all groups (Fig. 6.2). The notable exception is the loblolly-shortleaf pine (*Pinus taeda-P. echinata*) group, which has decreasing forest area with increasing forest age. Within the Central Hardwood Region this group is concentrated in Arkansas, where half the area is in plantations and half is in natural stands. The planted loblolly-shortleaf stands are predominantly less than 30 years old, and the natural stands are predominantly greater than 30 years old. For this forest type group intensive management in the last three decades has been highly influential in shaping the age class distribution through extensive establishment of forest plantations.

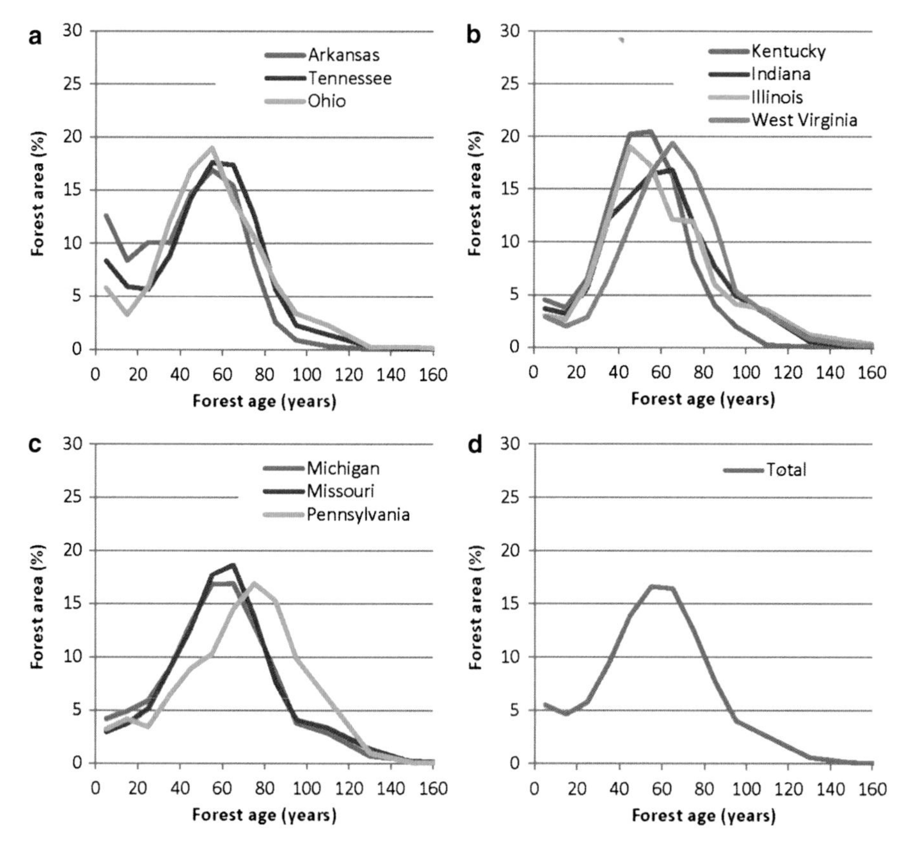

Fig. 6.2 Percent forest area by 10-year age classes for states in the Central Hardwood Region. Panels (**a**), (**b**), and (**c**) display groups of states that are similar in the shape of their age class distributions for forest less than 30 years old

In naturally regenerated forests the forest type groups and forest age classes are interrelated in two important ways. First, some groups are dominated by early successional tree species that are generally favored by a stand-initiating disturbance. Aspens (*Populus tremuloides*; *P. grandidentata*) and birches are relatively short-lived pioneer species that tend to increase in abundance on heavily disturbed sites within their range. As aspen-birch forests age, late successional species such as white spruce (*Picea glauca*), balsam fir (*Abies balsamea*), or sugar maple (*A. saccharum*) often increase in dominance. In the absence of disturbance those species can subsequently dominate the forest cover. Consequently, forests in the aspen-birch forest type group tend to be young relative to those in the maple-beech-birch or spruce-fir forest type groups. Figure 6.3b illustrates this forest-type-group by age progression from early successional aspen-birch and elm-ash-cottonwood (*Ulmus-Fraxinus-Populus*) groups with relatively short-lived species to the late successional oak-hickory and maple-beech groups with relatively long-lived species. Second, the silvicultural systems typically associated with a forest cover type affect the age structure. Intensive plantation management contributes to the abundance of young loblolly-shortleaf forest area.

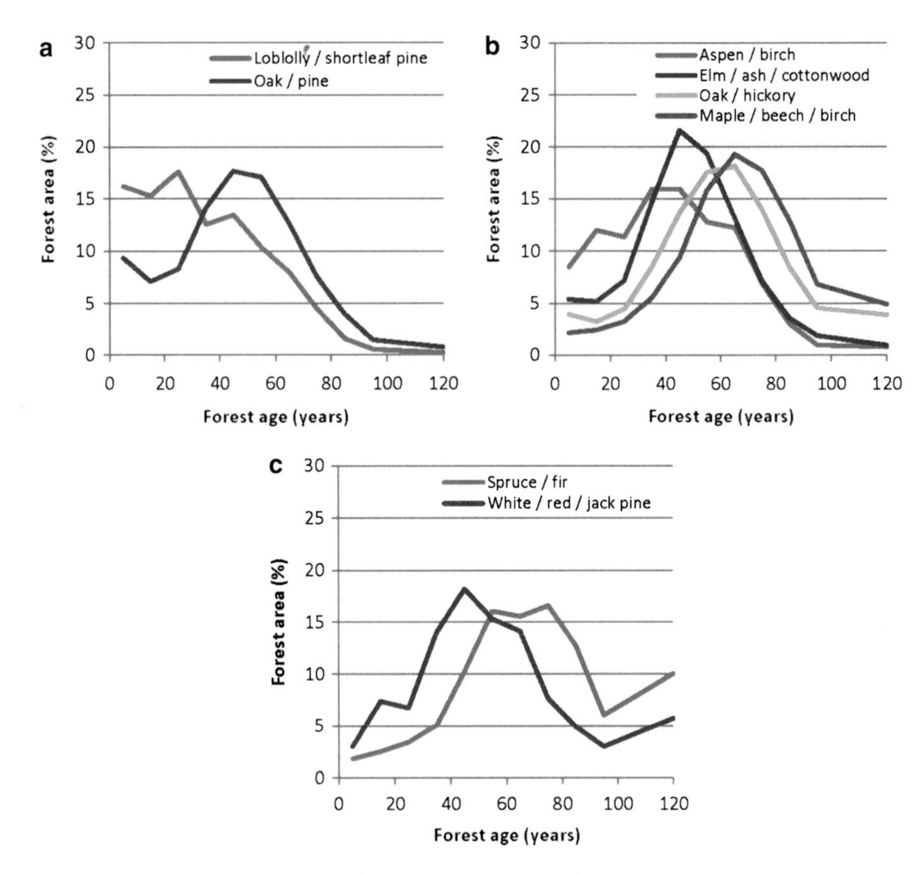

Fig. 6.3 Percent of total forest area by 10-year age class and forest type group for ten states in the Central Hardwood Region. Panel (**a**) displays the two forest type groups with the greatest relative area in the 0–10-year age class. Panel (**b**) illustrates increasing mean and peak forest age class for five hardwood forest type groups. Panel (**c**) shows age distributions for the remaining two conifer forest type groups. The composite age distribution for all forest type groups is identical to Fig. 6.2(**d**)

Short-rotation, even-aged silviculture with clearcutting is commonly used to regenerate aspen-birch forests. Thus, they tend to be relatively young compared to oak-hickory and maple-beech forests that are typically managed on longer rotations with crop tree management and harvest by individual-tree selection.

6.3.3 Forest Age Classes Over Time

The shape of an age class distribution (Figs. 6.1–6.3) reflects the accumulated history of past patterns of stand initiating disturbances for that population. For example, forests currently in the 60-year age class (young sawtimber) underwent a stand initiating disturbance 60 years ago and have not had a subsequent stand initiating disturbance since.

For some individual states it is possible to combine periodic forest inventories over a nearly 60 year period to directly observe changes over time in forest age-class distributions. Indiana lies in the center of the study region and is typical among states that have low abundance of early successional forests (Fig. 6.2). Statewide forest inventories for Indiana were conducted in 1950, 1969, 1986, 1998, 2003, and 2008 (USDA Forest Service 1953; Spencer 1969; Miles 2010). Early forest inventories reported values for timberland rather than forestland, so for consistency we also summarized area of timberland. Timberland excludes forested parks, wilderness areas, and other locations where timber harvest is restricted by policy or legislation. It also excludes areas that are so unproductive or physically inaccessible that commercial harvest operations are impractical. For Indiana statewide summaries this distinction is of little practical significance because 98% of all forest land in the state is classified as timberland. Between 1950 and 2008, total timberland area in Indiana increased from 1.6 to 1.9 million ha, largely due to natural forest regeneration on land that was previously used for agriculture. More than 95% of the state's timberland is in hardwood forest types. Inventories prior to 1998 recorded stand size class rather than age class. To maintain continuity across all inventory years we summarized Indiana forest area by three size classes:

- Seedling/sapling —stands with more than half the stocking in trees between 3 and 13 cm diameter at breast height (dbh). These stands are typically less than 30 years of age.
- Poletimber—stands with more than half the stocking in the combined poletimber (13–28 cm dbh) and sawtimber (>28 cm dbh) size classes with poletimber stocking exceeding sawtimber stocking. These stands are typically 30–60 years of age.
- Sawtimber—stands with more than half the stocking in the combined poletimber and sawtimber size classes with sawtimber stocking equal to or greater than poletimber stocking. These stands are typically at least 60 years old.

The time series from 1950 to 2008 reveals a steady increase in forest in the sawtimber size class (Fig. 6.4). The proportion of young forest represented by the seedling-sapling size class peaked at 24% in 1967, declined to 6% by 1998, and gradually increased to the current level of 8%. The poletimber size class varied in proportion over time as ageing forests in the seedling-sapling size class moved into the poletimber size class and aging forests in the poletimber size class moved into the sawtimber size class.

Abundance of several wildlife species in Indiana that are dependent on early successional habitats have declined precipitously and show a general correspondence with the decline of seedling-sapling forest from its peak in 1967 (Fig. 6.5). Prairie Warblers (*Dendroica discolor*) have steadily declined since monitoring by the North American Breeding Bird Survey began in 1966 (Sauer et al. 2008); American Woodcock (*Scolopax minor*) since monitoring by the Singing-ground Survey began in 1968 (Cooper and Parker 2010), and Ruffed Grouse (*Bonasa umbellus*) since monitoring by drumming routes began in 1979 (Backs 2010). The less than perfect correspondence among these curves could reflect that all are estimates and that habitat is not the sole determinant of bird abundance (also see Franzreb et al., Chap. 9).

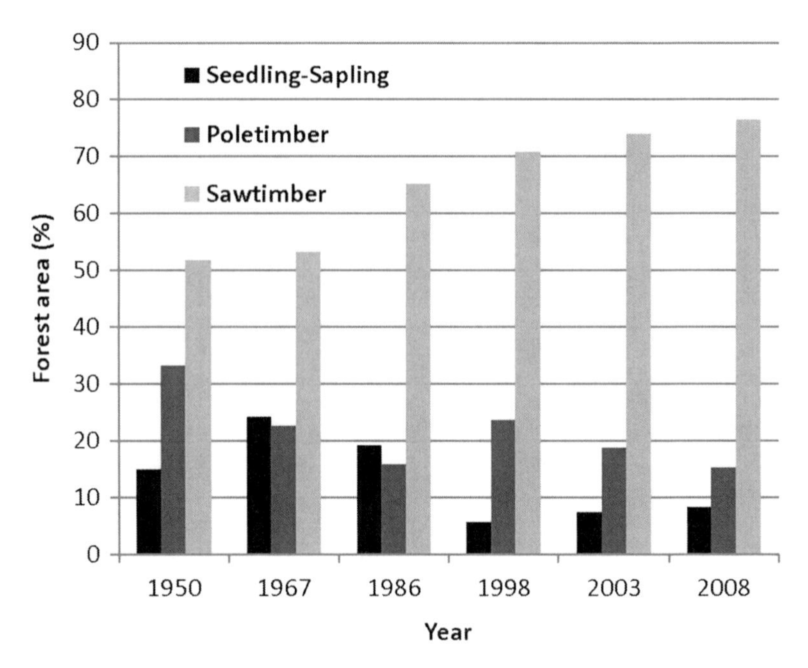

Fig. 6.4 Timberland area by stand size class and inventory year, Indiana. Forests in the seedling-sapling size class are stocked predominantly with trees less than 13 cm dbh. Forests in the poletimber size class are stocked predominantly with trees at least 13 cm and less than 28 cm dbh. Forests in the sawtimber size class are stocked predominantly with trees 28 cm dbh and larger

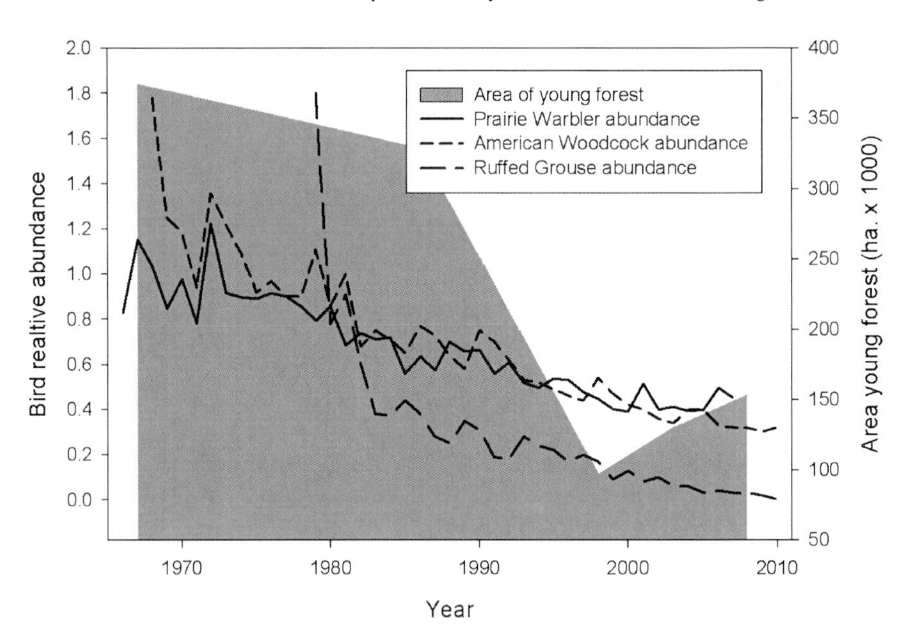

Fig. 6.5 Trends in Prairie Warbler, American Woodcock, and Ruffed Grouse from roadside surveys in Indiana and area of young forest (seedling-sapling class, Fig. 6.4)

6.3.4 Scenario Analyses

Early successional forest habitats are created by disturbances that regenerate a new forest stand, or they develop from open land that is allowed to succeed to forest cover. A given tract of early successional forest is transient; generally it will grow into mid-successional habitat within 20 years (see Greenberg et al., Chap. 1). Thus, sustaining early successional habitats on a forest landscape requires periodic disturbances that regenerate new forest stands and thereby recreate young forest habitats; these events can be forest management activities (e.g. tree harvest, prescribed fire) or natural events (e.g. wind, wild fire) (White et al., Chap. 3). Sustaining polesize forest habitats (e.g., 30–60-year-old forests in the stem exclusion stage of stand development) depends on a steady supply of seedling-sapling habitats that over time becomes polesize forest habitats. Likewise, older forest in the sawtimber size class requires a progression over time from undisturbed younger habitats. In stark contrast to old forest habitats, which require more than a century with minimal disturbance to develop, silvicultural treatments can quickly create young forest habitats via timber harvesting (see Loftis et al., Chap. 5), often with income from an associated timber sale.

Scenario analyses provide one way to examine rates of disturbances necessary to continuously replenish young forest habitats without depleting late successional forest habitats. Scenario outcomes can be compared with actual recent rates of stand initiation that are reflected in the current area of forest in the 0–20-year age class. Such comparisons can be used to estimate how much additional young forests should be created (or avoided) in order to sustain a stable total area of young forest over time.

The forest age class distributions (Figs. 6.1–6.3) reveal that both young forest habitats and old forest habitats are in short supply. Therefore, we constrained the scenarios to maintain a fixed proportion of the total forest area in age classes older than 120 years while balancing younger age classes to also sustain a constant proportion of forest habitats less than or equal to 10 years of age. Specific assumptions are as follows:

1. Dedicate 15% of total forest area to become old forest habitats (>120 years old, on average, but potentially varying by forest type, geographic location, and other factors).
2. Manage to evenly distribute the remaining forest area (85% within age classes less than 120 years) to (a) periodically replenish young forest habitats 10 years old or younger, (b) provide a uniform distribution of habitat in age classes 10–120 years, (c) maintain a pool of mature forest to serve as a replacement for old forest when it is lost to natural disturbances, and (d) increase forest habitat diversity (i.e., coarse scale age and structural diversity) compared to current forest conditions.
3. The current proportion of forest area 20 years of age or younger is a reasonable indicator of recent rates of early successional habitat creation by natural and anthropogenic disturbances over the past two decades.

4. The difference between (a) current rates of forest regeneration (assumption 3) and (b) equilibrium rates of forest regeneration derived from assumptions 1 and 2 is (c) the additional quantity of early successional habitats that should be created (or avoided) to move toward balanced forest age classes with a steady supply of young forest habitats over time.

The first two assumptions dictate that 15% of forest area is devoted to old forest habitats (older than 120 years) protected from harvest, and the remaining 85% of forest area is split evenly among twelve 10-year age classes from 0 to 120 years of age. Thus, keeping 7.1% of the forest area in young forest (up to 10 years old) and eventually in every 10-year age class up to 120 years would move the forest age distribution toward an equilibrium with greater age class diversity than currently exists.

Expressed more generally

$$Y_{eq} = (1 - R)/(0.1A_{min}) \qquad (6.1)$$

Where

Y_{eq} = the equilibrium proportion of total forest area to be maintained in each 10-year age class for the scenario. This is also the target proportion of young forests (i.e., forests less than or equal to 10 years old).

R = proportion of total forest area set aside in old forests or other protected forest areas that are excluded from management for early successional habitats.

A_{min} = minimum age for old forests in years

An alternate scenario with 20% old forests or other reserves and a 100 year minimum age for old forests results in a decadal target of 8% of forest area in young forest habitats. Similarly, assumptions of 30% old forests and other reserves with a 150 year minimum age for old forests indicate an equilibrium of 4.7% in young forest habitats per decade.

Comparison of the current proportion of young forest habitats for a given state or region to the target proportion indicated by a particular scenario indicates the deficit (or excess) of early successional habitats relative to that norm (Table 6.2). For the initial scenario with 15% of the forest area retained in old forest reserves and 120 years as the minimum age for old forests, roughly 3.6 million ha of young forests (up to 10 years old) should be established across the region each decade to maintain steady replenishment of young forest habitats. By extension, this assumes that roughly 3.6 million ha of forests will be maintained in each age class up through age 120 years. For this scenario the target by state for young forest area ranges from a low of 136,000 ha per decade in Indiana to a high of 568,000 ha per decade in Michigan (Table 6.2, column [f]). The difference by state between the target and current observed area of young forest habitats (0–10-year age class) ranges from a deficit of roughly 230,000 ha for Missouri, Pennsylvania and West Virginia to a surplus of about 250,000 ha for Arkansas (Table 6.2, column [h]). Area of young forests for Tennessee is currently very close to the target for this scenario.

Table 6.2 Young forest habitat target and observed surplus (or deficit) habitat by state for a scenario of 15% reserved forest (no harvest disturbance), a minimum old-growth habitat age threshold of 120 years, and the target area of young forest habitats (0–10 years old) computed as an equal proportion of the forest area 0–120 years old (see Eq. 6.1). Deficit or surplus area of young forest habitats is reported as a percent and as total area relative to the current area of young forest habitats. States are ordered from greatest percent deficit to greatest percent surplus of young forest habitats relative to the target (Column [e]). Under this scenario all states except Tennessee and Arkansas are presumed to be deficient in young forest habitats. Surplus or deficit area of young forest habitats on a county basis provides an indication of the annual management implications of addressing the surplus or deficit for the scenario if the forest regeneration treatments were distributed throughout the state on a county basis. Current forest area in the 0–10 and 11–20 year age classes are provided as points of reference indicating recent observed rates of stand initiating disturbances that create young forest habitats; column [g] is the mean of columns [k] and [l]

[a] State	[b] No of counties	[c] Target young forest area for the 0–10 year age class[a] (%)	[d] Observed young forest area for the 0–10 year age class[b] (%)	[e] Difference of observed and target young forest area in the 0–10 year age class (%)	[f] Target young forest area for the 0–10 year age class[a] (ha)	[g] Observed young forest area in the 0–10 year age class[b] (ha)	[h] Difference of observed and target young forest in the 0–10 year age class (ha)	[i] Deficit or surplus young forest habitat needed to meet target (ha/year)	[j] Additional young forest habitat needed annually per county to meet target (ha/year)	[k] Current area in 0–10 year age class (ha)	[l] Current area in 11–20 year age class (ha)
W. Virginia	55	7.1	2.4	–4.7	343,394	117,139	–226,255	–22,626	411	138,337	95,940
Illinois	102	7.1	2.8	–4.3	137,585	54,299	–83,286	8,329	82	57,687	50,911
Missouri	115	7.1	3.3	–3.8	441,543	207,047	–234,496	–23,449	204	182,213	231,881
Indiana	92	7.1	3.4	–3.7	136,051	65,433	–70,618	–7,062	77	70,070	60,796
Pennsylvania	67	7.1	3.7	–3.4	475,722	247,306	–228,416	–22,842	341	215,940	278,672
Kentucky	120	7.1	4.1	–3.0	354,518	206,560	–147,959	–14,796	123	225,105	188,014
Michigan	83	7.1	4.5	–2.5	568,426	364,067	–204,359	–20,436	246	334,746	393,388
Ohio	88	7.1	4.5	–2.5	230,362	147,706	–82,657	–8,266	94	189,930	105,482
Tennessee	95	7.1	7.1	0.0	397,741	400,094	2,353	235	0	467,696	332,492
Arkansas	75	7.1	10.5	3.4	531,098	785,440	254,343	25,434	0	942,566	628,315
Total	892	7.1	5.1	–2.0	3,616,440	2,595,091	–1,021,349	–102,135	1,236	2,824,290	2,365,891
Mean per State		7.1	4.6	–2.4	361,644	259,509	–102,135	–10,213	123	282,429	236,589

[a]Based on the scenario described in the table caption, 7.1% of the total forest area per state should be maintained as early successional forests in the 0–10 year age class

[b]Observed area of early successional habitats in the 0–10 year age class was computed as the average per decade observed for the 0–10 year and the 11–20 year age classes from columns [k] and [l], respectively. This provided a larger sample size than relying solely on data for the 0–10 year age class

On an absolute scale most states have large deficits in area of early successional forests relative to this scenario (Table 6.2) and, in fact, relative to any semblance of a balanced age class distribution (Figs. 6.1–6.3). However, when the deficits in early successional forest habitats are spread out across a state on an annual basis, they appear more manageable. For example, on a per county basis for an average county the total area of additional early successional habitats needed is about 100–400 ha annually (Table 6.2, column [j]). That quantity of early successional habitats can be achieved by the equivalent of a few to few dozen timber harvests per county each year. Obviously, in a practical implementation of this scenario, the total area of young forest habitats would be greater in areas with greater total forest cover. Nevertheless, the average proportion of forest area regenerated to young forest habitats in a given region for a given year would be about 0.007.

6.4 Landscape-Level Effects of Management and Disturbance

We used a landscape simulation model to demonstrate the landscape-scale cumulative effects of forest management and disturbances on forest age class distributions. We present results from simulations applied to a 71,142 ha portion of the Mark Twain National Forest, Missouri, in the Western Dry Subregion of the Central Hardwood Region (McNab, Chap. 2). Shifley et al. (2006) previously reported results for these simulations including species composition by forest size classes and dominant species group across the landscape. Here we present age-class distributions to illustrate the cumulative effects of alternative choices among silvicultural methods and management practices that can be used to used to create young forest habitat. We paid particular attention to the effects on patch size distributions for young forest age classes because for some species of conservation concern the size and spatial arrangement of young forest patches is as important as the total area of young forests.

We applied LANDIS version 3.6 (He and Mladenoff 1999; Mladenoff and He 1999; He et al. 2005) to simulate forest vegetation response to disturbance by timber harvest, wind, and fire. Details of our parameterization of LANDIS are provided in Shifley et al. (2006). Simulations were conducted for five management scenarios: uneven-aged management with group selection harvest affecting 5% and 10% of the landscape per decade, even-aged management with clearcut harvest affecting 5% and 10% of the landscape per decade, and no harvest. All scenarios included natural disturbance by wildfire (based on a 300-year mean wildfire return interval) and wind (based on an 800-year mean blowdown return interval) (Shifley et al. 2006). We reported results for year 200 of the scenarios because by that point in the simulation the age distributions and associated spatial patterns had equilibrated with respect to the assumptions of the scenario. We also reported the proportion of forest in 10-year age classes and the patch size distribution for forest 0–20 years old.

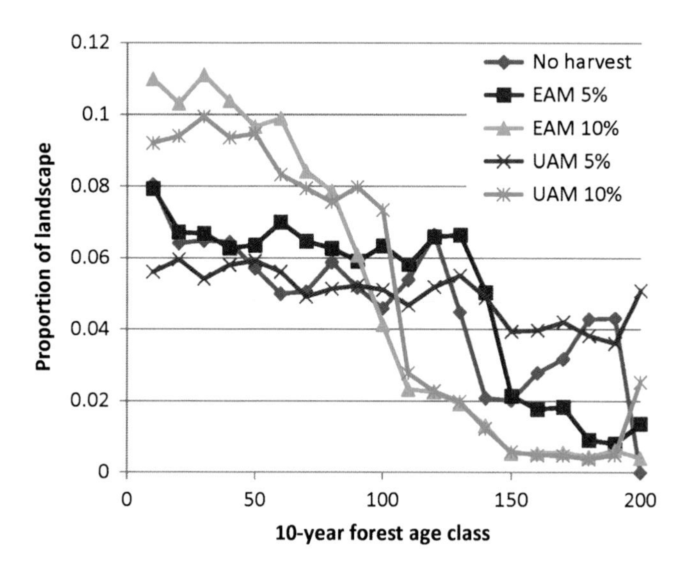

Fig. 6.6 Forest age class distributions following 200 years of simulated management and natural disturbance on a portion of the Mark Twain National Forest in the Missouri Ozarks. Age classes were summarized for each 0.01 ha site across the 71,142 ha modeled landscape. Scenarios represent no tree harvest and even-aged management (EAM) or uneven-aged management (UAM) that harvests 5% or 10% of the landscape per decade

The age class distributions resulting from our landscape simulation scenarios (Fig. 6.6) were much flatter than the current age class distributions observed by state and forest cover type (Figs. 6.1–6.3) because the scenarios implemented a consistent disturbance and management regime applied over a period of 200 years. This simulated pattern of future disturbance is very different than the actual history of timber harvest in the region over the last 200 years which was punctuated by widespread exploitive timber harvesting in the late 1800s and early 1900s.

Not surprisingly, 10% harvest per decade produces more young forest than 5% harvest per decade or no harvest (Fig. 6.6). Perhaps more unexpected is the similarity in age class distributions for no harvest and 5% even-aged or uneven-aged management. Two factors contribute to this similarity. First, the wildfire and wind disturbances modeled in the scenario affect, on average, 4.6% of the landscape per decade. Modeled wind and wildfire events in the scenarios do not always result in total loss of overstory with a stand initiating event, but over many decades the simulated wind and fire events forecast the creation of young forest habitats (see Spetich et al., Chap. 4; White et al., Chap. 3). The same is true in reality and such inevitable natural disturbances should be figured into management programs intended to maintain a specific proportion of the landscape in young forest habitats. Second, the initial conditions for the modeled scenarios are based on the observed forest age structure in the year 2000. Thus, the initial age distribution is similar to that for Missouri as shown in Fig. 6.2c. Over the 200 years of the modeled scenarios, tree species on some sites were predicted to suffer age dependent mortality when their typical

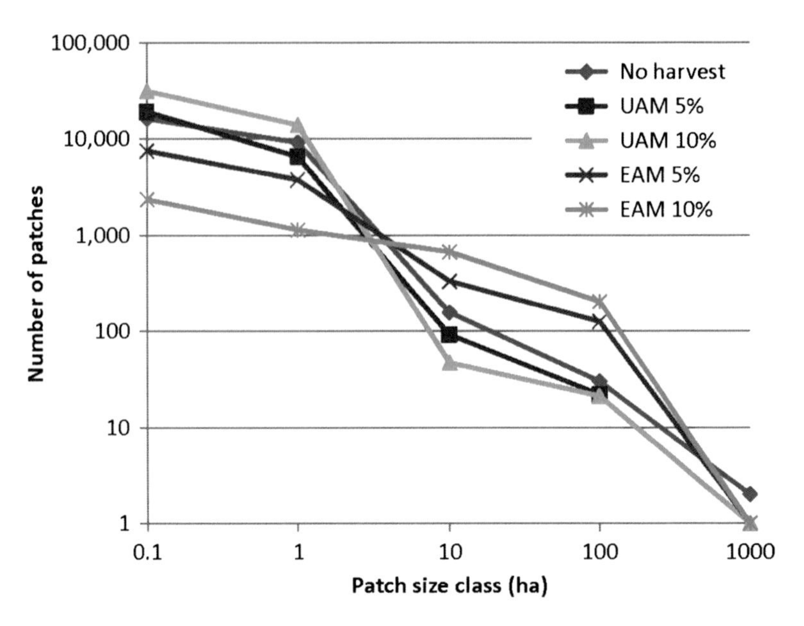

Fig. 6.7 Patch-size distribution for seedling-sapling size forest resulting from 200 years of simulated management and natural disturbance on a 71,142 ha portion of the Mark Twain National Forest in the Missouri Ozarks. Scenarios represent no tree harvest and even-aged management (EAM) or uneven-aged management (UAM) that harvests 5% or 10% of the landscape per decade

species longevity was exceeded. In certain situations this modeled natural mortality can result in initiation of a new, young forest stand.

Age-class distributions, however, do not indicate anything about spatial distribution of forests within estimation units, e.g., states. The simulated management regimes produce different patch-size distributions of young forests (Loftis et al., Chap. 5); these are evident when graphed (Fig. 6.7) or mapped (Fig. 6.8). Both the amount and spatial pattern of young forests affect habitat quality for early successional species such as the Prairie Warbler (Fig. 6.9). Higher intensity of harvest results in more young forests and generally results in higher overall levels of habitat suitability for this species. However, Prairie Warblers avoid small patches of young forests and favor large patches. Consequently, even-aged management with clearcutting results in larger patch sizes, providing substantially better Prairie Warbler habitat than a comparable total area of young forest distributed among many small patches as in the uneven-aged management regime with group selection harvesting. By contrast, the Hooded Warbler (*Wilsonia citrina*) inhabits older forests but uses small gaps in the forest; its habitat suitability is greater under uneven-aged management with group selection harvests that create small regeneration openings in the forest canopy. Estimated habitat suitability for the mature forest-associated Ovenbird (*Seiurus aurocapillus*) is largely unaffected by patch size. Similar habitat relationships exist for tree species that are characterized by their dependence on disturbance (Johnson et al. 2009).

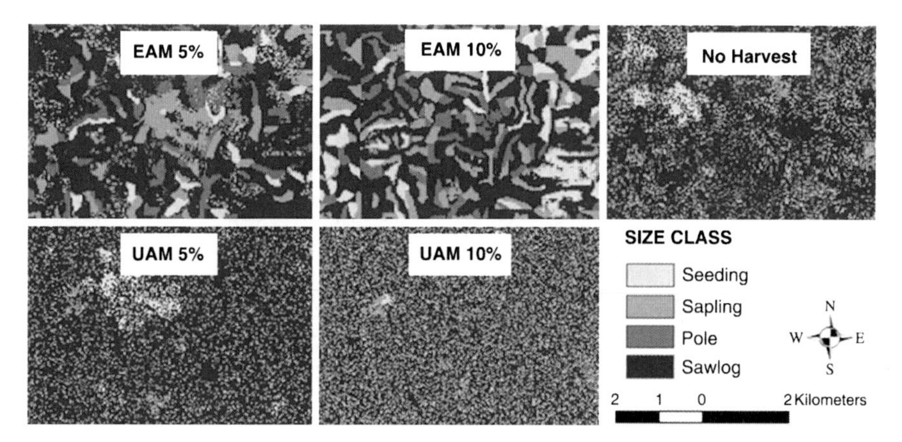

Fig. 6.8 Maps of forest patches by tree size class for landscapes resulting from 200 years of simulated management and natural disturbance on a 2,835 ha portion of the Mark Twain National Forest in the Missouri Ozarks. Scenarios represent no tree harvest and even-aged management (EAM) or uneven-aged management (UAM) that harvests 5% or 10% of the landscape per decade. The seedling size class includes forest up to 10 years old, sapling 11–30 years old, pole, 31–59 years old, and sawlog ≥60 years old (adapted from Shifley et al. 2006)

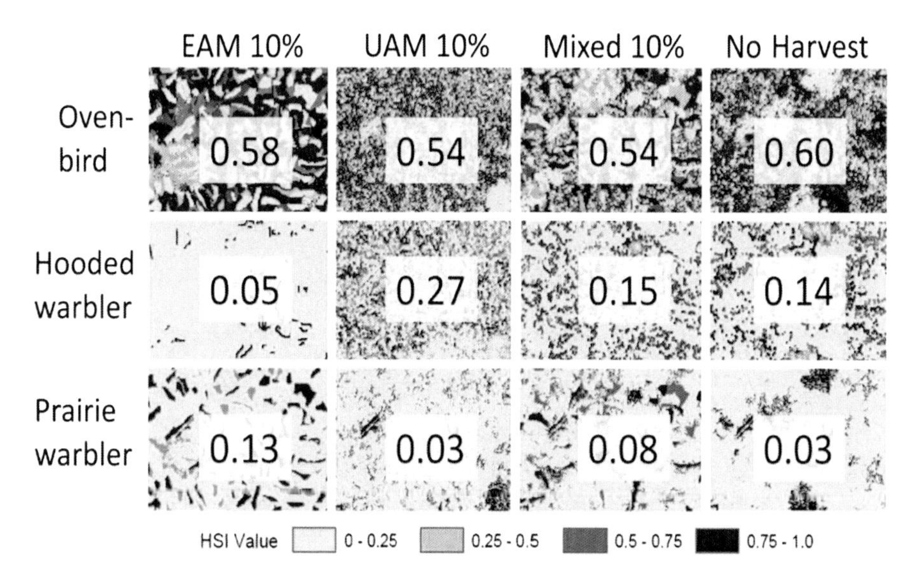

Fig. 6.9 Mean habitat suitability and maps of habitat suitability for three bird species from 200 years of simulated management and natural disturbance on a 2,835 ha portion of the Mark Twain National Forest in the Missouri Ozarks. Scenarios represent no tree harvest and even-aged management (EAM) or uneven-aged management (UAM) or an even mix of EAM and UAM (Mixed) that harvested 10% of the landscape per decade (adapted from Shifley et al. 2006)

6.5 Discussion and Conclusions

Summaries of forest age class distributions in the Central Hardwood Region reveal a paucity of young forests for most states and most forest types. More than half of the 51 million ha of forest area in the region is between 40 and 80 years of age. Only 25% of forest area is in stands between 0 and 40 years of age, and young forests up to 10 years of age constitute only 5.5% of the forest area. Nevertheless there is considerable variation in the proportion of young forests among states and forest types. Five states (Illinois, Indiana, Missouri, Pennsylvania, and West Virginia) have less than 4% of forest area in young forests while Arkansas has more than 10%. Consequently, any efforts to increase young forests will likely vary among states in magnitude and urgency.

No doubt managers will differ in opinions about how much young forest is desirable or necessary. However, it is relatively easy to examine scenarios with differing levels of old forest reserves and differing harvest rotation ages to explore alternatives for maintaining young forests. One scenario on which we focused suggested maintaining about 7% of total forest area as young forests in the 0–10 year age class. Under that scenario there is currently a region-wide young forest deficit of one million ha (Table 6.2). However that presumed deficit is unequally distributed among states and forest types.

The changes in management practices necessary to address a region-wide one million ha deficit of young forest can appear overwhelming in aggregate. However, when recast as the potential affected area per year distributed across all counties within a given state, the effort required to maintain young forest is less daunting. Given recent rates of forest disturbance and associated creation of young forests, management activities that created additional young forests in the amount of 200 ha per county per year would be sufficient, on average, to increase the region-wide proportion of these early successional habitats to more than 7% of total forest area. A potentially larger problem is continuing to periodically monitor the need for additional young forest habitats by region, state, and forest type, and annually creating additional young forest habitats where they are in deficit. Ultimately uncertainties about where, in what patch sizes, and in what forest types to create additional young forest habitats to maximize benefits to wildlife will present greater complexities than understanding how much total additional habitat is needed or the logistics of how to create it at a selected site.

Success or failure in sustaining young forests can be determined by the success or failure of wildlife species that depend upon these forests, although we acknowledge that factors other than habitat abundance also may be limiting some wildlife populations. Sustaining species that depend on young forests requires managers be cognizant of more than just the amount of young forests. We demonstrated through simulation modeling that different management practices and natural disturbance regimes create different landscape patterns of tree or stand size classes. Through simulation, management practices can be identified that provide beneficial patch and landscape characteristics for a particular target species. The reality is, however, that even among a small group of focal species habitat and landscape requirements will likely be varied enough that a diversity of practices will best meet the needs of all species. Desired distributions for patch size or landscape composition can be developed based on historic range of variation concepts (Landres et al. 1999), or at the coarser scale by simply ensuring that a diversity of management practices are used within and among landscapes.

Young forests are transient. Without repeated disturbance they are gone 10–20 years after they are created. Fortunately, it is easy to create additional young forests through timber harvest, often with the added economic benefit of a timber sale. Emerging markets for woody biomass may provide new opportunities for creating young forest habitats. Wildfire, weather events, and even insect or disease outbreaks also contribute to the replenishment of young forest habitats. Older forest habitats cannot be recreated as quickly. Consequently, an underlying assumption in our scenarios analyses is that an even distribution of forest area by age class (up to some specified threshold for "old" forests) is desirable. Strictly speaking, increasing the proportion of young forest area (0–10-years old) does not require that attention be given to the distribution of forest area among older age classes; young forest habitats can be recreated from forest of any age or condition. However, the broader underlying intent is to increase or sustain forest biodiversity, and young forest is only one component of a diverse forest age structure. Old forests and middle-aged forests can only be replenished over time via the aging of younger forests. Thus, forest landscapes characterized by non-uniform age class distributions (e.g., negative exponential, bimodal, irregular) present no problems for managers wishing to create additional young forest habitats, but over time perpetuation of non-uniform age class distributions will result in gaps in the quantity of older forest habitats.

The current age class distribution in the Central Hardwood Region is a product of the region's disturbance history. Exploitive timber harvesting that moved east to west was followed by woods burning, livestock grazing, and farming marginally productive lands. Much of the current abundance of forests in the 40–80-year age classes is the result of forest regeneration following abandonment of those practices. The region's current forest age structure is the cumulative effect of many decades of forest growth and disturbance. Increasing the area of young forests can be accomplished very quickly if forest owners and managers are motivated to do so. However altering the current forest age structure—which took many decades to create—to provide a more uniform distribution of forest area across all forest age classes would take many decades of purposeful management and monitoring to accomplish.

Forest ownership patterns in the Central Hardwood Region may prove to be a significant barrier to systematically increasing the area of young forest or to developing a more uniform distribution of forest area by age class (Wear and Huggett, Chap. 16). In the study area 31 million ha or 61% of the forest area is in family forests distributed among 3.5 million private owners. The mean ownership size is small and many owners likely will perceive creation of early successional habitats on their tracts to be at odds with their ownership goals which are weighted toward aesthetics, recreation, and protection. The most efficient options for creating young forest habitats on family woodlands may be in working with the 5% of family woodland owners with forest tracts at least 40 ha in size, because collectively they own 40% of all family forest area in the region (USDA Forest Service 2010).

The coarse filter approach we employed to examine forest age class structure, quantity of young forests, and temporal dynamics of young forests is simplistic but efficient for addressing large geographic extents over long timeframes. We assumed that all forest habitats can be categorized by age class and that there is a link between forest age and habitat characteristics. That may be realistic for young forest habitats 10 years old or younger, but for partially disturbed forests, woodlands, or savannahs the connection between forest age and habitat structure is more tenuous. Those limitations notwithstanding, the coarse filter analyses indicate the magnitude of the young forest resource, its spatial and temporal distributions, and the scope and complexity of some remedies where the amount of young forest habitats is considered to be deficient. The general approach can be used to help set and monitor regional goals for providing young forest habitats. It serves as a starting point for efforts to increase young forest habitats which, during implementation, must also address the more complex issues of where, when, and in what patch sizes to increase young forest habitats for the greatest benefit of different wildlife species that depend on them.

Acknowledgements Forest Inventory and Analysis personnel from the US Forest Service spent 60 years collecting the data we analyzed in this study. Bill Dijak was instrumental in the analysis and mapping of the landscape-scale disturbance scenarios and associated wildlife habitat suitability indices. Two anonymous reviewers provided valuable suggestions that improved an earlier version of this manuscript. We thank them all.

Literature Cited

Backs SE (2010) 2010 spring breeding indices of ruffed grouse. Wildlife management and research notes No. 998, Indiana Division Fish and Wildlife, Indianapolis

Baltimore County (2005) Baltimore county forest sustainability strategy, steering committee final draft. Baltimore County, 128 p. Accessed 23 Sept 2010

Carpenter C (2007) Forest sustainability assessment for the Northern United States. NA-TP-01-07CD. USDA Forest Service Northeastern Area State and Private Forestry, Newtown Square

Cooper TR, Parker K (2010) American woodcock population status, 2010. USDI Fish and Wildlife Service, Laurel

Eyre FH (ed) (1980) Forest cover types of the United States and Canada. Soc Amer For, Washington, DC

Forest Inventory and Analysis (2010a) Data and tools. USDA Forest Service, Washington, DC. http://www.fia.fs.fed.us/tools-data/default.asp. Accessed 23 Sept 2010

Forest Inventory and Analysis (2010b) Forest inventory and analysis national program. USDA Forest Service, Washington, DC. http://www.fia.fs.fed.us/. Accessed 23 Sept 2010

Haufler JB, Melh CA, Roloff GJ (1996) Using a coarse-filter approach with species assessment for ecosystem management. Wildl Soc Bull 24:200–208

He HS, Mladenoff DJ (1999) Spatially explicit and stochastic simulation of forest-landscape fire disturbance and succession. Ecology 80:81–99

He HS, Li W, Sturtevant BR, Yang J, Shang ZB, Gustafson EJ, Mladenoff DJ (2005) LANDIS 4.0 users guide. LANDIS: a spatially explicit model of forest landscape disturbance, management, and succession. Gen Tech Rep NC-263, USDA Forest Service North Central Research Station, St. Paul

Helms JA (1998) The dictionary of forestry. Soc Amer For, Bethesda
Johnson PS, Shifley SR, Rogers R (2009) The ecology and silviculture of oaks, 2nd edn. CABI Publishing, Wallingford/Oxfordshire
Landres PB, Morgan P, Swanson FJ (1999) Overview of the use of natural variability concepts in managing ecological systems. Ecol Appl 9:1179–1188
Mladenoff DJ, He HS (1999) Design and behavior of LANDIS, an object oriented model of forest landscape disturbance and succession. In: Mladenoff DJ, Baker WL (eds) Advances in spatial modeling of forest landscape change: approaches and applications. Cambridge University Press, Cambridge
Miles PD (2010) Forest inventory EVALIDator web-application version 4.01 beta. USDA Forest Service Northern Research Station, St. Paul. http://fiatools.fs.fed.us/Evalidator4/tmattribute.jsp. Accessed 1 Nov 2010
National Archives and Records Administration (2007) The president: executive order 13423 – strengthening federal environmental, energy, and transportation management. USA Federal Register, 26 Jan 2007, 3919–3923
Oliver CD, Larson BC (1990) Forest stand dynamics. McGraw-Hill, New York
Rittenhouse CD, Dijak WD, Thompson III FR, Millspaugh JJ (2007) Development of landscape-level habitat suitability models for ten wildlife species in the Central Hardwoods Region. Gen Tech Rep NRS-4, USDA Forest Service Northern Research Station, Newtown Square
Sauer JR, Hines JE, Fallon J (2008) The North American breeding bird survey, results and analysis 1966–2007. Version 15 May 2008. USGS Patuxent Wildlife Research Center, Laurel
Schulte LA, Mitchell RJ, Hunter ML Jr, Franklin JF, McIntyre RK, Palik BJ (2006) Evaluating the conceptual tools for forest biodiversity conservation and their implementation in the US. For Ecol Manage 232:1–11
Shifley SR, Thompson FR III, Dijak WD, Larson MA, Millspaugh JJ (2006) Simulated effects of forest management alternatives on landscape structure and habitat suitability in the Midwestern United States. For Ecol Manage 229:361–377
Smith WB, Miles PD, Perry CH, Pugh SA (2009) Forest resources of the United States, 2007. Gen Tech Rep GTR-WO-78, USDA Forest Service, Washington, DC
Spencer Jr JS (1969) Indiana's timber. Res Bull NC-7, USDA Forest Service North Central Forest Experiment Station, St. Paul
Tirpak JM, Jones-Farrand DT, Thompson III, Twedt DJ (2009) Multi-scale habitat suitability index models for priority landbirds in the Central Hardwoods and West Gulf Coastal Plain/Ouachitas bird conservation regions. Gen Tech Rep NRS-49, USDA Forest Service Northern Research Station, Newtown Square
Twedt DJ, Tirpak JM, Jones-Farrand DT, Thompson FT III, Uihlein WB III, Fitzgerald JA (2010) Change in avian abundance predicted from regional forest inventory data. For Ecol Manage 260:1241–1250
USDA Forest Service (1953) Forest statistics of Indiana. Forest survey release No. 15, USDA Forest Service Central States Forest Experiment Station, Columbus
USDA Forest Service (2004) National report on sustainable forests – 2003. FS-766, USDA Forest Service, Washington, DC
USDA Forest Service (2010) National woodland owner survey. USDA Forest Service, Washington. http://www.fia.fs.fed.us/nwos/. Accessed 31 Oct 2010
Woodall C, Hansen M, Brand G, McRoberts R, Gallion J, Jepsen E (2006) Indiana's forests 1999–2003 (Part B). Res Bull NC 253B, USDA Forest Service North Central Research Station, St. Paul
World Commission on Environment and Development (1987) Our common future. Oxford University Press, New York

Chapter 7
Herbaceous Response to Type and Severity of Disturbance

Katherine J. Elliott, Craig A. Harper, and Beverly Collins

Abstract The herbaceous layer varies with topographic heterogeneity and harbors the great majority of plant diversity in eastern deciduous forests. We described the interplay between disturbances, both natural and human-caused, and composition, dynamics, and diversity of herbaceous vegetation, especially those in early successional habitats. Management actions that create low to moderate disturbance intensity can promote early successional species and increase diversity and abundance in the herb layer, although sustaining communities such as open areas, savannahs, and woodlands may require intensive management to control invasive species or implement key disturbance types. A mixture of silvicultural practices along a gradient of disturbance intensity will maintain a range of stand structures and herbaceous diversity throughout the central hardwood forest.

7.1 Introduction

The herbaceous layer, made up of all herbaceous species and woody species under a meter height, harbors the great majority of plant diversity in eastern deciduous forests (Gilliam and Roberts 2003). In landscapes with significant topographic

K.J. Elliott (✉)
USDA Forest Service, Southern Research Station, Center for Forest Watershed Research, Coweeta Hydrologic Laboratory, Otto, NC, USA
e-mail: kelliott@fs.fed.us

C.A. Harper
Department of Forestry, Wildlife, and Fisheries, University of Tennessee, Knoxville, TN, USA
e-mail: charper@utk.edu

B. Collins
Department of Biology, Western Carolina University, Cullowhee, NC, USA
e-mail: collinsb@email.wcu.edu

C.H. Greenberg et al. (eds.), *Sustaining Young Forest Communities*,
Managing Forest Ecosystems 21, DOI 10.1007/978-94-007-1620-9_7,

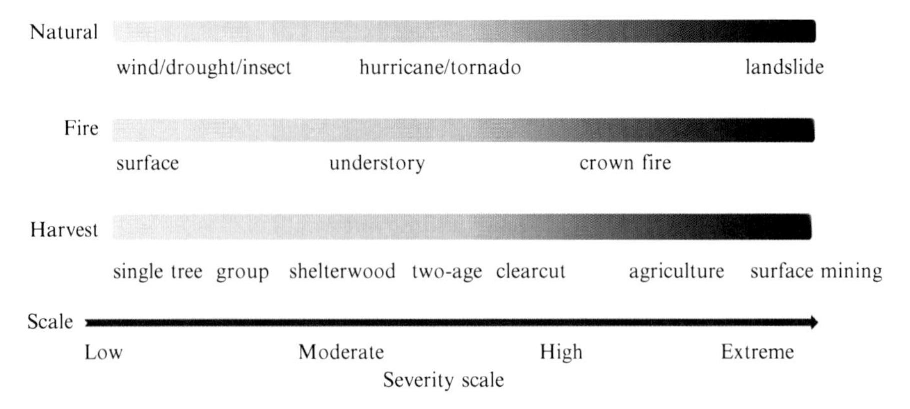

Fig. 7.1 Conceptual diagram of disturbance severity scale for natural and human-induced events

heterogeneity, herb layer composition and diversity vary with gradients of microclimate, soil moisture, and soil fertility (Hutchinson et al. 2005). Herb layer vegetation also is affected by natural and anthropogenic disturbances. Disturbances to the tree canopy, including individual tree falls, catastrophic wind events, catastrophic wildfire, and timber harvesting, result in moderate to large increases in resource availability (Small and McCarthy 2002; Roberts and Gilliam 2003). Low severity disturbances, such as surface fires, usually cause minor damage to overstory trees but affect herb layer vegetation directly by killing aboveground stems and indirectly by altering the forest floor and the availability of light, water, and nutrients (Elliott et al. 2004; Knoepp et al. 2009). At the highest end of a severity scale (Fig. 7.1), disturbances such as agriculture, landslides, and surface mining remove vegetation and till or entirely remove the soil, even down to bedrock. In this chapter, we examined the interplay between disturbance, both natural and human-caused, and composition, dynamics, and diversity of herbaceous vegetation. We briefly discuss herb layer contribution to early successional habitats in different communities, and then focus mostly on herbaceous layer response to specific types and severities of disturbance.

7.2 Early Successional Communities

The herb layer composition of open areas such as abandoned pastures, savannahs, and woodlands affects the quality of early successional habitats these communities provide to wildlife (Jones and Chamberlain 2004; Donner et al. 2010). Desirable plants provide protective cover and nutritious food sources, and allow travel, feeding, and loafing by wildlife within and under the cover. Conversely, undesirable plants provide suboptimal cover, seed, or forage that is not palatable or digestible and inhibit mobility of small animals. When undesirable plants dominate an area, usable space is limited and the abundance and species richness of wildlife may be

Fig. 7.2 An open field (**a**) and woodland (**b**) in eastern Tennessee with abundant native warm season grasses and forbs (photo by C.A. Harper)

relatively low. Management actions that increase diversity and abundance of desirable herb layer species can help sustain quality early successional habitats in these communities (Fig. 7.2).

In open areas and abandoned pastures, for example, eradicating non-native plant cover such as tall fescue (*Festuca elatior*) and bermudagrass (*Cynodon dactylon*), may be necessary before more desirable plant species can be established (Harper et al. 2007; Harper and Gruchy 2009). Tall fescue, which became the most important cultivated pasture grass in the Central Hardwood Region by the 1970s, develops a dense, sod-forming structure near the ground and deep thatch that restricts mobility of several birds (Harper and Gruchy 2009), including young Eastern Wild Turkey (*Meleagris gallopavo*), Northern Bobwhite (*Colinus virginianus*), Field Sparrows (*Spizella pusilla*) and Grasshopper Sparrows (*Ammodramus savannarum*). Its dense growth and thatch can suppress germination of more desirable ground layer plants such as broomsedge (*Andropogon virginicus*), big bluestem (*Andropogon gerardii*), little bluestem (*Schizachyrium scoparium*), blackberry (*Rubus* spp.), American pokeweed (*Phytolacca americana*), native lespedezas (*Lespedeza* spp.), ticktrefoil (*Desmodium* spp.), and partridge pea (*Chamaecrista fasiculata*). Desirable open areas have a mixture of native warm-season grasses and forbs with scattered patches of shrubs, such as wild plum (*Prunus* spp.), sumac (*Rhus* spp.), and crabapple (*Malus* spp.).

Prescribed fire, particularly growing season fires, may be necessary to reduce woody encroachment and maintain early successional grasses and forbs (Klaus et al. 2005; Harper 2007; Gruchy et al. 2009) in savannahs and woodlands. These communities are found throughout tropical and temperate portions of the world and are characterized by scattered overstory trees and a continuous herbaceous understory rich in grasses and forbs. Frequent fire, grazing, and periodic drought or relatively low annual precipitation maintain the open canopy of savannahs and woodlands (Brudvig and Asbjornsen 2008), and the vast majority of these communities in the eastern USA have been lost over the past century as a result of fire suppression, agriculture, and development (Scholes and Archer 1997; Abrams 2003; Spetich et al., Chap. 4). Management using late-dormant season fire at 3 year intervals led to dramatic increases in both richness and density of small mammals and songbirds, and provided more than adequate high-quality forage for white-tailed deer and elk in mature shortleaf pine (*Pinus echinata*) – bluestem sites (Masters 2007).

7.3 Disturbance and Forest Herb Layer Vegetation

In forests, herbaceous vegetation response depends on the type and severity of disturbances, which regulate supplies of resources such as light, soil nutrients, and moisture (Clinton 1995). Stand-replacing, high-severity disturbances (Fig. 7.1) create relatively homogeneous resource availability while low- to moderate-severity disturbances (Fig. 7.1) partially remove the canopy and generally result in greater resource heterogeneity (Gravel et al. 2010; White et al., Chap. 3). Silvicultural systems used in central hardwood forests represent a gradient of disturbance severity, from the least intense single-tree selection (harvesting individual selected trees from most of all size classes) to the most intense clear-cutting (complete removal of the stand in a single harvest) (Loftis et al., Chap. 5). In the following sections, we discuss human-caused and natural disturbances that commonly affect herbaceous vegetation in forests of the Central Hardwood Region.

7.3.1 Harvests

Herbaceous response to forest harvests differs among ecoregions within the Central Hardwood Region. In the Southern Appalachians and adjacent areas, high growth rates and nutrient concentrations of herbaceous plants result in faster recovery of aboveground biomass following clearcutting (Boring et al. 1988; Elliott et al. 2002a) compared to northern hardwood forests (Federer et al. 1989; Reiners 1992; Mou et al. 1993). For example, 1 year after harvest, aboveground biomass of herbs in clearcuts ranged from 0.18 to 0.40 Mg ha^{-1} in a hardwood watershed in western North Carolina (Elliott et al. 2002a) compared to only 0.09 Mg ha^{-1} in a northern hardwood forest in New Hampshire (Mou et al. 1993). However, herbaceous layer diversity in the harvested North Carolina watershed was lower than that in a nearby mature (≈70-years-old) forest (Table 7.1). In addition, it can take decades for herb layer diversity to recover from clearcut harvests. For example, flatter dominance diversity curves for reference and pre-harvest compared to post-harvest stands in two clearcut watersheds in the Coweeta Basin in western North Carolina show the herbaceous layer has not recovered diversity 30 years after disturbance (Fig. 7.3).

In contrast to the Southern Appalachians, all measures of herbaceous abundance and diversity in young (ca. 7 years old) clearcuts were greater than those in mature (more than 125 years old) stands in the Central Appalachians of Ohio (Small and McCarthy 2005), including mean cover (10.94% ± 1.42 versus 4.89 ± 0.57), richness, and H′ diversity (Table 7.1). Clearcut and mature forests shared high importance of several species, including white wood aster (*Aster divaricatus*), hog peanut (*Amphicarpaea bracteata*), whorled loosestrife (*Lysimachia quadrifolia*), Christmas-fern (*Polysticum acrostichoides*), and dooryard violet (*Viola sororia*). At the same time, younger stands showed greater importance of annual or shade-intolerant graminoids, such as sedges (*Carex digitalis*, *Carex laxiflora*), panic grass (*Panicum clandestinum*), and *Poa* spp., and non-native herbs (e.g., hoary bitter-cress (*Cardamine hirsuta*) and sulphur cinquefoil (*Potentilla recta*)), while mature stands showed greater importance of shade-tolerant perennials such as black cohosh (*Cimicifuga racemosa*), bland sweet cicely (*Osmorhiza claytonia*), Solomon's seal (*Polygonatum pubescens*), false Solomon's seal (*Smilacina racemosa*), and bellwort (*Uvularia perfoliata*) (Small and McCarthy 2005).

Other studies from sites within the Central Hardwood Region show diverse herb layer responses to forest harvests. Belote et al. (2009) used sites in Virginia and West Virginia to investigate how a gradient in disturbance intensity caused by different levels of timber harvesting influenced plant diversity through time and across spatial scales ranging from a square meter to 2 ha. The gradient of tree canopy removal and associated forest floor disturbance ranged from clearcut (95% basal area removed), leave-tree harvest (74% basal area removed leaving a few dominants), shelterwood harvest (56% basal area removed), understory herbicide (suppressed trees removed via basal application of herbicide), to uncut control. In the first year after disturbance, herbaceous species diversity increased at all spatial scales, but after 10 years of forest development shading by the canopy once again

Table 7.1 Mean (SE) herbaceous layer diversity (S = species richness; H′ = Shannon's diversity index; E′ = Pielou's evenness index) for different types and severities of disturbance for numerous studies across the Central Hardwood Region. S, H′ and E were calculated at the small plot (1.0 m^2) level with means and standard errors presented for each treatment and study location

Treatment	Location	Community	Time since last disturbance	Diversity[a] S	H′	E	References
Silviculture Rx							
Mature forest [precut forest]	Coweeta WS7	Mixed deciduous, low elevation, south-facing	50+ years	16.1 (1.3)	2.11 (0.10)	0.79 (0.02)	Elliott et al. 1997[b]
Clearcut	Coweeta WS7	Mixed deciduous, low elevation, south-facing	1 year	3.6 (0.3)	0.91 (0.09)	0.80 (0.02)	Elliott et al. 1997[b]
Clearcut	Coweeta WS7	Mixed deciduous, low elevation, south-facing	17 years	4.4 (0.4)	0.80 (0.08)	0.62 (0.04)	Elliott et al. 1997[b]
Clearcut	Coweeta WS7	Mixed deciduous, low elevation, south-facing	30 years	4.8 (0.5)	0.78 (0.08)	0.59 (0.04)	Elliott, unpublished
Mature forests [reference]	Allegheny Plateau, OH	Mixed-oak, low elevation (< 320 m)	>125 years	4.7 (0.4)	1.08 (0.09)		Small and McCarthy 2005
Clearcuts	Allegheny Plateau, OH	Mixed-oak, low elevation (< 320 m)	7 years	6.0 (0.4)	1.34 (0.07)		Small and McCarthy 2005
Mature forest [reference]	Wine Spring, NC	*Quercus rubra,* high elevation (> 1,200 m)	50+ years	15.0 (0.7)	2.12 (0.06)	0.79 (0.01)	Elliott and Knoepp 2005[b]
Two-age cut	Wine Spring, NC	*Quercus rubra,* high elevation (> 1,200 m)	2 years	14.9 (0.8)	2.12 (0.08)	0.80 (0.02)	Elliott and Knoepp 2005[b]

Group selection	Wine Spring, NC	*Quercus rubra,* high elevation (> 1,200 m)	2 years	16.8 (0.7)	2.25 (0.07)	0.81 (0.02)	Elliott and Knoepp 2005[b]
Shelterwood	Wine Spring, NC	*Quercus rubra,* high elevation (> 1,200 m)	2 years	18.8 (0.8)	2.40 (0.06)	0.82 (0.01)	Elliott and Knoepp 2005[b]
Mature forest [reference]	Fay Branch, NC	Mixed deciduous, mid elevation (850–950 m)	50+ years	21.0 (1.0)	1.89 (0.07)	0.63 (0.02)	Elliott, unpublished
Two-age cut	Fay Branch, NC	Mixed deciduous, mid elevation (850–950 m)	2 years	31.5 (1.7)	2.38 (0.06)	0.70 (0.01)	Elliott, unpublished
Abandoned pasture or old field							
Mature forest [reference]	Coweeta WS14	Mixed deciduous, low elevation, north-facing	70 years	12.1 (1.6)	1.51 (0.15)	0.70 (0.04)	Elliott et al. 1998[b]
Grass-to-forest	Coweeta WS6	Mixed deciduous, low elevation, north-facing	1 year	11.6 (0.58)	1.54 (0.08)	0.63 (0.03)	Elliott et al. 1998[b]
Grass-to-forest	Coweeta WS6	Mixed deciduous, low elevation, north-facing	28 years	11.6 (0.58)	1.54 (0.08)	0.63 (0.03)	Elliott et al. 1998[b]
Old Field	Wayne County, WV	Stream floodplain, bottomland hardwoods	20 years	13.4 (0.5)	2.21 (0.05)	0.86 (0.01)	Gilliam and Dick 2010
Pasture	Wayne County, WV	Stream floodplain,	< 1 year	5.9 (0.2)	1.38 (0.03)	0.78 (0.01)	Gilliam and Dick 2010
Fire							
Mature forest [pre-burn]	Wine Spring, NC	Pine-oak-heath	50+ years	5.8 (1.4)	1.02 (0.22)	0.35 (0.03)	Elliott et al. 1999[b]

(continued)

Table 7.1 (continued)

Treatment	Location	Community	Time since last disturbance	Diversity[a]			References
				S	H′	E	
Rx fire (single burn)	Wine Spring, NC	Pine-oak-heath	2 years	5.9 (0.3)	1.35 (0.05)	0.80 (0.02)	Elliott et al. 1999[b]
Rx fire (single burn)	Wine Spring, NC	Pine-oak-heath	10 years	7.0 (0.4)	1.53 (0.06)	0.81 (0.02)	Elliott et al. 2009[b]
Mature forest [reference]	Ocoee, TN	Shortleaf pine-mixed oak	50+ years	5.2 (0.4)	1.12 (0.10)	0.71 (0.04)	Elliott, unpublished
Rx fire (repeated 2X)	Ocoee, TN	Shortleaf pine-mixed oak	2 years	5.4 (0.5)	1.19 (0.10)	0.76 (0.03)	Elliott, unpublished
Cut+Rx fire (repeated 2X)	Ocoee, TN	Shortleaf pine-mixed oak	2 years	6.3 (0.4)	1.36 (0.08)	0.76 (0.02)	Elliott, unpublished
Cut+Rx fire (repeated 2X)	Ocoee, TN	Mesic, mixed oak-pine	2 years	11.8 (0.5)	1.90 (0.06)	0.78 (0.02)	Elliott, unpublished
Mature forest [reference]	Smoky Mountains, TN	Mixed deciduous	50+ years	22	2.4 (0.2)	0.74 (0.02)	Holzmueller et al. 2009
Wildfire (single burn)	Smoky Mountains, TN	Mixed deciduous	~20 years	27	2.5 (0.2)	0.75 (0.03)	Holzmueller et al. 2009
Wildfire (repeated 2X)	Smoky Mountains, TN	Mixed deciduous	~20 years	27	2.3 (0.3)	0.70 (0.03)	Holzmueller et al. 2009
Wildfire (repeated 3X)	Smoky Mountains, TN	Mixed deciduous	~20 years	27	2.0 (0.4)	0.62 (0.05)	Holzmueller et al. 2009
Mature forest [reference]	Allegheny Plateau, OH	Oak-hickory	50+ years	14 (0.8)	3.73 (0.04)	0.91 (0.01)	Hutchinson et al. 2005[c]

Rx fire (repeated 2X)	Allegheny Plateau, OH	Oak-hickory	2 years	17 (0.8)	3.82 (0.03)	0.92 (0.01)	Hutchinson et al. 2005[c]
Rx fire (annual 4X)	Allegheny Plateau, OH	Oak-hickory	1 year	17 (0.8)	3.81 (0.03)	0.91 (0.01)	Hutchinson et al. 2005[c]
Mature forest [reference]	Monongahela National Forest, WV	Mixed deciduous	50+ years		1.06 (0.11)		Royo et al. 2010
Rx fire + gap + grazing	Monongahela National Forest, WV	Mixed deciduous	5 years	5.6 (1.1)	1.31 (0.09)		Royo et al. 2010
Fire + gap	Monongahela National Forest, WV	Mixed deciduous	5 years	2.9 (0.7)			Royo et al. 2010
Post-burn, 2-year fire interval	Fort Benning, GA	Mixed oak-pine	postburn	4.7 (0.7)	3.63[e] (0.15)	0.49 (0.003)	Collins, unpublished[d]
Post-burn, 4-year fire interval	Fort Benning, GA	Mixed oak-pine	postburn	4.1 (0.3)	3.55[e] (0.09)	0.54 (0.01)	Collins, unpublished[d]
2-year fire interval	Fort Benning, GA	Mixed oak-pine	1 year	5.4 (0.9)	3.8 (0.17)	0.49 (0.04)	Collins, unpublished[d]
4-year fire interval	Fort Benning, GA	Mixed oak-pine	3 year	5.0 (0.54)	3.7 (0.11)	0.50 (0.02)	Collins, unpublished[d]
Wind disturbance							
Mature forest [reference]	Massac County, IL	Bottomland hardwoods	60+ years		0.92		Nelson et al. 2008[e]
Wind	Massac County, IL	Bottomland hardwoods	3 years		1.47		Nelson et al. 2008[e]
Wind + salvage	Massac County, IL	Bottomland hardwoods	3 years		1.72		Nelson et al. 2008[e]

(continued)

Table 7.1 (continued)

Treatment	Location	Community	Time since last disturbance	Diversity[a] S	H′	E	References
Mature forest [reference]	Coweeta	High elevation (> 1,100 m), Mixed-oak	70+ years	9.5 (1.0)	1.49 (0.14)	0.70 (0.04)	Elliott et al. 2002b[b]
Wind + salvage	Coweeta	High elevation (> 1,100 m), Mixed-oak	2 years	14.5 (0.7)	1.87 (0.07)	0.71 (0.02)	Elliott et al. 2002[b]
Mature forest [reference]	Northwestern, CO	Subalpine, high elevation (> 1,260 m), spruce-fir	>200 years	14.6 (1.5)	2.2 (0.2)	0.75 (0.04)	Rumbaitis del Rio 2006
Wind	Northwestern, CO	Subalpine, high elevation (> 1,260 m), spruce-fir	4 years	17.8 (1.6)	2.2 (0.1)	0.76 (0.02)	Rumbaitis del Rio 2006
Wind + salvage	Northwestern, CO	Subalpine, high elevation (> 1,260 m), spruce-fir	4 years	6.3 (0.7)	1.3 (0.1)	0.74 (0.02)	Rumbaitis del Rio 2006

[a]Shannon index was calculated as: $H' = pi \ln pi$, where pi = proportion of total percent cover or total aboveground biomass (g) of species i. Species evenness was calculated as: $E = H'/H'_{MAX}$, where H'_{MAX} = maximum level of diversity possible within a given population = ln(number of species) (Magurran 2004). Standard errors are in parentheses

[b]S, H, and E were re-calculated at a finer spatial scale (1.0 m^2 plot) than presented in the original manuscripts (≥100 m^2 plot) for comparisons among other studies listed in the table

[c]Hutchinson et al. (2005) used a 2.0 m^2 plot for herbaceous layer sampling

[d]Collins (unpublished) based calculations on twelve 12 m line intercept samples in each of 10 sites, significantly different ($p = 0.039$)

[e]Newman et al. (2008) found no significant differences ($p = 0.082$) among treatments

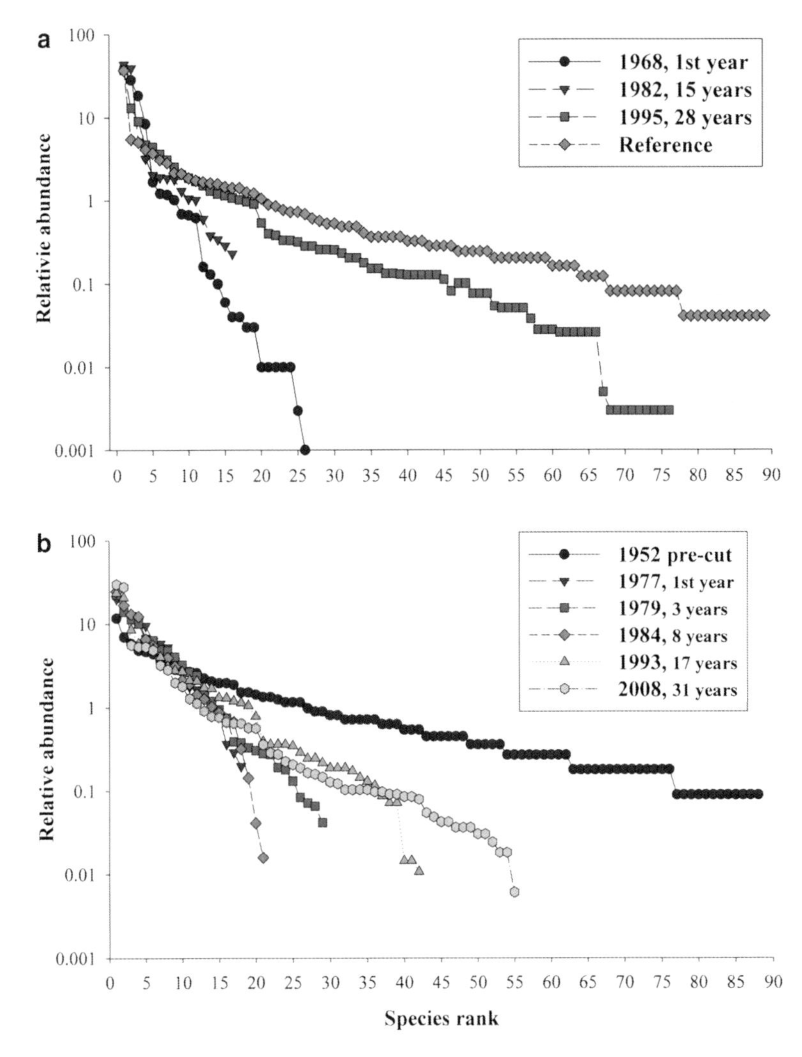

Fig. 7.3 Dominance-diversity curves for two clearcut watersheds, WS6 and WS7, in the Coweeta Basin, western North Carolina (Adapted from Elliott et al. 1997, 1998 and Elliott, unpublished). Curves were based on percent cover for (**a**) WS6 at 1, 15, and 28 years after the final disturbance and (**b**) WS7 prior to clearcutting in 1952, and 1, 3, 8, 17, and 31 years after cutting. Flatter curves represent high species diversity or low dominance by a few species; in contrast, steep curves represent low species diversity or a high degree of dominance (Whittaker 1965)

controlled diversity (Belote et al. 2009). Zenner et al. (2006) compared five harvest treatments in upland mixed oak hardwoods in the Missouri Ozarks. The harvest treatments caused overstory canopy reductions from 12.8% in controls to 83.6% in clearcuts. Herb layer vegetation showed a clear response that increased in proportion to harvest treatment intensity, with relative species composition and abundance of life forms increasing in proportion to harvest intensity. Dominance of legumes and tree seedlings decreased while woody vines, graminoids, and annuals/biennials increased along the harvest intensity gradient. Elliott and Knoepp (2005) found a similar pattern in herbaceous layer diversity in the Southern Appalachians; group selection (24% canopy reduction) and shelterwood harvests (68% canopy reduction) had higher species richness and diversity (Shannon's index of diversity, Magurran 2004) than the heavier two-age cut (80% canopy reduction) and reference forests (Table 7.1). In partial cuts, shade from the residual overstory trees created a mosaic of environmental conditions, which provided suitable microsites for a mix of shade-intolerant and shade-tolerant herbaceous species, and higher species richness and diversity than an undisturbed forest.

Taken together, the research shows that harvesting central hardwood forests affects diversity and species composition of herbaceous layer vegetation. Diversity can increase or decrease following harvest, then recovers, but the recovery can take decades to reach pre-harvest or reference values. In addition, harvests can increase abundance of shade-intolerant species associated with early successional habitats in proportion to intensity of the harvest treatment.

7.3.2 *Abandoned Agricultural Lands*

Abandoned agricultural land is common in the eastern USA (Parker and Merritt 1994; Bellemare et al. 2002), but is declining as oldfields shift to forest lands. In the Southern Appalachians, for example, agricultural lands have declined by an average 13% from the 1950s to the 1990s (Wear and Bolstad 1998). In fact, major portions of today's eastern National Forests were once abandoned agricultural land (Jenkins and Parker 2000; Thiemann et al. 2009).

Agricultural use has had a definite and severe effect on native plant communities (Flinn and Vellend 2005). Forests growing on former agricultural land often have lower frequencies of many native forest herbs than forests that were never cleared for agriculture. A leading explanation for this pattern is that many forest herbs are dispersal-limited, but environmental conditions can also hinder colonization (Fraterrigo et al. 2009a, b). Abandoned agricultural areas have a species composition that is highly variable and distinct from other disturbance types. For example, in southern Indiana, several typical disturbance species, such as blackberry (*Rubus* spp.) and northern groundcedar (*Lycopodium complanatum*), and many non-native species such as grass pink (*Dianthus armeria*), meadow fescue (*Festuca pratensis*), and oxeye daisy (*Leucanthemum vulgare*), were associated with abandoned agriculture plots (Jenkins and Parker 2000). Abandoned agriculture plots had significantly greater cover of giant ragweed (*Ambrosia trifida*, federally listed as a noxious-weed and common in oldfields) than four other stands types (Jenkins and Parker 2000).

In Great Smoky Mountains National Park, abandoned agricultural plots were associated with species normally found in dry and sub-mesic communities, including ebony spleenwort (*Asplenium platyneuron*), ribbed sedge (*Carex virescens*), poverty oatgrass (*Danthonia spicata*), hillside blueberry (*Vaccinium palladium*), and dwarf dandelion (*Krigia biflora*) (Thiemann et al. 2009). They also were associated with an influx of non-native and non-forest species such as northern ground-cedar (*Lycopodium complanatum*), Japanese honeysuckle (*Lonicera japonica*), and heart-leaved groundsel (*Senecio aureus*) (Thiemann et al. 2009). In addition, many other indicators of mesic forests, including star chickweed (*Stellaria pubera*), Canadian woodnettle (*Laportea canadensis*), bloodroot (*Sanguinaria canadensis*), celandine-poppy (*Stylophorum diphyllum*), and five-parted bitter-cress (*Cardamine concatenate*) were not found in the abandoned agriculture plots (Thiemann et al. 2009).

In general, abandoned agricultural fields can maintain early successional vegetation on the landscape from open site through young forest conditions. Early successional species that establish in the herbaceous layer can persist for several decades. At the same time, these sites may promote invasive species and have slow establishment of forest understory herbs.

7.3.3 *Surface Mining and Mountain-Top Removal*

Surface mining, particularly mountain-top removal, is the most severe disturbance type in the Central Hardwood Region, with the exception of landslides (Hales et al. 2009). Some coal surface mines have been reclaimed for more than 40 years, and reclamation has been mandated by USA federal law for almost 30 years (Surface Mining Control and Reclamation Act, Public Law 95–87 Federal Register 3 Aug 1977, 445–532). Coal surface mine reclamation practices are similar to those of other large-scale land reclamation projects: a few aggressive plant species are seeded or planted in an effort to achieve legal requirements for minimum ground cover and prevent soil erosion. Many mine reclamation efforts focus on establishing rapid-growing non-native species that control erosion but may slow or prevent the establishment of later-successional, native species (Holl 2002). Until recently, this seeded ground cover consisted of Kentucky-31 tall fescue (*Festuca elatior*), red clover (*Trifolium pratense*), sericea lespedeza (*Lespedeza cuneata*), and birdsfoot trefoil (*Lotus corniculatus*), all of which are non-native and dense.

Although efforts are underway to establish native species, many recently mined mountain tops are still hydro-seeded with a non-native mixture of species. Once these species are established, it can be difficult to reduce their cover and replace them with native species. In addition, these non-native plant communities may be susceptible to establishment of invasive woody species. For example, 50 years after being reclaimed with sericea lespedeza, red clover, and Kentucky-31 tall fescue beneath planted eastern white pine (*Pinus strobus*), autumn olive (*Eleagnus umbellata*), privet (*Ligustrum* spp.), and dying white pines made up a significant component of the woody understory and forest edge vegetation on a coal surface mine in eastern Kentucky (Collins, unpublished).

7.3.4 Fire

Prescribed burning is used by the USDA Forest Service, USDI National Park Service, The Nature Conservancy and other land owners to reduce fuel loads, improve wildlife habitat, and restore ecosystem structure and function. However, less is known about its effects on eastern hardwood ecosystems than on southern pine dominated ecosystems. There, prescribed fire has been used as a silvicultural tool for over 50 years (see review: Carter and Foster 2004). In general, vegetation is responsive to prescribed fire, but the magnitude of response depends on initial forest condition and fuel load, topography, and season and characteristics of the fire, among other factors (Spetich et al., Chap. 4). In the following sections, we discuss herbaceous vegetation response to fire in two major forest types of the Central Hardwood Region: oak forests and hardwood pine forests.

7.3.4.1 Fire in Oak Forests

Perennial herbs in oak forests usually emerge each season from rhizomes, but they are dormant during the spring and fall burning periods. Because little heat penetrates into the soil to the dormant rhizomes when leaf litter burns, resprouting usually is not affected by burning in either season. Any changes in herb layer species composition or abundance would more likely be due to indirect effects such as reduced competition with top-killed midstory shrubs, or consumption of the litter layer. Keyser et al. (2004) found plant cover and species richness in an oak-dominated forest increased following fire regardless of whether burning occurred in February, April, or August. However, the more intense spring and summer burns led to a shift toward herbaceous species, whereas the winter burn resulted in dominance by woody species (Keyser et al. 2004).

In some cases in central hardwood forests, prescribed fire resulted in increased cover and diversity of herbaceous layer species (Arthur et al. 1998; Elliott et al. 1999; Clinton and Vose 2000; Clendenin and Ross 2001). In mixed-oak communities, herbaceous layer species tend to be more diverse after moderate-severity fire (Elliott et al. 1999; Glasgow and Matlack 2007), partly due to removal of the litter layer, increased nutrient cycling rates, and increased light levels. However, low severity, dormant season fires often have little effect on plant community composition (McGee et al. 1995; Kuddes-Fischer and Arthur 2002), and in some cases they have little effect on diversity (Franklin et al. 2003; Dolan and Parker 2004; Hutchinson et al. 2005; Elliott et al. 2004; Phillips et al. 2007; Elliott and Vose 2010).

Although single prescribed burns may have little effect, repeated dormant-season fire may affect herbaceous layer diversity, particularly warm-season grasses and forbs (Holzmueller et al. 2009; Pyke et al. 2010) in oak forests. For example, Bowles et al. (2002) found a significant shift in herbaceous layer vegetation toward greater abundance of warm-season plants, without decline of cool-season plants, after 17 years of annual fires. They suggested repeated burning can increase forest herbaceous layer diversity in a predictable manner: repeated, annual burns reduce shrubs and saplings, which subsequently increases understory light levels. They also found a positive relationship between canopy light levels with warm-season herb cover and richness (Bowles et al. 2002).

7.3.4.2 Fire in Pine-Hardwood Forests

Mixed pine-hardwood forests on dry ridges are thought to be sustained by fire (Barden 2000; Lafon et al. 2007). Fire suppression and few natural fires in dry-to-xeric pine-hardwood forests have promoted dominance of hardwoods and decline of the pine component of these forests for the last three decades (Smith 1991; Vose et al. 1999; Elliott and Vose 2005). In addition, substantial drought-related insect populations (primarily southern pine beetle [*Dendroctonus frontalis*]) (Elliott et al. 1999; Elliott and Vose 2005) and previous forestry practices, such as high-grading, have contributed to changes such as a significant increase in acreage of stands with a dense understory of mountain laurel (*Kalmia latifolia*) on upper, drier slopes of the Southern Appalachians. Competition with mountain laurel inhibits reproduction and growth of woody and herbaceous vegetation, so changes in species composition and stand structure are likely to persist without management intervention.

Herbaceous species respond to direct and indirect effects of fire. An initial increase in nitrogen availability after fire can contribute to increased herbaceous cover (Elliott et al. 2004; Knoepp et al. 2009). In addition, low severity prescribed fires, coupled with dormant season ignition, allow the root systems and seed banks of herbaceous layer species to survive; thus, plants are able to re-emerge in the spring and summer after the burn treatments. The herbaceous layer includes several life forms that may respond differently to fire disturbance: tree seedlings, shrubs, forbs, ferns, and graminoids. In a Southern Appalachians pine-oak community, Elliott et al. (2009) found evergreen shrubs decreased, while deciduous shrubs, forbs, and grasses increased after a moderate-severity prescribed fire (Fig. 7.4). After 10 years, forbs and grasses were more abundant than they were before the prescribed fire treatment (Elliott et al. 2009).

In another site in the Southern Appalachians (Linville Gorge; Dumas et al. 2007), post-disturbance colonizers such as fireweed (*Erichtites hieracifolia*), daisy fleabane (*Erigeron annuus*), and white snakeroot (*Eupatorium rugosum*) were present only in burned plots, where they likely flushed from the seed bank. Greater diversity and abundance of herbs and tree seedlings in the first post-fire growing season were likely a response to the combination of forest floor removal by fire and increased penetration of light associated with the loss of the mountain laurel. These findings are consistent with Reilly et al. (2006), who argued that changes in species diversity after the Linville Gorge fire were the result of local scale phenomena, and not long distance dispersal. Fire would favor seed bank species and species able to propagate from protected meristems.

Dilustro et al. (2002, 2006) and Collins et al. (2006a, b) examined herb layer response to prescribed fire and land use (military) in pine and mixed pine-hardwood

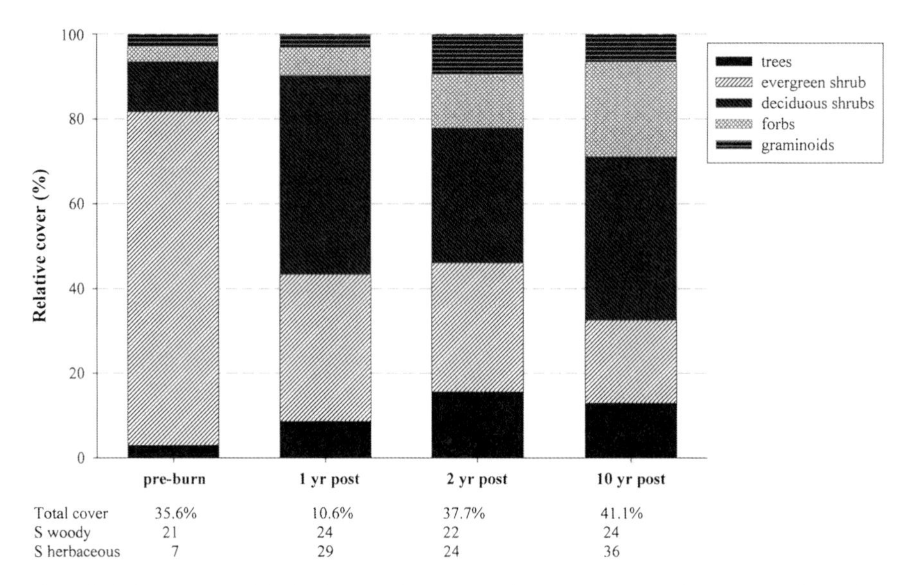

Fig. 7.4 Relative cover of the herbaceous layer (all herbaceous species and woody species <1.0 m height) by growth forms for Wine Spring Creek, western North Carolina; a dormant season, moderate-to-high intensity prescribed burn. Total cover (%), S_{woody} (number of woody species), $S_{herbaceous}$ (number of herbaceous species) for pre-burn (1994), 1-year post-burn (1995), 2-years post-burn (1996), and 10-years post-burn (2006) (adapted from Elliott et al. 2009)

forests at Fort Benning, GA. In a subset of these sites with a significant hardwood component, neither species richness nor evenness differed between 2 year and 4 year fire treatments, either in the post-burn season or after one (2-year treatment) or three (4-year treatment) years. Overall diversity (H′) was higher in 2-year burn treatments in the post-burn season, but this difference was not apparent 1–3 years(s) post-burn (Table 7.1). Across all sites, however, fire, harvests, and disturbances associated with mechanized military training favor pine dominance and maintain early successional or fire-tolerant species in the ground layer (Dilustro et al. 2002).

Positive response of some herb layer species provides evidence that growing season fire is an important part of the natural disturbance regime in pine-hardwood forests. However, what is best for one species may not be for all; other species respond more to dormant-season than growing-season burns (Sparks et al. 1998; Hiers et al. 2000; Liu and Menges 2005). In addition, many species do not appear to be influenced by burning season. For example, in a shortleaf pine-grassland community in Arkansas, fewer than 10% of 150 plant species evaluated for response to late growing-season (September–October) and late dormant-season (March–April) burns were differentially affected by burning season (Sparks et al. 1998). The variable response of understory species to fire season suggests a heterogeneous fire regime (including variation in the seasonal timing of fire) may help conserve biodiversity (Hiers et al. 2000; Liu et al. 2005) and maintain early successional stages of pine-hardwood forests on the landscape.

7.3.5 Drought

Canopy gaps, created by wind or death of canopy trees, are widely known to influence woody seedling and sapling species recruitment and abundance through their effect on resource availability and heterogeneity (Clinton et al. 1993; Elliott and Swank 1994; Kneeshaw and Bergeron 1998; Kloeppel et al. 2003; Gravel et al. 2010). Less is known about the effects of gaps created by drought on the herbaceous layer in temperate forested ecosystems (Roberts and Gilliam 2003; Neufeld and Young 2003). Information is especially lacking on how interactions among drought-induced canopy gaps and other disturbances, such as herbivory and fire, affect herbaceous vegetation (sensu Royo et al. 2010).

One long-term study conducted in the Southern Appalachians provides an example of the complex interactions among disturbances. Webster et al. (2008) investigated effects of Japanese stilt grass (*Microstegium vimineum*), an invasive grass, and deer herbivory on native herbaceous layer species in Cades Cove, Great Smoky Mountains National Park. A severe drought occurred in 2000, partway through their 10 year study (1997–2006). With deer herbivory, Japanese stilt grass populations rebounded quickly following drought and native herbaceous and woody species were unable to capitalize on the ephemeral release of growing space. In contrast, in the absence of deer herbivory (i.e., in exclosure plots), there was an increase in cover of woody plants and native species richness (Webster et al. 2008).

7.3.6 Windthrow and Salvage Logging

Canopy gaps caused by windthrow have different consequences for herb layer vegetation than gaps caused by drought. Windthrow uproots trees and breaks or kills surrounding trees, which, in turn, creates pit and mound topography (Clinton and Baker 2000) and generally creates larger canopy openings (Greenberg and McNab 1998; Peterson 2000; Elliott et al. 2002b; Peterson and Leach 2008) than drought-created gaps. Elliott et al. (2002b) reported a greater number of both early and late successional herb species in forests with windthrow and subsequent salvage logging than in an undisturbed forest (Table 7.1). In addition, some late successional species that were found in both forests were more abundant in the disturbed forest; these included Jack-in-the-pulpit (*Arisaema triphyllum*), black cohosh (*Cimicifuga racemosa*), wild licorice (*Galium lanceolatum*), common yellow wood-sorrel (*Oxalis stricta*), and violets (*Viola* spp.).

In a bottomland hardwood forest in southern Illinois, Nelson et al. (2008) investigated differences in vegetation composition and diversity among undisturbed, wind disturbed, and wind + salvage areas. They found species diversity (H′) generally increased as a function of soil disturbance (based on soil disturbance severity classes ranging from undisturbed < compressed < ruts < churned), with no significant differences between wind and wind + salvage areas (Table 7.1). Significantly less herbaceous cover in undisturbed and transition areas was attributed to having at least partial canopy cover in these sites versus wind and wind + salvage areas. Nelson et al. (2008) argued that large yearly variation in herbaceous cover among soil disturbance classes was due to creation of ruts, berms, pits, and mounds, which led to variation in moisture availability on a fine spatial scale. Three years after the wind disturbance, herbaceous cover in all soil disturbance classes declined rapidly as the canopy closed.

In subalpine forests of northwestern Colorado, Rumbaitis (2006) compared windthrow, windthrow + salvage logging, and undisturbed forests. Species richness and diversity were lower in the wind + salvage logged areas than the windthrow or undisturbed areas (Table 7.1). Species growing in the wind + salvage logged areas primarily were early successional specialists, whereas mixtures of early and late successional species grew in the windthrow only areas (Rumbaitis 2006). In contrast to the results of Elliott et al. (2002b), few shade-tolerant forbs were found in the wind + salvage logged areas. Rumbaitis (2006) concluded differences in understory disturbance severity were likely responsible for the observed differences in species diversity and composition between the windthrow only and wind + salvage logged areas.

In general, windthrow generates microsite heterogeneity that can facilitate species diversity and abundance in the herb layer. For example, pits and mounds associated with treefalls can have higher species diversity and greater herb cover than adjacent undisturbed areas (Peterson and Campbell 1993). Changes in light quality and quantity associated with gaps generate the greatest responses in understory herbs because many species are light limited (Whigham 2004). Woodland herbs often show greater growth and reproduction in response to increased light (Collins and Pickett 1988; Neufeld and Young 2003); however, positive responses may depend on gap size (Collins and Pickett 1988) and negative impacts associated with competition (Hughes 1992). Overall, windthrow gaps can increase herb layer species diversity and abundance, but may increase abundance of early successional or light-demanding species only when the canopy is removed and there is considerable soil disturbance.

7.4 Summary

Over the landscape, open areas, savannahs, and woodlands can provide early successional habitats for numerous wildlife species, but maintaining or restoring these vegetation types can require intensive management, such as removing invasive grasses with herbicide applications, increasing fire, and mechanical disturbance (e.g., disking). Herb layer response to disturbance varies with the type and severity of the disturbance, but also among ecoregions and forest types within the Central Hardwood Region. Low to moderate fire severity can increase herb cover and diversity and promote emergence from the seed bank and protected meristems in oak and pine-hardwood forests. Windthrow, at the low end of a canopy and soil disturbance gradient, can promote diversity of native species in the understory. At the other end of the spectrum, abandoned agricultural land and surface mining, especially mountain top removal, create early successional communities, but can also promote non-native species, especially if initially seeded with these species. Although herbaceous response differs over ecoregions, a mixture of silvicultural practices along a gradient of disturbance severity will maintain a range of stand ages and structures, and subsequently maximize landscape level herbaceous diversity.

Literature Cited

Abrams MD (2003) Where has all the white oak gone? Bioscience 53:927–939

Arthur MA, Paratley RD, Blankenship BA (1998) Single and repeated fires affect survival and regeneration of woody and herbaceous species in an oak-pine forest. J Torrey Bot Soc 125:225–236

Barden LS (2000) Population maintenance of *Pinus pungens* Lam. (table mountain pine) after a century without fire. Nat Areas J 20:227–233

Bellemare J, Motzkin G, Foster DR (2002) Legacies of the agricultural past in the forested present: an assessment of historical land-use effects on rich mesic forests. J Biogeogr 29:1401–1420

Belote RT, Sanders NJ, Jones RH (2009) Disturbance alters local–regional richness relationships in Appalachian forests. Ecology 90:2940–2947

Boring LR, Swank WT, Monk CD (1988) Dynamics of early successional forest structure and processes in the Coweeta Basin. In: Swank WT, Crossley DA Jr (eds) Forest hydrology and ecology at Coweeta, ecological studies 66. Springer, New York, pp 161–180

Bowles ML, Jacobs KA, Mengler JL (2002) Long-term changes in an oak forest's woody understory and herb layer with repeated burning. J Torrey Bot Soc 134:223–237

Brudvig LA, Asbjornsen H (2008) Patterns of oak regeneration in a Midwestern savanna restoration experiment. For Ecol Manage 255:3019–3025

Carter MC, Foster CD (2004) Prescribed burning and productivity in southern pine forests: a review. For Ecol Manage 191:93–109

Clendenin MA, Ross WG (2001) Effects of cool season prescribed fire on understory vegetation in a mixed pine hardwood forest of east Texas. Tex J Sci 53:65–78

Clinton BD (1995) Temporal variation in photosynthetically active radiation (PAR) in mesic Southern Appalachian hardwood forests with and without Rhododendron understories. In: Gottschalk KW, Fosbroke SL (eds) Proceedings of the 10th Central Hardwood forest conference, 5–8 Mar 1995, Morgantown WV. Gen Tech Rep NE-197, USDA Forest Service Northeastern Forest Experiment Station, Radnor, pp 534–540

Clinton BD, Baker CR (2000) Catastrophic windthrow in the Southern Appalachians: characteristics of pits and mounds and initial vegetation responses. For Ecol Manage 126:51–60

Clinton B D, Vose JM (2000) Plant succession and community restoration following felling and burning in the Southern Appalachians. In: Moser WK, Moser CF (eds) Fire and forest ecology: innovative silviculture and vegetation management, Proceedings of the 21st tall timbers fire ecology conference. Tall Timbers Research Station, Tallahassee, pp 22–29

Clinton BD, Boring LR, Swank WT (1993) Canopy gap characteristics and drought influences in oak forests of the Coweeta Basin. Ecology 74:1551–1558

Collins BS, Pickett STA (1988) Response of herb layer cover to experimental canopy gaps. Am Midl Nat 119:282–290

Collins B, Minchin P, Dilustro J, Duncan L (2006a) Land use effects on groundlayer composition and regeneration of mixed pine hardwood forests in the Fall Line Sandhills, S.E. USA. For Ecol Manage 226:181–188

Collins B, Sharitz R, Madden K, Dilustro J (2006b) Comparison of sandhills and mixed pine hardwood communities at Fort Benning, Georgia. SE Nat 51:92–102

Dilustro J, Collins B, Duncan L, Sharitz R (2002) Soil Texture, land use intensity, and vegetation of Fort Benning upland forest sites. J Torrey Bot Soc 129:280–297
Dilustro J, Collins B, Duncan L (2006) Land use history effects in mixed pine hardwood forests at Fort Benning. J Torrey Bot Soc 133:460–467
Dolan B J, Parker GR (2004) Understory response to disturbance: an investigation of prescribed burning and understory removal treatments. In: Spetich MA (ed) Upland oak ecology symposium: history, current conditions, and sustainability. Gen Tech Rep SRS-73, USDA Forest Service Southern Research Station, Asheville, pp 285–291
Donner DM, Ribic CA, Probst JR (2010) Patch dynamics and the timing of colonization–abandonment events by male Kirtland's Warblers in an early succession habitat. Cons Biol 143:1159–1167
Dumus S, Neufeld HS, Fisk MC (2007) Fire in a thermic oak-pine forest in Linville Gorge wilderness area, North Carolina: importance of the shrub layer to ecosystem response. Castanea 72:92–104
Elliott KJ, Knoepp JD (2005) The effects of three regeneration harvest methods on plant diversity and soil characteristics in the Southern Appalachians. For Ecol Manage 211:296–317
Elliott KJ, Swank WT (1994) Impacts of drought on tree mortality and growth in a mixed hardwood forest. J Veg Sci 5:229–236
Elliott KJ, Vose JM (2005) Effects of understory burning on shortleaf pine (*Pinus echinata* Mill.) /mixed-hardwood forests. J Torrey Bot Soc 132:236–251
Elliott KJ, Vose JM (2010) Short-term effects of prescribed fire on mixed oak forests in the Southern Appalachians: vegetation response. J Torrey Bot Soc 137:49–66
Elliott KJ, Boring LR, Swank WT, Haines BR (1997) Successional changes in diversity and composition in a clearcut watershed in Coweeta Basin, North Carolina. For Ecol Manage 92:67–85
Elliott KJ, Boring LR, Swank WT (1998) Changes in vegetation diversity following grass-to-forest succession. Am Midl Nat 140:219–232
Elliott KJ, Hendrick RL, Major AE, Vose JM, Swank WT (1999) Vegetation dynamics after a prescribed fire in the Southern Appalachians. For Ecol Manage 114:1–15
Elliott KJ, Boring LR, Swank WT (2002a) Aboveground biomass and nutrient pools in a Southern Appalachian watershed 20 years after clearcutting. Can J For Res 32:667–683
Elliott KJ, Hitchcock SL, Krueger L (2002b) Vegetation response to large scale disturbance in a Southern Appalachian forest: hurricane opal and salvage logging. J Torrey Bot Soc 129:48–59
Elliott KJ, Vose JM, Clinton BD, Knoepp JD (2004) Effects of understory burning in a mesic mixed-oak forest in the Southern Appalachians. In: Engstrom RT, Galley KEM, de Groot WJ (eds) Proceedings of the 22nd tall timbers fire ecology conference: fire in temperate, boreal, and montane ecosystems. Tall Timbers Research Station, Tallahassee, pp 272–283
Elliott KJ, Vose JM, Hendrick RL (2009) Long-term effects of prescribed wildland fire on vegetation dynamics in the Southern Appalachians. Fire Ecol 5:66–85
Federer CA, Hornbeck JW, Tritton LM, Martin CW, Pierce RS, Smith CT (1989) Long-term depletion of calcium and other nutrients in eastern US forests. Environ Manage 13:593–601
Flinn KM, Vellend M (2005) Recovery of forest plant communities in post-agricultural landscapes. Front Ecol Environ 3:243–250
Franklin SB, Robertson PA, Fralish JS (2003) Prescribed burning effects on upland *Quercus* forest structure and function. For Ecol Manage 184:315–335
Fraterrigo JM, Pearson SM, Turner MG (2009a) The response of understory herbaceous plants to nitrogen fertilization in forests of different land-use history. For Ecol Manage 257:2182–2188
Fraterrigo JM, Turner MG, Pearson SM (2009b) Interactions between past land use, life-history traits and understory spatial heterogeneity. Landsc Ecol 21:777–790
Gilliam FS, Dick DA (2010) Spatial heterogeneity of soil nutrients and plant species in herb-dominated communities of contrasting land use. Plant Ecol 209:83–94
Gilliam FS, Roberts MR (2003) Conceptual framework for studies of the herbaceous layer. In: Gilliam FS, Roberts MR (eds) The herbaceous layer in forests of Eastern North America. Oxford University Press, Oxford, pp 3–11
Glasgow LS, Matlack GR (2007) Prescribed burning and understory composition in a temperate deciduous forest, Ohio, USA. For Ecol Manage 238:54–64

Gravel D, Canham CD, Beaudet M, Messier C (2010) Shade tolerance, canopy gaps and mechanisms of coexistence of forest trees. Oikos 119:475–484

Greenberg CH, McNab WH (1998) Forest disturbance in hurricane-related downbursts in the Appalachian mountains of North Carolina. For Ecol Manage 104:179–191

Gruchy JP, Harper, Gray MA (2009) Methods for controlling woody invasion into CRP fields in Tennessee. Proceedings of the Gamebird 2006: Quail VI and Perdix XII 6: 315–321

Hales TC, Ford CR, Hwang T, Vose JM, Band LE (2009) Topographic and ecologic controls on root reinforcement. J Geophys Res 114:F03013. doi:10.1029/2008JF001168

Harper CA (2007) Strategies for managing early succession habitat for wildlife. Weed Tech 21:932–937

Harper CA, Gruchy JP (2009) Conservation practices to promote quality early successional habitat. In: Burger LW, Evans KO (eds) Managing working lands for northern bobwhite: the USDA NRCS Bobwhite Restoration Project, Washington, DC, pp 87–114

Harper CA, Bates GE, Hansbrough MP, Gudlin MJ, Gruchy JP, Keyser PD (2007) Native warm-season grasses: identification, establishment, and management for wildlife and forage production in the Mid-South. UT Extension, PB 1752, Knoxville, 189 pp

Hiers JK, Wyatt R, Mitchell RJ (2000) The effects of fire regime on legume reproduction in longleaf pine savannas: is a season selective? Oecologia 125:521–530

Holl KD (2002) Long-term vegetation recovery on reclaimed coal surface mines in the Eastern USA. J Appl Ecol 39:960–970

Holzmueller EJ, Jose S, Jenkins MA (2009) The response of understory species composition, diversity, and seedling regeneration to repeated burning in Southern Appalachian oak-hickory forests. Nat Areas J 29:255–262

Hughes JW (1992) Effects of removal of co-occurring species on distribution and abundance of *Erythronium americanum* (Liliaceae), a spring ephemeral. Am J Bot 790:1329–1336

Hutchinson TF, Boerner REJ, Sutherland S, Sutherland EK, Ortt M, Iverson LR (2005) Prescribed fire effects on the herbaceous layer of mixed-oak forests. Can J For Res 35:877–890

Jenkins MA, Parker GR (2000) The response of herbaceous-layer vegetation to anthropogenic disturbance in intermittent stream bottomland forests of southern Indiana, USA. Plant Ecol 151:223–237

Jones JDJ, Chamberlain MJ (2004) Efficacy of herbicides and fire to improve vegetative conditions for northern bobwhites in mature pine forests. Wildl Soc Bull 32:1077–1084

Keyser PD, Sausville DJ, Ford WM, Schwab DJ, Brose PH (2004) Prescribed fire impacts to amphibians and reptiles in shelterwood-harvested oak-dominated forests. Va J Sci 55:159–168

Klaus NA, Buehler DA, Saxton AM (2005) Forest management alternatives and songbird breeding habitat on the Cherokee National Forest, Tennessee. J Wildl Manage 69:222–234

Kloeppel BD, Clinton BD, Vose JM, Cooper AR (2003) Drought impacts on tree growth and mortality of Southern Appalachian forests. In: Greenland D, Goodin DG, Smith RC (eds) Climate variability and ecosystem response at long-term ecological research sites. Oxford University Press, New York, pp 43–55

Kneeshaw DD, Bergeron Y (1998) Canopy gap characteristics and tree replacement in the southeastern boreal forest. Ecology 79:783–794

Knoepp JD, Elliott KJ, Vose JM, Clinton BD (2009) Effects of prescribed fire in mixed-oak forests of the Southern Appalachians: forest floor, soil, and soil solution nitrogen responses. J Torrey Bot Soc 136:380–391

Kuddes-Fischer LM, Arthur MA (2002) Response of understory vegetation and tree regeneration to a single prescribed fire in oakpine forests. Nat Areas J 22:43–52

Lafon CW, Waldron JD, Cairns DM, Tchakerian MD, Coulson RN, Klepzig KD (2007) Modeling the effects of fire on the long-term dynamics and restoration of yellow pine and oak forests in the Southern Appalachian Mountains. Rest Ecol 15:400–411

Liu H, Menges ES (2005) Winter fires promote greater vital rates in the Florida keys than summer fires. Ecology 86:1483–1495

Liu H, Menges ES, Snyder JR, Koptur S, Ross MS (2005) Effects of fire intensity on vital rates of an endemic herb of the Florida keys, USA. Nat Areas J 25:71–76

Magguran AE (2004) Measuring biological diversity. Blackwell, Oxford
Masters RE (2007) The importance of shortleaf pine for wildlife and diversity in mixed oak-pine forests and pine-grassland woodlands. In: Kabrick JM, Dey DC, Gwaze D (eds) Shortleaf pine restoration and ecology in the Ozarks. Gen Tech Rep NRS-P-15, USDA Forest Service Northern Research Station, Newtown Square, pp 35–46
McGee GG, Leopold DJ, Nyland RD (1995) Understory response to springtime prescribed fire in two New York transition oak forests. For Ecol Manage 76:149–168
Mou P, Fahey TJ, Hughes JW (1993) Effects of soil disturbance on vegetation recovery and nutrient accumulation following whole-tree harvest of a Northern Hardwood ecosystem. J Appl Ecol 30:661–675
Nelson JL, Groninger JW, Battaglia LL, Ruffner CM (2008) Bottomland hardwood forest recovery following tornado disturbance and salvage logging. For Ecol Manage 256:388–395
Neufeld HS, Young DR (2003) Ecophysiology of the herbaceous layer in temperate deciduous forests. In: Gilliam FS, Roberts MR (eds) The herbaceous layer in forests of eastern North America. Oxford University Press, Oxford, pp 38–90
Parker GR, Merritt C (1994) The central region. In: Barren JW (ed) Regional silviculture of the United States. Wiley, New York, pp 129–172
Peterson CJ (2000) Catastrophic wind damage to North American forests and the potential impact of climate change. Sci Total Environ 262:287–311
Peterson CJ, Campbell JE (1993) Microsite differences and temporal change in plant communities of treefall pits and mounds in an old-growth forest. Bull Torrey Bot Club 120:451–460
Peterson CJ, Leach AD (2008) Limited salvage logging effects on forest regeneration after moderate-severity windthrow. Ecol Appl 18:407–420
Phillips R, Hutchinson T, Brudnak L, Waldrop T (2007) Fire and fire surrogate treatments in mixed-Oak forests: effects on herbaceous layer vegetation. In: Butler BW, Cook W (comps) The fire environment-innovations, management, and policy. Gen Tech Rep RMRS-P-46, USDA Forest Service Rocky Mountain Research Station, Fort Collins, pp 1–11
Pyke DA, Brooks ML, D'Antonio C (2010) Fire as a restoration tool: a decision framework for predicting the control or enhancement of plants using fire. Restor Ecol 18:274–284
Reilly MJ, Wimberly MC, Newell CL (2006) Wildfire effects on plant species richness at multiple spatial scales in forest communities of the Southern Appalachians. J Ecol 94:118–130
Reiners WA (1992) Twenty years of ecosystem reorganization following experimental deforestation and regrowth suppression. Ecol Monogr 62:503–523
Roberts MR, Gilliam FS (2003) Response of the herbaceous layer to disturbance in Eastern forests. In: Gilliam FS, Roberts MR (eds) The herbaceous layer in forests of eastern North America. Oxford University Press, Oxford, pp 302–320
Royo AA, Collins R, Adams MB, Kirschbaum C, Carson WP (2010) Pervasive interactions between ungulate browsers and disturbance regimes promote temperate forest herbaceous diversity. Ecology 91:93–105
Rumbaitis del Rio CM (2006) Changes in understory composition following catastrophic windthrow and salvage logging in a subalpine forest ecosystem. Can J For Res 36:2943–2954
Scholes RJ, Archer SR (1997) Tree-grass interactions in savannas. Annu Rev Ecol Syst 28:517–544
Small CJ, McCarthy BC (2002) Spatial and temporal variation in the response of understory vegetation to disturbance in a Central Appalachian oak forest. J Torrey Bot Soc 129:136–153
Small CJ, McCarthy BC (2005) Relationship of understory diversity to soil nitrogen, topographic variation, and stand age in an eastern oak forest, USA. For Ecol Manage 217:229–243
Smith RN (1991) Species composition, stand structure, and woody detrital dynamics associated with pine mortality in the Southern Appalachians. Thesis, University of Georgia, Athens
Sparks JC, Masters RE, Engle DM, Palmer MW, Bukenhofer GA (1998) Effects of late growing-season and late dormant-season prescribed fire on herbaceous vegetation in restored pine-grassland communities. J Veg Sci 9:133–142
Thiemann JA, Webster CR, Jenkins MA, Hurley PM, Rock JH, White PS (2009) Herbaceous-layer impoverishment in a post-agricultural Southern Appalachian landscape. Am Midl Nat 162:148–168

Vose JM, Swank WT, Clinton BD, Knoepp JD, Swift LW Jr (1999) Using prescribed fire to restore Southern Appalachian pine-hardwood ecosystems: effects on mass, carbon, and nutrients. For Ecol Manage 114:215–226

Wear DN, Bolstad P (1998) Land-use changes in Southern Appalachian landscapes: spatial analysis and forecast evaluation. Ecosystems 1:575–594

Webster CR, Rock JH, Froese RE, Jenkins MA (2008) Drought-herbivory interaction disrupts competitive displacement of native plants by *Microstegium vimineum*, 10-year results. Oecologia 157:497–508

Whigham DF (2004) Ecology of woodland herbs in temperate deciduous forests. Annu Rev Ecol Evol Syst 35:583–621

Whittaker RH (1965) Dominance and diversity in land plant communities. Science 147:250–260

Zenner EK, Kabrick JM, Jensen RG, Peck JE, Grabner JK (2006) Responses of ground flora to a gradient of harvest intensity in the Missouri Ozarks. For Ecol Manage 222:326–334

Chapter 8
The Role of Young, Recently Disturbed Upland Hardwood Forest as High Quality Food Patches

Cathryn H. Greenberg, Roger W. Perry, Craig A. Harper, Douglas J. Levey, and John M. McCord

Abstract Young (1–10 year post-disturbance) upland hardwood forests function as high-quality food patches by providing abundant fruit, and nutritious foliage and flowers that attract pollinating and foliar arthropods and support high populations of small mammals that, in turn, are prey for numerous vertebrate predators. Reductions in basal area increase light penetration to the forest floor, which stimulates vegetative growth and promotes fruiting. Fruit biomass (dry edible pulp) can be 5 to nearly 50 times greater in young forest than mature forest as "pioneer" species, such as pokeweed and blackberry, ericaceous shrubs, various forbs and grasses, and stump sprouts of many tree species produce fruit. Forage production can increase substantially after disturbances that significantly reduce overstory basal area, such

C.H. Greenberg (✉)
Upland Hardwood Ecology and Management Research Work Unit, USDA Forest Service, Southern Research Station, Bent Creek Experimental Forest, 1577 Brevard Rd, Asheville, NC 28806, USA
e-mail: kgreenberg@fs.fed.us

R.W. Perry
Southern Pine Ecology and Management Research Work Unit, USDA Forest Service, Southern Research Station, 1270, Hot Springs, AR 71902, USA
e-mail: rperry03@fs.fed.us

C.A. Harper • J.M. McCord
Department of Forestry, Wildlife and Fisheries, University of Tennessee, Knoxville, TN 37996, USA
e-mail: charper@utk.edu; jmccord3@utk.edu

D.J. Levey
Department of Zoology, University of Florida, 118525, Gainesville, FL 32611-8525, USA
e-mail: dlevey@zoo.ufl.edu

C.H. Greenberg et al. (eds.), *Sustaining Young Forest Communities*, Managing Forest Ecosystems 21, DOI 10.1007/978-94-007-1620-9_8,

as timber harvests, heavy thinning, or intense prescribed fire. Hard mast (nut) production can be sustained in young forests if some mature, good mast-producing oak, hickory, or beech trees are retained. Balancing the creation of young, recently disturbed upland hardwood forests with the desired amount and distribution of other forest age-classes will sustain high-quality food patches for wildlife within a landscape context.

8.1 Introduction

Deciduous forest of the Central Hardwood Region is a patchwork of stand ages and structures that result from natural small-scale disturbance, such as death of individual trees, and larger-scale events, including fire, ice, wind, and insect outbreaks (White et al., Chap. 3). Many forest management activities, such as timber harvest, thinning, and controlled burning, also create disturbances. Varied types and intensities of disturbances result in an assortment of structural features that complicate a simple definition of young upland hardwood forest. Yet, all share similar attributes, including a well-developed groundcover or shrub and young tree component, and absence of or discontinuous mature tree canopy (Greenberg et al., Chap. 1).

Abundant light and reduced competition created by reductions in overstory tree density coupled with soil perturbation and scarification from disturbances promote germination, foliar growth, flowering, and fruiting by many plant species on the forest floor. Disturbance also promotes colonization by disturbance-adapted plants, such as blackberry (*Rubus* spp.) and pokeweed (*Phytolacca americana*), that produce prodigious amounts of fruit (Greenberg et al. 2007). Open, recently disturbed forests provide an abundance of native fruits, woody browse, nutritious foliage and flowers that attract arthropods and high densities of small mammals that serve as prey for numerous snake, bird, and mammalian predators. Thus, these young forests function as high-quality food patches for many wildlife species. The important role of young hardwood forests in supporting wildlife is becoming increasingly recognized by natural resource professionals. In this chapter, we synthesize results of our own research and other studies on fleshy fruit, hard mast, browse, and arthropod and small mammal (as prey) production in young (less than 10 years post-disturbance) upland hardwood forests of the Central Hardwood Region of the USA (see Fig. 1.1).

8.2 Fleshy Fruit

Fleshy fruit (soft mast) is a key food resource for many game and nongame wildlife species (Martin et al. 1951). Most species of birds and mammals consume fruit at least occasionally (Martin et al. 1951; Willson 1986). Fruit consumption has been linked to mammalian survival and reproductive success (e.g., Rogers 1976; Eiler et al. 1989). Fruit choice is a complex interplay between the nutritional composition of fruit, changing nutritional needs, availability of alternative food sources, and seasonal patterns of fruit and consumer abundance (Levey and Martinez del Rio 2001). Some studies suggest birds consume high-lipid fruits more rapidly than "low-quality" (low-lipid) fruits in fall (White and Stiles 1992), but others indicate nutritional quality is not an important determinant of fruit selection by birds (Borowicz 1988; Fuentes 1994; Jordano 2000; Whelan and Willson 1994). Further, digestive abilities may differ among avian species (Fuentes 1994; Martinez del Rio and Restrepo 1993). For example, Cedar Waxwings (*Bombycilla cedrorum*) specialize in sugary fruits, whereas thrushes specialize in lipid-rich fruits (Witmer and Van Soest 1998). American Robins (*Turdus migratorius*) produce low levels of the enzyme sucrase, and thus cannot digest high-sucrose fruits (Martinez del Rio and Restrepo 1993).

Abundant fruit in young forests may be a particularly important high-energy food source for neotropical migratory birds during fall migration (Parrish 1997). During winter, soft mast is important to many vertebrates when other food resources are scarce (e.g., McCarty et al. 2002; Greenberg and Forrest 2003; Whitehead 2003). For example, the local distribution of Hermit Thrushes (*Catharus guttatus*) and Yellow-rumped Warblers (*Dendroica coronata*) during winter may be influenced by fruit availability (Kwit et al. 2004; Borgmann et al. 2004). The open conditions in young forests provide greater abundance of fruit, and also facilitate discovery by fruit-eating vertebrates. Fruit removal rates may be more rapid in gaps and along forest edges than under closed-canopy forests (Thompson and Willson 1978).

Fruit availability and abundance vary spatially and temporally across heterogeneous landscapes comprised of different forest age classes and site quality. This variation in fruit abundance results from differences in the composition of fruiting species, fruiting phenology, and the dynamic process of colonization and recovery of fruiting plants in young, recently disturbed forests. At local scales, fruit production is dictated by the composition of plant species, many of which are patchy in their occurrence.

Fruit production per hectare is inversely related to the residual density or basal area (BA) of overstory trees (shade) remaining after a disturbance, and declines over time with canopy closure (Perry et al. 1999, Fig. 8.1). Fruit production is much greater in forest openings than in closed canopy conditions, regardless of whether openings are caused by natural disturbance (e.g., Thompson and Willson 1978; Blake and Hoppes 1986) or by silvicultural disturbance, such as timber harvest (e.g., Lay 1966; Halls and Alcaniz 1968; Johnson and Landers 1978; Campo and Hurst 1980; Stransky and Roese 1984; Perry et al. 1999; Mitchell and Powell 2003; Greenberg et al. 2007). For example, Blake and Hoppes (1986) reported 44 fruits/80 m^2 in single-tree gaps, but only 2 fruits/80 m^2 in adjacent closed canopy forest in Illinois. Perry et al. (1999; Perry, unpublished) found that in the Interior Highlands of Arkansas and Oklahoma, production of dry edible fruit pulp biomass ($\leq$2 m height) 5 years post-harvest was about three times greater in group selection matrix (the forest surrounding group openings) and eight times greater in single-tree selection harvests where BA reduction was minor and light increased only slightly, compared to mature (>50 years old), closed-canopy forest. However, dry edible fruit pulp biomass production ($\leq$2 m height) in their study area was 31 times greater in clearcuts,

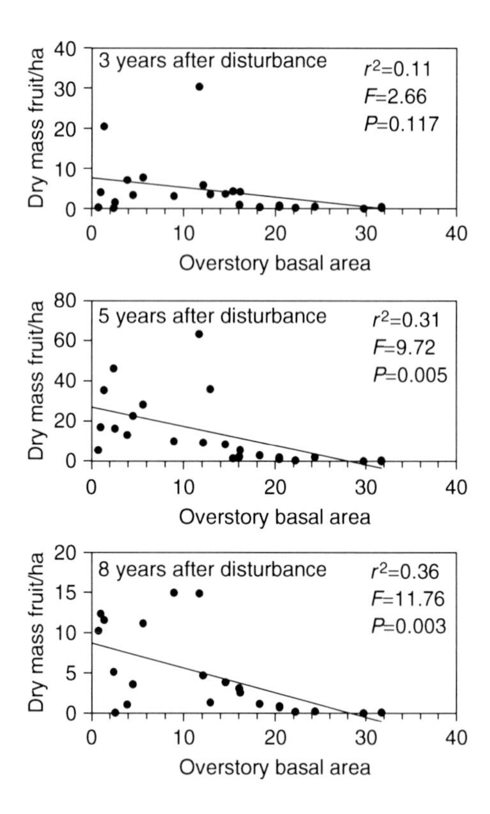

Fig. 8.1 Relationship between total fruit production (dry edible pulp biomass; kg/ha ≤ 2 m high) and overstory BA (m^2/ha) in forest stands thinned/harvested to various densities in the Interior Highlands of Arkansas and Oklahoma. Production was greatest in stands with lowest BA, but the relationship changed over time since disturbance with the strongest relationship at 5 and 8 years after disturbance (data from Perry et al. 1999; Perry, unpublished)

46 times greater in group openings, and 49 times greater in shelterwood harvests than in mature forest (Figs. 8.2, 8.3). In the Southern Appalachians, production of dry edible fruit pulp biomass was 5–20 times greater in shelterwood harvests (with about 15% BA retention) beginning 3–5 years post-harvest than in mature forest (Fig. 8.3) (Greenberg et al. 2007). Increases in fruit production are generally less in small openings, such those created by single-tree selection or gaps compared to larger openings, such clearcut or shelterwood harvests, because smaller openings are typically shaded more by surrounding forest than larger openings (Perry et al. 1999).

Fruit production in young forests can be affected by the type of disturbance and prior land uses. In areas subjected to timber harvest, site preparation after harvest can affect the length of time plants take to establish fruiting or overall long-term fruit production. After logging, sites not subjected to site preparation or sites only burned after harvest may produce more fruit from woody shrubs than sites subjected to site preparation methods, such as mechanical chopping or blading, which destroy the roots of pre-established plants (Stransky and Halls 1980). However, more intense site preparation can potentially facilitate establishment of disturbance-adapted herbaceous plants from seed, such as pokeweed and blackberry. Seeds of these "pioneer" species are dispersed by vertebrates that eat the fruits, and can be abundant in seed banks prior to disturbances (T. Keyser, unpublished). In reforested areas subjected to timber harvest, lands that were previously cleared and farmed produce

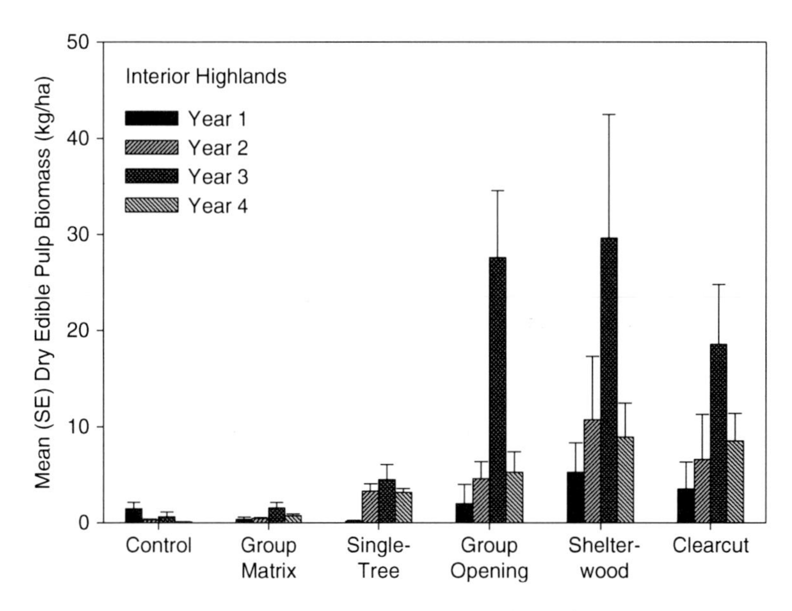

Fig. 8.2 Mean total dry biomass (kg/ha) of edible pulp from fleshy fruit ($\leq$2 m high) produced 1, 3, 5, and 8 years after harvest in different silvicultural treatments with different average retained BAs in the Interior Highlands of Arkansas and Oklahoma. Treatments are unharvested forests (control; 29.4 m^2/ha BA), the forested matrix surrounding group openings in group-selection stands (group matrix; 20.3 m^2/ha BA), single-tree selection stands (15.5 m^2/ha BA), group openings (4.1 m^2/ha BA), shelterwood stands (11.4 m^2/ha BA), and clearcuts (1.4 m^2/ha BA) (data from Perry et al. 1999; Perry, unpublished)

substantially less fruit because of sparse seed beds and fewer pre-established root systems (Stransky and Halls 1980).

High-intensity (hot) fires in upland hardwood forests can create open, structurally diverse conditions by killing overstory and midstory trees. Burning in upland hardwood forests may reduce fruit production immediately following the fire, but may eventually result in increased production if light to the forest floor is increased and top-killed plants resprout, or disturbance-adapted species colonize or germinate from the seedbank, and fruit (Jackson et al. 2007, J Michael McCord, unpublished). More commonly, prescribed fires in upland hardwood forests are low-intensity with minimal disturbance or increases in light reaching the understory (Jackson et al. 2007). Post-burn increases in fruit production generally correspond with reductions in canopy cover and increased light to the forest floor, and thus are greater following high-intensity burns that kill trees. Post-burn fruit production may be spatially patchy (Jackson et al. 2007), reflecting the mosaic of light and disturbance conditions created by the patchy burn patterns typical in upland hardwood forests.

Burning at 7-year intervals or less in young forests may impede canopy closure and stimulate the development of herbaceous groundcover (Masters et al. 1993), thereby also prolonging young forest conditions that promote abundant fruit production.

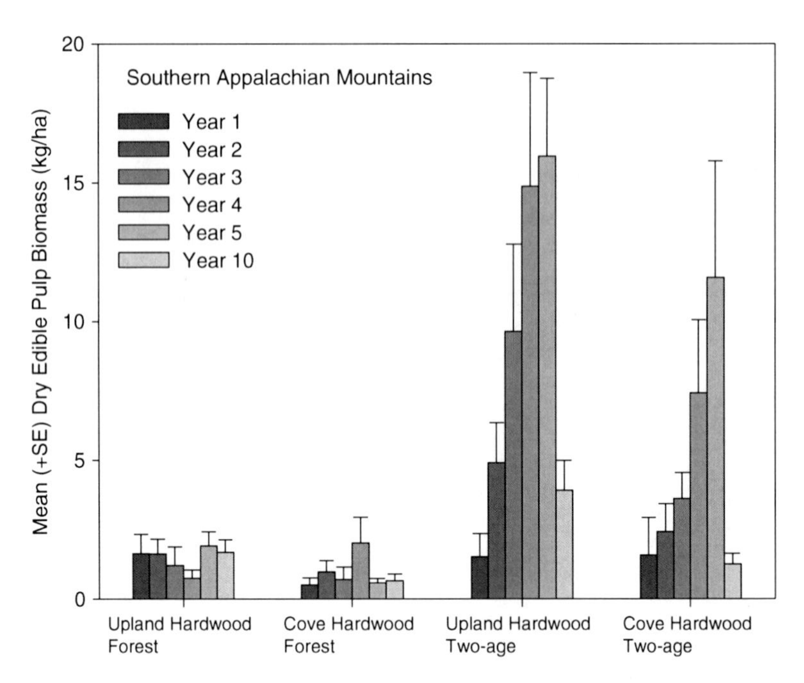

Fig. 8.3 Mean total fruit production (dry edible pulp biomass; kg/ha) produced 1, 2, 3, 4, 5, and 10 years after two-age harvests and in mature forest in upland hardwood and cove hardwood forests of the Southern Appalachians (data from Greenberg et al. 2007)

Prescribed fire may create opportunities for germination, establishment, and (or) growth for non-native invasive plant species, such as Russian and autumn olive (*Eleagnus* spp.) or oriental bittersweet (*Celastrus orbiculatus*), but it may also cause mortality or otherwise suppress population growth of many non-native species (D'Antonio 2000). A pre-fire inventory of non-native invasive plants and pre- or post-fire removal of highly invasive species may help to reduce the possibility of proliferation by some non-native species.

Total fruit production levels are typically tied more closely to stand age than to topographic position, and fruit production is generally highest in young forests (Reynolds-Hogland et al. 2006; Greenberg et al. 2007). In the Southern Appalachians, dry pulp biomass of fleshy fruit in young, recently harvested (using a low-leave shelterwood harvest where about 15% of the overstory BA was retained) stands is similar to that in mature forests during the first 2 years after harvest, but increases 5–20-fold by the third year after harvest (Greenberg et al. 2007; Fig. 8.3). Fruit production peaks around 5–8 years after harvest and remains high for several years before declining. By about the tenth year post-harvest, fruit production in young forests may be similar to production in mature, unthinned forests as growth of regenerating trees creates a fully shaded environment (Fig. 8.3). Reynolds-Hogland et al. (2006) found production of berries (*Gaylussacia* spp., *Vaccinium* spp., and *Rubus* spp.) was highest in 2–9 year old stands.

The length of time fruit production remains high in young forests varies with the growth rate of regenerating trees that eventually shade the forest understory. Woody plant growth rates are influenced by moisture or site quality, which is dictated by topographic position, soils, and geography (Elliott et al., Chap. 7; Loftis et al., Chap. 5). Thus, moist, high-quality sites may reach canopy closure and reduced fruit production more rapidly than xeric upland hardwood forests. Further, the occurrence and relative abundance of many fruit-producing species are influenced by site quality. For example, blackgum (*Nyssa sylvatica*) and ericaceous shrubs are most common on dry, lower quality sites in the Southern Appalachians, whereas spicebush and many herbaceous species are generally associated with moist, high-quality cove hardwood forests.

The disturbance-associated species pokeweed and blackberry are prodigious fruit producers in recently disturbed hardwood forests throughout the Central Hardwood Region, including the Southern Appalachians, Interior Highlands, Ridge and Valley, and upper Coastal Plain (Perry et al. 1999; Greenberg et al. 2007; Greenberg et al. in review; McCord and Harper in press). A "relay" between pokeweed and blackberry sustains high levels of fruit production in young hardwood forests for several years. Pokeweed dominates fruit production for the first few years after disturbance, but generally is shaded out by the fourth or fifth year. In contrast, blackberry is usually present, but takes 3 or 4 years before it produces substantial amounts of fruit. Sumac (*Rhus* spp.) is an ephemeral pioneer shrub that occurs throughout the Central Hardwood Region and produces prodigious amounts of fruit, but may occur less frequently in young forest patches than pokeweed and blackberry (Greenberg et al. 2007). In southern portions of the Central Hardwood Region, American beautyberry (*Callicarpa americana*) is also an important fruit producer in young forests.

Many species not typically associated with disturbance also produce abundant fruit in young forests – often more fruit than in mature forests. In the Southern Appalachians, flowering dogwood (*Cornus florida*), American holly (*Ilex americana*), Fraser magnolia (*Magnolia fraseri*), black cherry (*Prunus serotina*), sassafras (*Sassafras albidum*), and blackgum all produce fruit from stump sprouts within 1–3 years post-harvest. In the Interior Highlands, flowering dogwood, black cherry, sassafras, blackgum, serviceberry (*Amelanchier arborea*), and muscadine grapes (*Vitis rotundifolia*) are species not associated with disturbance that can produce great amounts of soft mast in both older (7+years old) openings and in mature forests (Segelquist and Green 1968; Rogers et al. 1990; Perry et al. 1999). Several herbaceous species that are generally associated with mature cove hardwood forests, including Jack-in-the-pulpit (*Arisaema triphyllum*), mandarin (*Disporum lanuginosum*), Solomon's seal (*Polygonatum biflorum*), and *Trillium* spp., also produce more fruit in recently-harvested forests than in mature forests (Greenberg et al. 2007).

Ericaceous shrubs, including huckleberry (*Gaylussacia* spp.) and blueberry (*Vaccinium* spp.), produce abundant fruit within a year after disturbance, but also produce a large proportion of the total fruit in mature forests. Dominant species include huckleberry in the Southern Appalachians, and deerberry (*V. stamineum*), which is widespread throughout the Central Hardwood Region. The relative abundance of huckleberry and blueberry species (and their fruit) varies with topography and geography. Huckleberry tends to be most abundant on dry, lower-quality sites. Blueberries produce minor amounts of fruit compared to huckleberry in the Southern Appalachians (Greenberg et al. 2007), though this may vary with location. They are the dominant ericaceous, fruit-producing species in the upper Coastal Plain and the Interior Highlands (Perry et al. 1999; Greenberg et al. in review).

Only a handful of native plant species in upland hardwood forests produce or retain fruit during winter. American holly, greenbriar (*Smilax* spp.), and sumac are important winter fruits throughout the Central Hardwood Region. Sumac is limited to recently disturbed forests, whereas holly and greenbriar produce fruit in all forest age-classes. Several species of non-native, invasive plants, including oriental bittersweet, Chinese privet (*Ligustrum sinense*), and multiflora rose (*Rosa multiflora*), produce or retain fruit during winter (Greenberg and Walter 2010) and can invade disturbed, or sometimes undisturbed, forests when these stands are near seed sources. Whereas these non-native plant species may provide food for wildlife, animals did not historically rely on those food sources and they are not part of the ecological balance that evolved between native animals and food sources in the Central Hardwood Region. Further, consumption of non-native fruits by birds and vertebrates promotes widespread dispersal and establishment of non-native plants across the landscape where they compromise native plant communities.

8.3 Hard Mast

Nuts produced by oak (*Quercus* spp.), hickory (*Carya* spp.) and beech (*Fagus grandifolia*) trees provide a valuable food resource to many wildlife species (Martin et al. 1951) and influence the distribution, recruitment and survival, and behavior of wildlife, ranging from migratory birds to black bear (*Ursus americanus*) (McShea and Healy 2002; Rodewald 2003; Clark 2004). Acorns are considered a "keystone" to biological diversity because their nourishment affects abundance of rodents that are an important prey base for raptors and carnivores, and affects populations of white-tailed deer (*Odocoileus virginianus*) that in turn alter forest structure and composition through browsing (Feldhamer 2002). Hard mast production may be reduced in young forests when mature oak, hickory, or other nut-producing trees are removed or killed. Thus, retention of some hard mast production when creating young forest stands through silviculture should be considered.

The age at which regenerating trees begin to produce mast varies; most oak species produce acorns by age 20–25 and reach full production potential around age 50 (Burns and Honkala 1990). Age of hard mast production, however, likely differs between trees that originate from seedlings versus stump sprouts (coppice) from rootstocks of mature, harvested trees. For example, coppice scarlet oaks and white oaks in the Appalachians produce abundant acorns within 25 years after harvest (Greenberg and Parresol 2002). Oak trees grown from seed in open conditions, such as nurseries, can produce acorns within 10 years (Scott Schlaurbaum, unpublished).

Some hard mast production can be sustained if mature, mast-producing trees are retained, such as in partial harvest techniques like shelterwood, single-tree selection, and group-selection harvests. The level of potential hard mast production depends partly on the number of mature mast-producing trees remaining after the disturbance, but is also affected by the selection of individual trees. Production of hard mast by retained trees in recently disturbed forests is confounded by various factors that affect nut production by individual trees, including tree size, genetics, and site quality. The influence of tree size (diameter at breast height; dbh) on acorn production is largely a function of crown area (Rose et al. in review). Larger-diameter oak trees generally have bigger crowns (Bechtold 2003) and thus can potentially produce more acorns than smaller-diameter trees. However, the influence of oak dbh on acorn density per unit of crown area is negligible (Greenberg and Parresol 2002; Lashley et al. 2010).

Generally <50% of individual oaks of any given species are "good" producers, yet the majority of the total acorn crop at a site may be produced by these trees (Greenberg and Parresol 2002; Lashley et al. 2010). Thus, high acorn production levels could be potentially sustained with the removal of ≥ 50% of individual oaks if good producers could be identified for retention (Lashley et al. 2010). Unfortunately, no measurable parameter can predict whether an individual oak is a good producer or a poor producer other than observation of individual trees over several years.

Any sustained post-harvest increase in acorn production by residual oaks or hickories is difficult to detect with confidence because of variation in hard mast production among individual trees and years. However, studies have established a clear relationship between forest density and seed production in pines (e.g., Croker 1952; Bilan 1960; Godman 1962), and foresters often thin pine stands to promote seed production.

Although few studies have evaluated the effects of stand density on mast production by oaks and hickories, some research suggests heavy thinning may increase hard mast production by individual trees (Paugh 1970; Healy 1997; Perry and Thill 2003). However, these reductions in tree density may reduce overall net production within a stand (Harlow and Eikum 1963; Minckler and McDermott 1960). Residual oaks and hickories may increase their production of nuts after thinning or timber harvests, likely a result of decreased competition, increased light to tree crowns, and possible increases in crown size over time (Perry and Thill 2003; Perry et al. 2004). Thus, reducing the BA of forests may increase production by the individual hard mast-producing trees that are left (Perry and Thill 2003; Fig. 8.4). Areas with reduced BA could potentially maintain similar hard mast production indices to areas of mature, unthinned forest because of the greater output by individual residual trees (Perry and Thill 2003), while at the same time promoting soft mast and forage production in the understory. Reduced hard mast production in individual harvested forest stands that comprise a small proportion of a forested landscape may be relatively inconsequential, and may be offset by a large increase in fleshy fruit production.

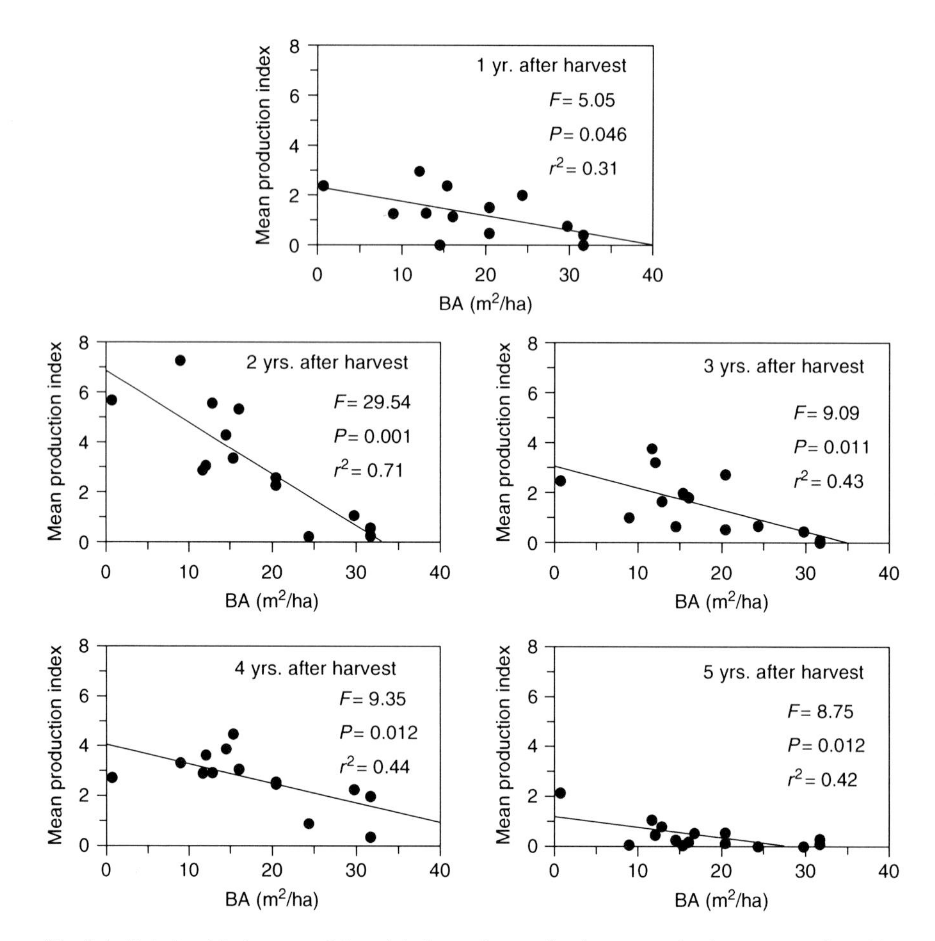

Fig. 8.4 Relationship between BA and indices of mean hard mast production per tree for white oaks (*Quercus alba*) in 13–15 forest stands differing in total BA in the Interior Highlands 1, 2, 3, 4, and 5 years after partial harvest and thinning (data from Perry et al. 2004). Annual differences in overall mast production among all areas demonstrate annual variation in mast production regardless of BA, which can be great

8.4 Herbaceous Forage and Woody Browse

Many wildlife species, including white-tailed deer, rabbits (*Sylvilagus* spp.), Ruffed Grouse (*Bonasa umbellus*), Bobwhite Quail (*Colinus virginianus*), black bear, Eastern Wild Turkey (*Meleagris gallopavo),* voles (*Microtus* spp.), and cotton rats (*Sigmodon* spp.), use various forbs, grasses, brambles, and browse (leaves and twigs of woody species ≤ about 1.4 m above the ground) to meet nutritional demands, and many other species require this low vegetative stratum for nesting, food (soft mast, seeds, and arthropods), and cover from predators.

Several studies within the Central Hardwood Region have evaluated forage availability following thinning and timber harvests (Morriss 1954; Ripley and Campbell 1960; Patton and McGinnes 1964; Della-Bianca and Johnson 1965; Moore and Downing 1965; Moore and Johnson 1967; Harlow and Downing 1969, 1970; Beck and Harlow 1981; Ford et al. 1993; Tilghman 1989; Johnson et al. 1995; Lashley et al. in press). Substantial reductions in BA significantly increase light to the forest floor and stimulate growth and development of the understory (Morriss 1954; Ford et al. 1993; Loftis et al., Chap. 5; Elliott et al., Chap. 7). In Texas pine-hardwood forest, forage production (herbaceous and woody vegetation < 1.5 m height) increased eightfold to twelvefold – from 309–383 dry kg/ha (preharvest) to 1,983–3,774 dry kg/ha – within 1–4 years after clearcutting and site preparation (Stransky and Halls 1978). In the Tennessee Ridge and Valley, forage availability (2008), dominated by tree species, was five times greater in shelterwood harvests (2001) followed by prescribed fire (2005) compared to mature forest controls (722 dry kg/ha versus 129 kg/ha, respectively), and more than seven times greater in retention cuts followed by multiple burns (2001, 2005, 2007) compared to controls (940 dry kg/ha versus 129 kg/ha, respectively) (Lashley 2009). In the pine-hardwood forest of the Ouachita Mountains in Oklahoma, total mean forage availability was 16–24 times greater in stands where pine timber was harvested, hardwoods thinned, and winter prescribed burns conducted at 1, 2, 3, or 4-year intervals (2–4 times) compared to mature forest controls (2,832–4,123 dry kg/ha versus 171 kg/ha, respectively); grasses composed the majority, whereas woody vegetation composed a small fraction of total forage (Masters et al. 1993). Forage availability in young forests declines appreciably after canopy closure (within 7–10 years), when sunlight no longer reaches the forest floor, but gradually increases, albeit to a relatively lower level, as stands mature (Johnson et al. 1995; Beck 1983).

Site quality can have a significant effect on forage availability (Beck 1983). Herbaceous plant diversity and quantity may be greater on mesic, high-quality sites than on dry, low-quality sites (Elliott et al., Chap. 7). In contrast, woody sprouts generally dominate on dry, poor-quality sites after heavy thinning (Beck 1983; Crawford 1971). Post-disturbance production of grasses and forbs may vary geographically, and with disturbance types and frequency (such as fire) (Spetich et al., Chap. 4).

In the Southern Appalachians nutritional quality of leaves from five woody browse species was similar between recent clearcuts and mature forest (Ford et al. 1994). However, forage quality may be greater in young forest than in mature forest because of increased diversity of forbs and other shade intolerant plant species (Elliott et al., Chap. 7). In addition, a high density of stump sprouts or seedlings in young forest increases browse availability from species such as blackgum, red maple (*Acer rubrum*), yellow-poplar (*Liriodendron tulipifera*), sassafras, oak, and hickory (Harlow and Hooper 1972; Warren and Hurst 1981; Beck and Harlow 1981; Ford et al. 1994; Loftis et al., Chap. 5). Forage quality for a given species, whether herbaceous or woody, is related to stage of growth. New growth of any plant is more digestible than older growth; as plants mature, cell walls thicken and lignin content, which is relatively indigestible, increases (Ball et al. 2002). Thus, greater forage quality and nutritional carrying capacity of young forests compared to mature forests is related to increased plant diversity, young foliar growth, and higher biomass resulting from increased available sunlight.

Without periodic disturbances, woody vegetation grows into the midstory (Jackson et al. 2007), reducing forage availability and a thick understory structure that benefits several forest songbirds and other wildlife species (Della-Bianca and Johnson 1965; Jackson et al. 2007; Thatcher et al. 2007; Lashley 2009, Shifley and Thompson, Chap. 6; Franzreb et al., Chap. 9). Intense fire that kills trees, or timber stand improvement treatments, such as heavy thinning or retention cuts, can create or perpetuate open-canopy conditions typical of young forests. Low-intensity prescribed fire in hardwood stands with an incomplete canopy cover can also maintain a diverse understory structure for various wildlife species without harming the residual overstory (Jackson et al. 2007). Repeated low-intensity prescribed fire within a 7-year period following canopy reduction will also sustain greater forage production by impeding canopy closure (Lashley et al. in press). Without a reduction in canopy closure and an increase in available sunlight, low-intensity prescribed fire is relatively ineffective in maintaining high forage production and diverse understory structure (Jackson et al. 2007; Shaw et al. 2010; Lashley et al. in press).

8.5 Arthropods

Arthropods are an important food source for many vertebrates. Most bird species in temperate deciduous forests are primarily insectivorous during the breeding season, and reproductive output may be limited by low food abundance (Holmes et al. 1986). Small, litter-dwelling arthropods are important for terrestrial salamanders (Duellman and Trueb 1986), whereas larger ground-dwelling arthropods are consumed by many birds, mammals, and larger amphibians and reptiles (Martin et al. 1951). Flying and foliar arthropods, such as Lepidoptera and Diptera, are important for many species of insectivorous birds and bats (Rodenhouse and Holmes 1992; Kurta et al. 1990; Loeb and O'Keefe, Chap. 10). Soil arthropods, such as larval beetles, are important components of skunk (*Mephitis* spp.) and shrew (Soricidae) diets (Martin et al. 1951). Forest condition and microclimate requirements differ among orders, families, and even species of arthropods (Greenberg and Forrest 2003). Therefore, forest disturbances that create open-canopy conditions have different effects on arthropod guilds, or groups, according to their habitat requirements. Disturbances that increase protective cover may benefit vertebrates that forage for arthropods and thus functionally increase invertebrate availability (Jackson et al. 2007).

Results of studies on arthropod response to forest disturbances have been inconsistent. Discrepancies may result from differences in sampling methodologies, site quality, season or month(s) studied, and timing of disturbance. For example, litter extraction methods sample the abundance of litter-dwelling arthropods as a snapshot in time, whereas pitfalls and other trapping methods sample a combination of both arthropod abundance and activity levels (Swengel 2001). Efficiency of sweep net sampling, area sampled, and forest strata sampled may differ among vegetation types because of differences in vegetation structure, thus biasing results (Harper and Guynn 1998). Insect activity periods differ among orders and species and studies conducted during different months may not be directly comparable. Disturbances that occur during peak activity periods or affect arthropod food sources could have a greater impact than disturbances during the non-growing season. Despite these types of inconsistencies, some general themes emerge, with overall responses to high-intensity disturbances and young forest conditions differing among litter-dwelling, ground-dwelling, and flying/foliar arthropods.

Forest disturbances that reduce canopy cover, increase light, and increase temperature at the forest floor, may result in decreased depth, cover, and moisture of leaf litter and cause declines in the biomass of litter- and ground-dwelling arthropods. Post-harvest reductions in leaf litter depth reported in the literature range from 14% to 70% (Buckner and Shure 1985; Ashe 1995), and may vary with site quality and the size and type of disturbance. However, the rapid growth of stump sprouts and other vegetation (Loftis et al., Chap. 5; Elliott et al., Chap. 7), and residual mature trees provide shade and replenish the leaf litter through leaf fall within 1–2 years post-disturbance (e.g., Greenberg and Waldrop 2008; Greenberg et al. 2010).

Ground- and litter-dwelling arthropod abundance and biomass is positively associated with leaf litter depth and moisture. For example, when compared to mature forests, arthropod abundance or biomass is lower in large forest gaps created by wind disturbance (Greenberg and Forrest 2003), on unpaved roads and up to 100 m into the adjacent mature forest (Haskell 2000), in managed and unmanaged forest openings (Harper et al. 2001), and recently harvested cove- and upland hardwood forest stands (Whitehead 2003). Several orders, such as Carabidae, Julida, Scolopendromorpha, and Spirobolidae, may be more abundant in mature forests where leaf litter depth and cover are greater (e.g., Greenberg and Forrest 2003), but other orders, such as Orthoptera and Homoptera may be more abundant in disturbed forests with greater cover of forbs and young foliage (e.g., Hollifield and Dimmick 1995).

Burning may have short-term negative impacts on litter- and ground-dwelling macroarthropod communities by direct mortality, or indirectly by altering forest floor conditions. Impacts of burning also correspond with the intensity and patchiness of burns, the availability of refugia, such as coarse woody debris, and the timing of burns in relation to taxon-specific life history traits (Swengel 2001). For example, burning during winter may affect ground-dwelling macroarthropods less because most of these species (including eggs and larvae) are underground and activity levels are generally low (Greenberg and Forrest 2003). Thus, life history traits, mobility, and behavior can mitigate direct effects of burning on arthropods.

Early spring burns may have little detectable impact on the relative abundance of ground dwelling arthropods (Coleman and Rieske 2006; Greenberg et al. 2010). However, Kalisz and Powell (2000) reported a 36% reduction in total dry biomass of forest floor and soil invertebrates after a March burn on the Cumberland Plateau in Kentucky, primarily from reductions in Coleopterans and Coleopteran larvae. Litter-dwelling arthropods, such as springtails, may be more sensitive to prescribed fire than ground-dwelling arthropods. For example, single and multiple prescribed burns in early spring reduced litter-dwelling arthropod abundance by 83% the first year and 48% the second year after burning in upland forests on the Cumberland Plateau in southeastern Kentucky (Coleman and Rieske 2006). Dress and Boerner (2004) reported lower relative abundance of microarthropods in an annually burned watershed where leaf litter mass was reduced, compared to periodically burned and unburned watersheds in southern Ohio. However, reduced leaf litter cover may increase arthropod availability to predators (Harper et al. 2001). Nevertheless, post-burn recovery of leaf litter arthropods is rapid and corresponds with leaf litter replenishment the following year (Greenberg et al. 2010).

Abundance and species richness of flying/foliar arthropods is often associated with plant species richness and herbaceous groundcover because many of these arthropods feed on foliage of specific plants, pollen, or nectar of woody and herbaceous plants. For example, abundance or species richness of foliage- or floral-feeding arthropods tends to be lower in pasture monocultures (Hollifield and Dimmick 1995; Burford et al. 1999; Harper et al. 2001; Fettinger et al. 2002; Dodd et al. 2008). In the Central Hardwood Region, disturbance does not usually change species richness of woody plants (Loftis et al., Chap. 5), but may increase richness of herbaceous plant species (Elliott et al., Chap. 7) or stimulate flowering and fruiting. In the southern Appalachians, the abundance of floral-visiting insects increased following hot prescribed fires that killed trees and increased herbaceous cover (Campbell et al. 2007). In the Ozark Mountains, moth occurrence was correlated with density and richness of woody plants, though abundance was similar among forest age classes (Dodd et al. 2008). Species richness and diversity of butterflies and their food- and host plants was higher in South Carolina early successional utility rights-of-way (Lanham et al., Chap. 12). In contrast, the abundance or species richness of foliage- or floral-feeding arthropods tends to be lower where forest stands are converted to pasture dominated by graminoids of homogeneous composition (Hollifield and Dimmick 1995; Burford et al. 1999; Harper et al. 2001; Fettinger et al. 2002; Dodd et al. 2008).

Site quality may influence arthropod availability because of potential differences in herbaceous cover and richness, leaf litter depth, and moisture that are associated with topographic position (Harper et al. 2001). In one study, three times more invertebrates occurred in mesic than xeric forest types, which in turn corresponded with herbaceous cover (Healy 1985). Other studies indicate that stand age is most important in determining arthropod abundance. In the Southern Appalachians, mature upland- and cove hardwood forests had more litter-dwelling arthropods and fewer flying/foliar arthropods than young upland- or cove hardwood forests (Whitehead 2003).

8.6 Small Mammals

Terrestrial small mammals (rats, mice, voles, shrews, squirrels, and rabbits) are the primary prey base for many species of vertebrates, including snakes, hawks, owls, and mammalian carnivores. For example, small mammals comprised 63% of Red-tailed

Hawk (*Buteo jamaicensis*) diets in hardwood forests of Pennsylvania (Sutton 1928), 76% of copperhead (*Agkistrodon contortrix*) diets in hardwood forests of Tennessee (Garton and Dimmick 1969), and occurred in 13% of coyote (*Canis latrans*) stomachs (Gipson 1974) and 65% of bobcat (*Lynx rufus*) stomachs (Fritz and Sealander 1978) examined in Arkansas. Consequently, small mammals may be considered a food source, and their abundance may be viewed as food biomass for many predators.

Species of small mammals respond differently to young, open-canopy conditions created by forest disturbance throughout the Central Hardwood Region. In deciduous forests, some groups of small mammals (e.g., *Peromyscus*) may decline after intense disturbances (Kirkland 1990); however, overall abundance of small mammals as a group (with the exceptions of squirrels [*Sciurus* and *Tamiasciurus*]) is generally much greater in young, recently disturbed, open-canopy forests compared to mature, closed-canopy forests. For example, Kirkland (1990) evaluated 21 studies on effects of clearcutting on small mammals (rodents and sorcids) in North America and found a significant pattern of increased relative abundance of all species combined after clearcutting. Furthermore, he found three out of four studies examining small mammal density reported increases after clearcutting. In hardwood forests of West Virginia, captures rates of all small mammals combined were 50% greater in 8–9 year-old stands than in mature (>100 years old) stands (Healy and Brooks 1988). In the Interior Highlands of Arkansas and Oklahoma, overall abundance of small mammals is low in mature, closed-canopy forests (Perry and Thill 2005); however, reductions in BA via timber harvest can dramatically increase overall abundance. Capture rates of all small mammals combined in areas harvested via single-tree selection, group selection, shelterwood cuts, and clearcuts can be 4–7 times greater than in closed-canopy, mature forest (Perry and Thill 2005).

Young forests provide the necessary habitat features for many species of small mammals. Abundance of some small mammals is correlated with coarse woody debris and logs (e.g., Loeb 1999; McCay 2000), and abundant woody debris often results from natural disturbances, such as windstorms or fires, or by logging and its associated slash. Shrub cover is also an important habitat component for many small mammals (e.g., Healy and Brooks 1988; Carey and Johnson 1995; Bellows et al. 2001), and shrub cover is characteristically much greater in young forests than mature, closed-canopy forests. Increased food supply typically results in increased vertebrate density (Boutin 1989), and abundant hard mast, soft mast, and grass/weed seeds in young, recently disturbed forest may provide substantially more food for small mammals than in surrounding mature forests. Furthermore, many small mammals, including voles, rabbits, and cotton rats, are primarily herbivores, and young, recently disturbed, open-canopy forest may provide substantially more herbaceous vegetation than mature forests. Declines in rabbit numbers in the eastern United States are attributable to changing land practices that reduced habitat, such as young forests, which provide critical cover for winter survival and predator evasion (Litvaitis 2001). Consequently, young hardwood forests provide abundant structural components and the necessary foods to support relative large densities of small mammals.

Many species of small mammals are associated with grasslands or hayfields (e.g., Hamilton and Whitaker 1979; Sealander and Heidt 1990), and in their earliest stages of development, young forests may provide habitat similar to grasslands (abundant herbaceous vegetation) and attract species such as hispid cotton rats (*S. hispidus*) and deer mice (*P. maniculatus*). A portion of the increase in small mammal abundance in young, recently disturbed forest may be attributable to exploitation of these sites by non-forest small mammals, such as jumping mice (*Zapus* spp.) and voles (Kirkland 1990).

Predator activity may be greatest in areas with the most prey (e.g., Ozoga and Harger 1966; Litvaitis and Shaw 1980), and predators of small mammals are often abundant in early successional habitat where they take advantage of abundant prey and cover. Many predators of small mammals, including gray fox (*Urocyon cinereoargenteus*), bobcats, and many snakes use young forests, shrubby areas, or areas with dense understories for cover (e.g., Hamilton 1982; Haroldson and Fritzell 1984; Kjoss and Litvaitis 2001; Perry et al. 2009) or avoid open areas with little cover (e.g.,Weatherhead and Prior 1992). For example, bobcats often prefer brushy areas or regenerating clearcuts where prey is most abundant (e.g., Hamilton 1982; Rolley and Warde 1985; Chamberlain et al. 2003). Furthermore, abundant burrows created by small mammals in areas of high small mammal abundance may provide habitat for predators such as snakes (Perry et al. 2009). Thus, young forest and other early successional or shrubby areas provide habitat for many predators of small mammals.

8.7 Conclusion

Young upland hardwood forests of the Central Hardwood Region provide a number of functions important to many wildlife species. These young forests provide habitat necessary for many species, including dense cover, abundant shrubs for shrub-nesting birds, and open areas for aerial predators, and also function as high-quality food patches that generally provide greater levels of many food resources than mature forests. Food resources abundant in young upland hardwood forests include fleshy fruit, forbs and grasses , browse, arthropods, and small mammals. Continuous creation of young forest patches through natural and silvicultural disturbance creates a shifting mosaic of age-classes and patch-sizes across the forested landscape. Partial reductions in tree density or canopy cover created by windstorms, hot fires, or partial timber harvests can provide a complex, heterogeneous forested landscape. Reductions in overstory tree density, while retaining some hard mast-producing trees, can promote production of fleshy fruit, foliage and flowers, and increase densities of arthropods and small mammals, while maintaining some level of hard mast production. Over time, young stands mature and provide other important features, such as high stem densities for grouse (Jones et al. 2008), or mature forest conditions that provide habitat for "forest interior" bird species (Greenberg and Lanham 2001). Balancing the creation of young, recently disturbed forest areas with the desired amount and distribution of other forest age classes will sustain high-quality food patches for wildlife within a landscape context.

Literature Cited

Ashe AN (1995) Effects of clear-cutting on litter parameters in the southern Blue Ridge Mountains. Castanea 60:89–97
Ball DM, Hoveland CS, Lacefield GD (2002) Southern forages, 3rd edn. Potash and Phosphate Institute, Norcross
Bechtold WA (2003) Crown-diameter prediction models for 87 species of stand-grown trees in the eastern United States. South J Appl For 27:269–278
Beck DE (1983) Thinning increases forage production in Southern Appalachian cove hardwoods. South J Appl For 7:53–57
Beck DE, Harlow RF (1981) Understory forage production following thinning in Southern Appalachian cove hardwoods. Proc Annu Conf Southeast Assoc Fish Wildl Agen 35: 185–196
Bellows AS, Pagels JF, Mitchell JC (2001) Macrohabitat and microhabitat affinities of small mammals in a fragmented landscape on the upper coastal plain of Virginia. Am Midl Nat 146:345–360
Bilan VM (1960) Stimulation of cone and seed production in pole-sized loblolly pine. For Sci 6:207–220
Blake JG, Hoppes WG (1986) Resource abundance and microhabitat use by birds in an isolated east-central Illinois woodlot. Auk 103:328–340
Borgmann KL, Pearson SF, Levey DJ, Greenberg CH (2004) Wintering yellow-rumped warblers (*Dendroica coronata*) track manipulated abundance of *Myrica cerifera* fruits. Auk 121:74–87
Borowicz VA (1988) Fruit consumption by birds in relation to fat content of pulp. Am Midl Nat 119:121–127
Boutin S (1989) Food supplementation experiments with terrestrial vertebrates: patterns, problems, and the future. Can J Zool 68:203–220
Buckner CA, Shure DJ (1985) Response of *Peromyscus* to forest opening size in the Southern Appalachian Mountains. J Mammal 66:299–307
Burford LS, Lacki MJ, Covell CV Jr (1999) Occurrence of moths among habitats in a mixed mesophytic forest: implications for management of forest bats. For Sci 45:323–332
Burns RM, Honkala BH (1990) Silvics of North America. In: Hardwoods, vol 2, Agri Handbook 654, USDA Forest Service, Washington, DC
Campbell JW, Hanula JL, Waldrop TA (2007) Effects of prescribed fire and fire surrogates on floral visiting insects of the Blue Ridge province in North Carolina. Biol Conserv 134:393–404
Campo JJ, Hurst GA (1980) Soft mast production in young loblolly plantations. Proc Annu Conf Southeast Assoc Fish Wildl Agen 34:470–475
Carey AB, Johnson ML (1995) Small mammals in managed, naturally young, and old growth forests. Ecol Appl 5:336–352
Chamberlain MJ, Leopold BD, Conner LM (2003) Space use, movements and habitat selection of adult bobcats (*Lynx rufus*) in central Mississippi. Am Midl Nat 149:395–405
Clark JD (2004) Oak-black bear relationships in southeastern uplands. In: Spetich MA (ed) Upland oak ecology symposium: history, current conditions, and sustainability. Gen Tech Rep SRS-73, USDA Forest Service, Southern Research Station, Asheville
Coleman TW, Rieske LK (2006) Arthropod response to prescription burning at the soil-litter interface in oak pine forests. For Ecol Manage 233:52 60
Crawford HS Jr (1971) Wildlife habitat changes after intermediate cutting for even-aged oak management. J Wildl Manage 35:275–286
Croker TC (1952) Early release stimulates cone production. Res Note 79, USDA Forest Service, Washington, DC
D'Antonio CM (2000) Fire, plant invasions, and global changes. In: Mooney HA, Hobbs RJ (eds) Invasive species in a changing world. Island Press, Washington, DC
Della-Bianca L, Johnson FM (1965) Effect of an intensive clearing on deer-browse production in the Southern Appalachians. J Wildl Manage 29:729–733

Dodd LE, Lacki MJ, Rieske LK (2008) Variation in moth occurrence and implications for foraging habitat of Ozark big-eared bats. For Ecol Manage 255:3866–3872
Dress WJ, Boerner REJ (2004) Patterns of microarthropod abundance in oak-hickory forest ecosystems in relation to prescribed fire and landscape position. Pedobiologia 48:1–8
Duellman WE, Trueb L (1986) Biology of amphibians. McGraw-Hill, New York
Eiler JH, Wathen WG, Pelton MR (1989) Reproduction in black bears in the Southern Appalachian Mountains. J Wildl Manage 53:353–360
Feldhamer GA (2002) Acorns and white-tailed deer: interrelationships in forest ecosystems. In: McShea WJ, Healy WM (eds) Oak forest ecosystems: ecology and management for wildlife. Johns Hopkins Press, Baltimore, pp 215–223
Fettinger JL, Harper CA, Dixon CE (2002) Invertebrate availability for upland game birds in tall fescue and native warm-season grass fields. J Tenn Acad Sci 77:83–87
Ford WM, Johnson AS, Hale PE, Wentworth JM (1993) Availability and use of spring and summer woody browse by deer in clearcut and uncut forests of the Southern Appalachians. South J Appl For 17:116–119
Ford WM, Johnson AS, Hale PE (1994) Nutritional quality of deer browse in Southern Appalachian clearcuts and mature forests. For Ecol Manage 67:149–157
Fritz SH, Sealander JA (1978) Diets of bobcats in Arkansas with special reference to age and sex differences. J Wildl Manage 42:553–539
Fuentes M (1994) Diets of fruit-eating birds: what are the causes of interspecific differences? Oecologia 97:134–142
Garton JS, Dimmick RW (1969) Food habits of the copperhead in middle Tennessee. J Tenn Acad Sci 44:113–117
Gipson PS (1974) Food habits of coyotes in Arkansas. J Wildl Manage 38:848–853
Godman RM (1962) Red pine cone production stimulated by heavy thinning. Tech Note 628, USDA Forest Service, Washington, DC
Greenberg CH, Levey DJ, Kwit C, McCarty JP, Pearson SF, Sargent S, Kilgo J (in review) Fruit production in five habitat types of the South Carolina Piedmont. J Wildl Manage (revision submitted)
Greenberg CH, Forrest TG (2003) Seasonal abundance of ground-occurring macroarthropods in forest and canopy gaps in the Southern Appalachians. Southeast Nat 2:591–608
Greenberg CH, Lanham DJ (2001) Breeding bird assemblages of hurricane-created gaps and adjacent closed canopy forest in the Southern Appalachians. For Ecol Manage 153:251–260
Greenberg CH, Parresol BR (2002) Dynamics of acorn production by five species of Southern Appalachian oaks. In: McShea WJ, Healy WM (eds) Oak forest ecosystems: ecology and management for wildlife. Johns Hopkins University Press, Baltimore
Greenberg CH, Waldrop TA (2008) Short-term response of reptiles and amphibians to prescribed fire and mechanical fuel reduction in a Southern Appalachian upland hardwood forest. For Ecol Manage 255:2883–2893
Greenberg CH, Walter ST (2010) Fleshy fruit removal and nutritional composition of winter-fruiting plants: a comparison of non-native and native invasive and native species. Nat Areas J 30:312–321
Greenberg CH, Levey DJ, Loftis DL (2007) Fruit production in mature and recently regenerated upland and cove hardwood forests of the Southern Appalachians. J Wildl Manage 71:321–329
Greenberg CH, Forrest TG, Waldrop TA (2010) Short-term response of ground-dwelling macroarthropods to prescribed fire and mechanical fuel reduction in a Southern Appalachian upland hardwood forest. For Sci 59:112–121
Halls LK, Alcaniz R (1968) Browse plants yield best in forest openings. J Wildl Manage 32:185–186
Hamilton DA (1982) Ecology of bobcat in Missouri. MS thesis, University of Missouri, Columbia
Hamilton WJ Jr, Whitaker JO Jr (1979) Mammals of the eastern United States, 2nd edn. Cornell University Press, Ithaca
Harlow RF, Downing RL (1969) The effects of size and intensity of cut on production and utilization of some deer foods in the Southern Appalachians. Northeast Fish Wildl Conf Trans 26:45–55
Harlow RF, Downing RL (1970) Deer browsing and hardwood regeneration in the Southern Appalachians. J For 68:298–300

Harlow RF, Eikum RL (1963) The effects of stand density on the acorn production of turkey oaks. In: Proceedings of the 17th annual conference Southeastern Association Game and Fish Commissioner, Hot Springs, September 1963, pp 126–133
Harlow RF, Hooper RG (1972) Forages eaten by deer in the Southeast. Proc Annu Conf Southeast Assoc Fish Wildl Agen 25:18–46
Haroldson KJ, Fritzell EK (1984) Home ranges, activity, and habitat use by gray foxes in an oak-hickory forest. J Wildl Manage 48:222–227
Harper CA, Guynn DC Jr (1998) A terrestrial vacuum sampler for macroinvertebrates. Wildl Soc Bull 26:302–306
Harper CA, Guynn DC Jr, Knox JK, Davis JR, Williams JG (2001) Invertebrate availability for wild turkey poults in the Southern Appalachians. In: Proceedings of the eighth National Wild Turkey Symposium, National Wild Turkey Federation, Edgefield, pp 145–156
Haskell DG (2000) Effects of forest roads on macroinvertebrate soil fauna of the Southern Appalachian Mountains. Conserv Biol 14:57–63
Healy WM (1985) Turkey poult feeding activity, invertebrate abundance, and vegetation structure. J Wildl Manage 49:466–472
Healy WM (1997) Thinning New England oak stands to enhance acorn production. North J Appl For 14:152–156
Healy WM, Brooks RT (1988) Small mammal abundance in northern hardwood forest stands in West Virginia. J Wildl Manage 52:491–496
Hollifield BK, Dimmick RW (1995) Arthropod abundance relative to forest management practices benefiting ruffed grouse in the Southern Appalachians. Wildl Soc Bull 23:756–764
Holmes RT, Sherry TW, Sturges FW (1986) Bird community dynamics in a temperate deciduous forest: long-term trends at Hubbard Brook. Ecol Monogr 56:201–220
Jackson SW, Basinger RG, Gordon DS, Harper CA, Buckley DS, Buehler DA (2007) Influence of silvicultural treatments on eastern wild turkey habitat characteristics in eastern Tennessee. In: Proceedings of the ninth National Wild Turkey Symposium, National Wild Turkey Federation, Edgefield, pp 190–198
Johnson AS, Landers JL (1978) Fruit production in slash pine plantations in Georgia. J Wildl Manage 42:606–613
Johnson AS, Hale PE, Ford WM, Wentworth JM, French JR, Anderson OF, Pullen GB (1995) White-tailed deer foraging in relation to successional stage, overstory type and management of Southern Appalachian forests. Am Midl Nat 133:18–35
Jones BC, Kleitch JL, Harper CA, Buehler DA (2008) Ruffed grouse brood habitat use in a mixed hardwood forest: implications for forest management in the Appalachians. For Ecol Manage 255:3580–3588
Jordano P (2000) Fruits and frugivory. In: Fenner M (ed) Seeds: the ecology of regeneration in plant communities. CAB International, New York, pp 111–124
Kalisz PJ, Powell JE (2000) Effects of prescribed fire on soil invertebrates in upland forests on the Cumberland Plateau of Kentucky, USA. Nat Areas J 20:336–341
Kirkland GL Jr (1990) Patterns of small mammal community change after clearcutting in temperate North American forests. Oikos 59:313–320
Kjoss VA, Litvaitis JA (2001) Community structure of snakes in a human-dominated landscape. Biol Conserv 98:285–292
Kurta A, Kunz TH, Nagy KA (1990) Energetics and water flux of free-ranging big brown bats (*Eptesicus fuscus*) during pregnancy and lactation. J Mammal 71:59–65
Kwit C, Levey DJ, Greenberg CH, Pearson SF, McCarty JP, Sargent S, Mumme RL (2004) Fruit abundance and local distribution of wintering hermit thrushes (*Catharus guttatus*) and yellow-rumped warblers (*Dendroica coronata*) in South Carolina. Auk 121:46–57
Lashley MA (2009) Deer forage available following silvicultural treatments in upland hardwood forests and warm-season plantings. Thesis, University of Tennessee, Knoxville
Lashley MA, McCord JM, Greenberg CH, Harper CA (2010) Masting characteristics of white oaks: implications for management. Proc Annu Conf Southeast Assoc Fish Wildl Agen 63:21–26
Lashley MA, Harper CA, Bates GE, Keyser PD (in press) Forage availability for white-tailed deer following silvicultural treatments. J Wildl Manage

Lay DL (1966) Forest clearings and for browse and fruit plantings. J For 64:680–683
Levey DJ, Martinez del Rio C (2001) It takes guts (and more) to eat fruit: lessons from avian nutritional ecology. Auk 118:819–831
Litvaitis JA (2001) Importance of early successional habitats to mammals in eastern forests. Wildl Soc Bull 29:466–473
Litvaitis JA, Shaw JH (1980) Coyote movements, habitat use, and food habits in southwestern Oklahoma. J Wildl Manage 44:62–68
Loeb SC (1999) Responses of small mammals to coarse woody debris in a southeastern pine forest. J Mammal 80:460–471
Martin AC, Zim HS, Nelson AL (1951) American wildlife and plants: a guide to wildlife food habits. Dover, New York
Martinez del Rio C, Restrepo C (1993) Ecological and behavioral consequences of digestion in frugivorous animals. Vegetatio 107:205–216
Masters RE, Lochmiller RL, Engle DM (1993) Effect of timber harvest and prescribed fire on white-tailed deer forage production. Wildl Soc Bull 21:401–411
McCarty JP, Levey DJ, Greenberg CH, Sargent S (2002) Spatial and temporal variation in fruit use by wildlife in a forested landscape. For Ecol Manage 164:277–291
McCay TS (2000) Use of woody debris by cotton mice (*Peromyscus gossypinus*) in a southeastern pine forest. J Mammal 81:527–535
McCord JM, Harper CH (in press) Brood habitat following canopy reduction understory herbicide application, and fire in mature upland hardwoods. Proc Nat Wild Turkey Symp 10 (In press)
McShea WJ, Healy WM (eds) (2002) Oak forest ecosystems: ecology and management for wildlife. Johns Hopkins University Press, Baltimore
Minckler LS, McDermott RE (1960) Pin oak acorn production and regeneration as affected by stand density, structure, and flooding. Agriculture Experiment Station, Res Bull 750, University of Missouri, Columbia
Mitchell MS, Powell RA (2003) Response of black bears to forest management in the Southern Appalachian Mountains. J Wildl Manage 67:692–705
Moore WH, Downing RL (1965) Some multiple-use benefits of even-aged management in the Southern Appalachians. Proc Soc Am Forest 1965:227–229
Moore WH, Johnson FM (1967) Nature of deer browsing on hardwood seedlings and sprouts. J Wildl Manage 31:351–353
Morriss DJ (1954) Correlation of wildlife management with other uses on the Pisgah National Forest. J For 52:419–422
Ozoga JJ, Harger EM (1966) Winter activities and feeding habits of northern Michigan coyotes. J Wildl Manage 30:809–818
Parrish JD (1997) Patterns of frugivory and energetic condition in Nearctic landbirds during autumn migration. Condor 99:681–697
Patton DR, McGinnes BS (1964) Deer browse relative to age and intensity of timber harvest. J Wildl Manage 28:458–463
Paugh JH (1970) Effects of thinning on acorn production on the West Virginia University Forest. Thesis, West Virginia University, Morgantown
Perry RW, Thill RE (2003) Effects of reproduction cutting treatments on residual hard mast production in the Ouachita Mountains. South J Appl For 27:253–258
Perry RW, Thill RE (2005) Small-mammal responses to pine regeneration treatments in the Ouachita Mountains of Arkansas and Oklahoma, USA. For Ecol Manage 219:81–94
Perry RW, Thill RE, Peitz DG, Tappe PA (1999) Effects of different silvicultural systems on soft mast production. Wildl Soc Bull 27:915–923
Perry RW, Thill RE, Tappe PA, Peitz DG (2004) The relationship between basal area and hard mast production in the Ouachita Mountains. In: Guldin JM (tech comp) Ouachita and Ozark Mountains symposium: ecosystem management research. Gen Tech Rep SRS-74, USDA Forest Service, Southern Research Station, Asheville, pp 55–59
Perry RW, Rudolph DC, Thill RE (2009) Reptile and amphibian responses to restoration of fire-maintained pine woodlands. Restor Ecol 17:917–927

Reynolds-Hogland MJ, Mitchell MS, Powell RA (2006) Spatio-temporal availability of soft mast in clearcuts in the Southern Appalachians. For Ecol Manage 237:103–114
Ripley TH, Campbell RA (1960) Browsing and stand regeneration in clear- and selectively-cut hardwoods. Trans North Am Wildl Conf 25:407–415
Rodenhouse NL, Holmes RT (1992) Result of experimental and natural food reductions for breeding Black-throated Blue Warblers. Ecology 73:357–372
Rodewald AD (2003) Decline of oak forests and implications for forest wildlife conservation. Nat Areas J 23:368–371
Rogers LL (1976) Effects of mast and berry crop failures on survival, growth and reproductive success of black bears. Trans North Am Wildl Nat Res Confer 41:431–438
Rogers MJ, Halls LK, Dickson JG (1990) Deer habitats in the Ozark forests of Arkansas. Gen Tech Rep SO-259, USDA Forest Service, Southern Research Station, New Orleans
Rolley RE, Warde WD (1985) Bobcat habitat use in southeastern Oklahoma. J Wildl Manage 49:913–920
Rose, A., Greenberg CH, Fearer T (in review) Acorn production prediction models for 5 oak species of the eastern United States. J Wildl Manage (submitted Feb. 2011)
Sealander JA, Heidt GA (1990) Arkansas mammals: their natural history, classification, and distribution. University of Arkansas Press, Fayetteville
Segelquist CA, Green WE (1968) Deer food yields in four Ozark forest types. J Wildl Manage 32:330–337
Shaw CE, Harper CA, Black MW, Houston AE (2010) The effects of prescribed burning and understory fertilization on browse production in closed- canopy hardwood stands. J Fish Wildl Manage 1:64–72
Stransky JJ, Halls LK (1978) Forage yield increase by clearcutting and site preparation. Proc Ann Conf Southeast Assoc Fish Wildl Agen 32:38–41
Stransky JJ, Halls LK (1980) Fruiting of woody plants affected by site preparation and prior land use. J Wildl Manage 44:258–263
Stransky JJ, Roese JH (1984) Promoting soft mast for wildlife in intensively managed forests. Wildl Soc Bull 12:234–240
Sutton GM (1928) Notes on a collection of hawks from Schuylkill County, Pennsylvania. Wilson Bull 40:84–95
Swengel AB (2001) A literature review of insect responses to fire, compared to other conservation management of open habitat. Biodivers Conserv 10:1141–1169
Thatcher BS, Buehler DA, Martin PD, Wheat RM (2007) Forest management to improve breeding habitat for priority songbirds in upland oak-hickory forests. In: Buckley DS, Clatterbuck WK (eds) Proceedings of the 15th central hardwood forest conference. Gen Tech Rep SRS-101, USDA Forest Service, Southeastern Research Station, Asheville
Thompson JN, Willson MF (1978) Disturbance and dispersal of fleshy fruits. Science 200:1161–1163
Tilghman NG (1989) Impacts of white-tailed deer on forests regeneration in northwestern Pennsylvania. J Wildl Manage 53:524–532
Warren RC, Hurst GA (1981) Ratings of plants in pine plantations as white-tailed deer food. Inform Bull 18, Mississipi Agricultural and Forestry Experiment Station, Mississippi state
Weatherhead PJ, Prior KA (1992) Preliminary observations of habitat use and movements of the eastern massasauga rattlesnake (*Sistrurus c. catenatus*). J Herp 26:447–452
Whelan CJ, Willson MF (1994) Fruit choice in North American migrant birds: aviary and field experiments. Oikos 71:137–151
White WW, Stiles EW (1992) Bird dispersal of fruits of species introduced into eastern North America. Can J Bot 70:1689–1696
Whitehead MA (2003) Seasonal variation in food resource availability and avian communities in four habitat types in the Southern Appalachian Mountains. Dissertation, Clemson University, Clemson
Willson MF (1986) Avian frugivory and seed dispersal in eastern North America. Curr Ornith 3:223–279
Witmer MC, Van Soest PJ (1998) Contrasting digestive strategies of fruit-eating birds. Funct Ecol 12:728–741

Chapter 9
Population Trends for Eastern Scrub-Shrub Birds Related to Availability of Small-Diameter Upland Hardwood Forests

Kathleen E. Franzreb, Sonja N. Oswalt, and David A. Buehler

Abstract Early successional habitats are an important part of the forest landscape for supporting avian communities. As the frequency and extent of the anthropogenic disturbances have declined, suitable habitat for scrub-shrub bird species also has decreased, resulting in significant declines for many species. We related changes in the proportion and distribution of small-diameter upland hardwood forest throughout the eastern USA (US Forest Service Forest Inventory and Analysis data) with North American Breeding Bird Survey data (US Geological Survey) on population trends of 11 species that use early successional hardwood forest. The availability of small-diameter upland hardwood forest has changed over the past four decades, with the biggest differences seen as declines from the 1990s to the 2000s. Most scrub-shrub species also declined since the inception of the Breeding Bird Survey in 1966. The declines in most of the bird species, however, did not closely track the changes in small-diameter forest availability. Scrub-shrub birds use a variety of habitats that

K.E. Franzreb (✉)
Research Wildlife Biologist with the Upland Hardwood Ecology and Management Research Work Unit, USDA Forest Service, Southern Research Station,
Southern Appalachian Mountains Cooperative Ecosystems Studies Unit, Department of Forestry, Wildlife, and Fisheries,
University of Tennessee, Knoxville, TN 37996, USA
e-mail: franzreb@utk.edu

S.N. Oswalt
Forester with the Resource Analysis Team, USDA Forest Service, Southern Research Station, Forest Inventory and Analysis,
4700 Old Kingston Pike, Knoxville, TN 37919, USA
e-mail: soswalt@fs.fed.us

D.A. Buehler
Department of Forestry, Wildlife and Fisheries, University of Tennessee,
Knoxville, TN 37996, USA
e-mail: DBuehler@utk.edu

C.H. Greenberg et al. (eds.), *Sustaining Young Forest Communities*,
Managing Forest Ecosystems 21, DOI 10.1007/978-94-007-1620-9_9,

originate from a diverse array of disturbance sources. The total availability of these habitats across the region apparently limits the populations for these species. A comprehensive management strategy across all of these types is required to conserve these species.

9.1 Introduction

Conservation biologists have become increasingly aware of the plight of wildlife species that require the early stages of forest succession for habitat. Two journals recently dedicated sections on this topic (Thompson et al. 2001; Litvaitis 2003). Historically, disturbances in forest ecosystems from natural and anthropogenic sources created a mosaic of habitats ranging from the earliest stages of succession through old growth conditions (see Greenberg et al., Chap. 1; White et al., Chap. 3). A multitude of wildlife species are adapted to take advantage of young forest habitats created by these disturbances and populations of many are declining as abandoned farmland and pastures return to forest and recently harvested or disturbed forests re-grow (Greenberg et al., Chap. 1). For example, populations of many avian species that breed in small-diameter forested habitats are declining throughout the eastern United States (Askins 2001; Brawn et al. 2001; Hunter et al. 2001; Dettmers 2003), as are some that breed in mature forests but use small-diameter forested habitats during the post-breeding season (Marshall et al. 2003; Bulluck and Buehler 2006; Vitz and Rodewald 2006).

Early successional habitats arise from a variety of natural and anthropogenic disturbance sources, including catastrophic weather (tornados, hurricanes, severe ice storms, flooding), wild fire, grazing, clearing of land for agriculture and subsequent abandonment, insect outbreaks, creation and management of utility rights-of-way, roadside edges, mining, and forest management (Greenberg et al., Chap. 1). Numerous studies have documented avian response to various types of forest management at the stand scale (e.g., Annand and Thompson 1997; Krementz and Christie 2000; Pagen et al. 2000; Marshall et al. 2003; Rodewald and Vitz 2005; Vitz and Rodewald 2006; Campbell et al. 2007) and at the landscape scale (e.g., Thompson et al. 1992; Bourque and Villard 2001; Rodewald and Yahner 2001a; Rodewald and Yahner 2001b; Gram et al. 2003). The effect of clearcutting on birds at the stand scale in eastern forests has received the most research attention (Sallabanks et al. 2000). Studies on avian response to other sources of disturbance are available but less numerous (e.g., King and Byers 2002; Tingley et al. 2002; Confer and Pascoe 2003; Lacki et al. 2004; Bulluck and Buehler 2006). Maintaining a mosaic of different stand age classes (i.e., differing years post-harvest) in a forested landscape can provide habitat for a diversity of avifauna, especially when the requirements of regional species of concern, patch size, and landscape context are considered (King et al. 1998; Krementz and Christie 2000; King et al. 2001; Rodewald and Yahner 2001a; Rodewald and Yahner 2001b; Gram et al. 2003).

Our goals were to (1) summarize the changes in availability of small-diameter upland hardwood forests in the eastern USA over time based on analysis of US Forest Service Forest Inventory and Analysis (FIA) data, and (2) examine population trends for scrub-shrub avian species that use these small-diameter upland hardwood habitats based on US Geological Survey North American Breeding Bird Survey (BBS) data analyses. Finally, we evaluate how well the avian population trends track documented changes in small-diameter hardwood forest availability in the region. FIA data represent the only source of stand-level data collected over the entire area of upland hardwood forest in the eastern USA with a statistically-sound sampling design and standardized data collection protocols (Bechtold and Patterson 2005). We analyzed changes in small-diameter forests, rather than forest stand age, because birds respond to changes in the structural properties of forests (Raphael et al. 1987; Diaz et al. 2005), and those properties may vary considerably within the young age class depending on tree species composition and site productivity (Moran et al. 2000). We believe that tracking forests of the structure required by scrub-shrub birds would be a better fit than using age as the classification criterion. Even so, small-diameter forests in the FIA database represent a subset of the potential available habitat for many eastern scrub-shrub birds. Hence, we are assessing the relationship between FIA small-diameter forests and population trends for this suite of bird species.

9.2 Approach

We conducted analyses at three spatial scales: (1) the upland hardwood forest area of the eastern USA as defined by three Bird Conservation Regions (BCRs), (2) within three BCRs, and (3) within BCR-state intersections (Fig. 9.1). Bird Conservation Region boundaries are described on the BBS website (www.mbr-pwrc.usgs.gov/bbs/) and were designed to provide a spatial framework for avian conservation planning under the North American Bird Conservation Initiative (NABCI). Data from ten states within three BCRs (Central Hardwoods, Appalachian Mountains, and Piedmont; Fig. 9.1) are included in the analysis and largely overlap the Central Hardwood Region considered in this book (see Greenberg et al., Chap. 1). The Central Hardwoods BCR includes the Ozark Mountains on the west and extends eastward including the Interior Low Plateau with the entire area being dominated by oak-hickory (*Quercus-Carya*) deciduous forest. The Appalachian Mountains BCR contains the Blue Ridge, Ridge and Valley, Cumberland Plateau, Ohio Hills and the Allegheny Plateau. This area is characterized at lower elevations by oak-hickory and other deciduous forest types and at higher elevations by various combinations of pine (*Pinus* spp.), hemlock (*Tsuga canadensis*), spruce (*Picea* spp.), fir (*Abies* spp.), northern hardwoods, and northern red oak (*Q. rubra*). The Piedmont BCR is considered to be transitional between the rugged, mountainous Appalachians dominated by hardwoods and the relatively flat Coastal Plain dominated by pines and mixed southern hardwoods. For detailed descriptions of these upland hardwood forest types see Chap. 2 (McNab).

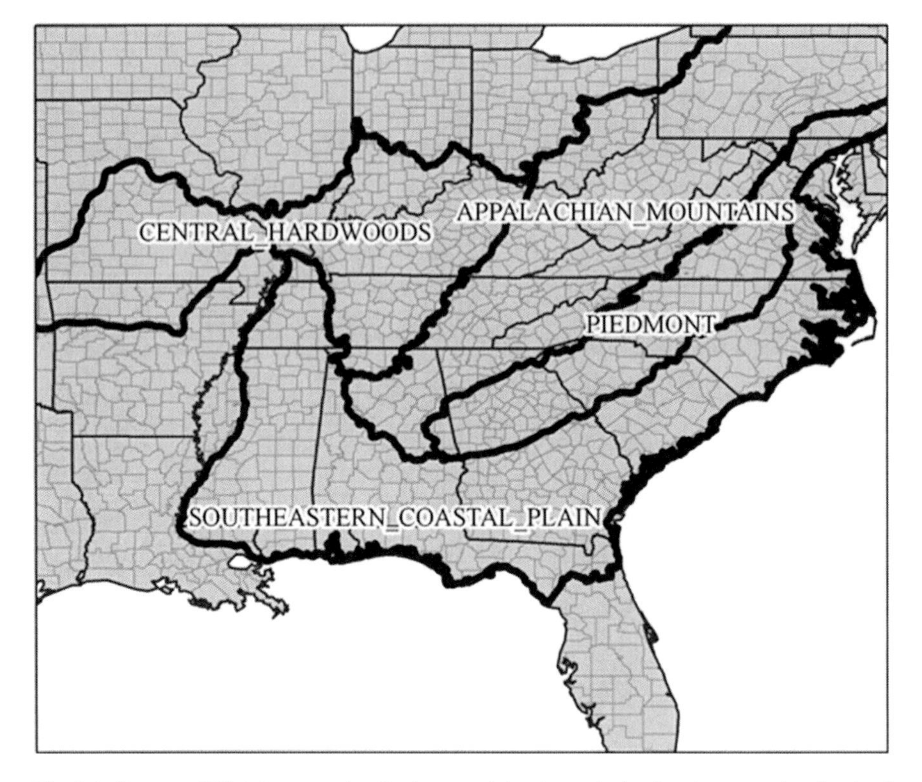

Fig. 9.1 States and Bird Conservation Regions used for the analysis of early successional upland hardwood forests and avian population trends

In addition, we examined data for ten states individually that were within the three referenced BCRs. Only FIA plots that fell inside the boundaries of the BCRs of interest were included in the state totals, thus the numbers do not represent complete state-level coverage. The states included were Alabama, Arkansas, Georgia, Kentucky, Missouri, North Carolina, South Carolina, Tennessee, Virginia, and West Virginia.

9.2.1 Forest Inventory and Analysis (FIA)

We used FIA data to identify small-diameter hardwood forests, based on dominance at the stand level by small-diameter hardwood trees. We examined trends in availability of small-diameter hardwood forests across four decadal time periods (1970s, 1980s, 1990s, and 2000s) within the three BCRs of interest. The sample population was defined by intersecting the outline of BCRs with FIA plot locations in ten states using ESRI ArcGIS (Fig. 9.1). FIA plots were located on the map using actual coordinates collected in the field, with the exception of plot locations in Missouri

Table 9.1 States, years, and number of plots within three Bird Conservation Regions in the eastern USA used to analyze trends in availability of small-diameter hardwood forests by decade

	Year in decade			
State	1970s	1980s	1990s	2000s
Alabama	1972	1982	1990	2008
Arkansas	1978	1988	1995	2007
Georgia	1972	1989	1997	2008
Kentucky		1988		2007
Missouri		1989		2008
North Carolina	1974	1984	1990	2007
South Carolina	1976	1986	1993	2007
Tennessee	1980	1989	1999	2007
Virginia	1977	1985	1992	2008
West Virginia		1989		2006
Total Number of Plots	12,479	29,926	22,074	25,603

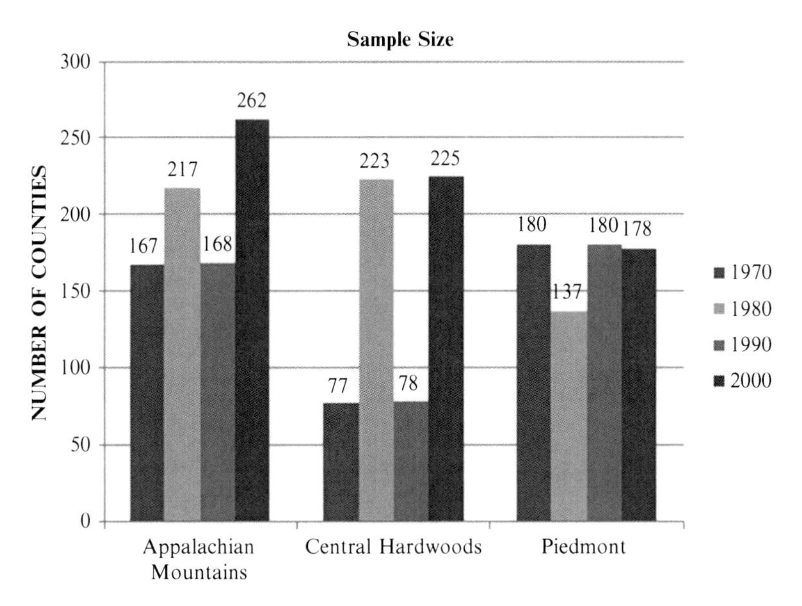

Fig. 9.2 Number of sample units (county aggregates of plots, n) used in statistical analysis by year and Bird Conservation Region

and West Virginia, where FIA "perturbed and swapped" locations were based on availability at the time of analysis (Bechtold and Patterson 2005). Survey periods and numbers of plots used in this analysis varied by state (Table 9.1).

We analyzed county aggregates of selected plots as the sample unit (Fei and Steiner 2007; Oswalt and Turner 2009; Fig. 9.2). We calculated metrics based on timberland areas within BCRs. Timberland is defined by FIA as "forest land that is producing or capable of producing in excess of 20 cubic feet per acre per year of wood at culmination of mean annual increment." This definition excludes reserved forest land and "unproductive" forest land. Until recently, FIA collected individual tree metrics only on timberland, thus, for trend analysis utilizing specific plot and tree metrics, timberland must be used (Bechtold and Patterson 2005). The total timberland area in hectares (TTA), total hardwood timberland area (THA), and total small-diameter hardwood timberland area (TSD) were calculated for each Decade-State-BCR-County combination. Some states were not sampled in some decades (for example, Kentucky only was sampled in two of the four decades). Thus, sample size and area differed through time. Therefore, to facilitate comparison among decades, area estimates were normalized for analysis by converting raw numbers to proportions, yielding the proportion of total timberland area that was hardwood (PTTA), the proportion of total timberland area that was small-diameter hardwood (PTSD), and the proportion of total hardwood timberland that was small-diameter (PTHA). Concerns that the use of proportions might produce erroneous results with regards to changes in avian habitat if raw TTA and raw TSD both experienced declines but PTSD remained stable were relieved by Smith et al. (2009), who showed that in the regions encompassing the BCRs of interest, timberland area has remained stable or increased since the mid-1970s. We were unable to use discrete area numbers because not all states were sampled in each decade. Thus, the sample area was not the same and discrete area values would reflect the differences in sample area instead of true differences in forest acreage. Hardwood stands were identified as those falling within a pre-selected set of FIA forest-type groups containing primarily hardwood species (Table 9.2). Small-diameter stands were identified using the FIA variable STDSZCD, which defines small-diameter stands as "stands with an all live stocking value of at least 10 (base 100) on which at least 50% of the stocking is trees less than 12.7 cm in diameter" (USDA Forest Service 2009).

Analyses of variance were used to determine changes in PTTA, PTSD, and PTHA over time across the whole study area, by BCR, and by state. Generalized least square means were compared among decades for each ANOVA. We also provide data from the latest publication of the nationwide USDA Forest Service, Forest Resources of the United States report (Smith et al. 2009) for comparison with localized results.

9.2.2 Breeding Bird Survey (BBS) Analyses

We used data analyses from the North American BBS to examine avian population trends covering the time periods 1966–1979 and 1980–2007, and the combined period of 1966–2007 (Sauer et al. 2008). We examined population trends for scrub-shrub species to determine which species were undergoing changes and the direction (increasing or decreasing) of change in the three BCRs and the ten aforementioned states. We used the species group designations of Sauer et al. (2008) to identify scrub-shrub species. In addition, we provide detailed analyses on 11 representative scrub-shrub species of eastern upland hardwood forest.

Table 9.2 Forest Inventory and Analysis forest types used in the analysis of trends in availability of small-diameter hardwood trees in three Bird Conservation Regions in the eastern USA

FIA code	Forest type	FIA code	Forest type
400	Oak (*Quercus* spp.)/Pine (*Pinus* spp.) group	510	Scarlet oak (*Q. coccinea*)
401	E. white pine (*P. strobus*)/n. red oak (*Q. rubra*)/white ash (*Fraxinus americana*)	511	Yellow-poplar
		512	Black walnut (*Juglans nigra*)
402	Eastern red cedar (*Juniperus virginiana*)/hardwood	513	Black locust (*Robinia pseudoacacia*)
403	Longleaf pine (*P. palustris*)/oak	514	Southern scrub oaks
404	Shortleaf pine (*P. echinata*)/oak	515	Chestnut oak (*Q. prinus*)/black oak (*Q. velutina*)/scarlet oak
405	Virginia pine (*P .virginiana*) /southern red oak (*Q. falcata*)	516	Cherry (*Prunus* spp.)/white ash /yellow-poplar
406	Loblolly pine (*P. taeda*) /hardwood	517	Elm (*Ulmus* spp.)/ash/black locust
407	Slash pine (*P. elliottii*)/hardwood	519	Red maple (*Acer rubrum*)/oak
409	Other pine/hardwood	520	Mixed upland hardwoods
500	Oak/hickory (*Carya* spp.) group	800	Maple (*Acer* spp.)/American beech (*Fagus grandifolia*)/birch (*Betula* spp.) group
501	Post oak (*Q. stellata*)/blackjack oak (*Q. marilandica*)		
502	Chestnut oak (*Q. prinus*)	801	Sugar maple (*A. saccharum*)/beech yellow/birch (*Betula alleghaniensis*)
503	White oak (*Q. alba*)/red oak (*Q. rubra*)/hickory		
504	White oak	802	Black cherry(*Prunus serotina*)
505	Northern red oak	805	Hard maple (*A. nigrum*)/basswood (*Tilia* spp.)
506	Yellow-poplar (*Liriodendron tulipifera*)/white oak/n. red oak	809	Red maple/upland
507	Sassafras (*Sassafras albidum*) /persimmon (*Diospyros virginiana*)	905	Pin cherry (*P. pensylvanica*)
		962	Other hardwoods
508	Sweetgum (*Liquidambar styraciflua*)/yellow-poplar	971	Deciduous oak woodland
509	Bur oak (*Q. macrocarpa*)	976	Misc. woodland hardwoods

Beginning in 1966, the BBS has been conducted annually and provides the only long-term database on breeding birds in North America. During the survey, observers collect data along a series of 24.5 mile routes using the point count method of recording all birds heard or seen within 0.25 miles of the point over a three-minute period. Points are established every 0.5 mile along the routes and data are collected using standardized collection protocols. The data are then forwarded to the US Geological Survey (USGS) for analysis by BBS staff using the route-regression procedure (Geissler and Sauer 1990) and modified through the use of estimating equations (Link and Sauer 1994). Their null hypothesis is that there has been no population change for the time period with a significance level of $P<0.10$. BBS data do not categorize vegetation or stand type at the point or route level. Hence, the data presented here are not restricted to only situations where the species occurred in early successional habitats.

Considerable controversy exists regarding the methods used to collect and analyze BBS data and, hence, the conclusions derived from it. Limitations of the methodology have been discussed in various venues (Sauer and Droege 1990; Peterjohn et al. 1995; James et al. 1996; Thomas and Martin 1996) and will not be discussed further here. In spite of this controversy, the different methods usually yield similar results, although the estimated rates of change may differ (Peterjohn et al. 1997).

9.2.3 Bird-Habitat Change Analyses

We regressed annual bird population change on annual change in availability of small-diameter hardwood forests using simple linear regression to test hypotheses that bird population trends were related to changes in small-diameter forest habitat availability for the 11 focal avian species. The percent change per year in small-diameter upland hardwood forest for each state ($n=10$ states) was the independent variable. The annualized change in the index of relative abundance for a given avian species was the response variable. We measured total change in small-diameter forest hectares per time period as described above and then calculated a percent change per year index. We used the earliest forest inventory date and the latest forest inventory date for each state to determine the number of years in the time period. We then used that same time period for calculating the percent change/year in bird relative abundances based on analysis tools provided by Sauer et al. (2008). The number of years used in the analyses varied depending on when the first forest inventory was completed in a given state (range = 17 years from 1989 to 2006 for West Virginia to 36 years from 1972 to 2007 for Georgia). Regression assumptions include (1) linearity of the relationship between dependent and independent variables; (2) independence of the errors (no serial correlation); (3) homoscedasticity; and (4) normality of the error distribution. We evaluated regression models for compliance with these assumptions with plots of residuals versus predicted values and normal probability plots of residuals. In general, the individual regressions met assumptions, thus no transformations were required. The regression assumption of measurement of the x and y variables without error was generally not met because data used in the regression were averaged values.

9.3 Results and Discussion

9.3.1 *Availability of Small-Diameter Upland Hardwood Forests*

Hardwood area trends, as a proportion of total timberland, varied by BCR and time. In the Appalachian Mountains BCR, PTTA increased between the 1970s and 1990s, and then increased again between the 1990s and 2000s ($P=0.0075$; Fig. 9.3). In the Central Hardwood BCR, PTTA remained stable across all four decades ($P=0.0810$). The PTTA increased in the Piedmont BCR between the 1980s and 1990s ($P<0.0001$). Timberland in the Appalachian Mountains and Central Hardwoods BCRs was predominantly hardwood, and contained the highest proportion of hardwood to softwood timberland in the study (91.2 ± 4.1 and $86.2 \pm 0.9\%$ in the 2000s, respectively). In comparison, the Piedmont BCR sample area was composed of approximately $60.8 \pm 1.4\%$ hardwoods in the 2000s.

Proportionally, the area of small-diameter hardwood timberland across the entire sample of interest remained stable from the 1970s to the 1980s (27.0 ± 0.7 and $26.8 \pm 0.7\%$, respectively), increased in the 1990s to $32.3 \pm 0.8\%$, then declined in the 2000s to $21.7 \pm 0.6\%$ ($P<0.0001$; Fig. 9.4). In the Appalachian Mountains BCR, no differences occurred from the 1970s to the 1980s (18.0 ± 1.3 and $16.0 \pm 0.9\%$, respectively), but small-diameter area increased in the 1990s to $19.6 \pm 1.4\%$ of hardwood timberland before declining precipitously to $11.7 \pm 0.9\%$ in the 2000s ($P<0.0001$). Small-diameter hardwood area was stable in the Central Hardwoods BCR from the 1970s through the 1990s (23.8 ± 2.2, 21.5 ± 1.2, and $21.8 \pm 1.8\%$,

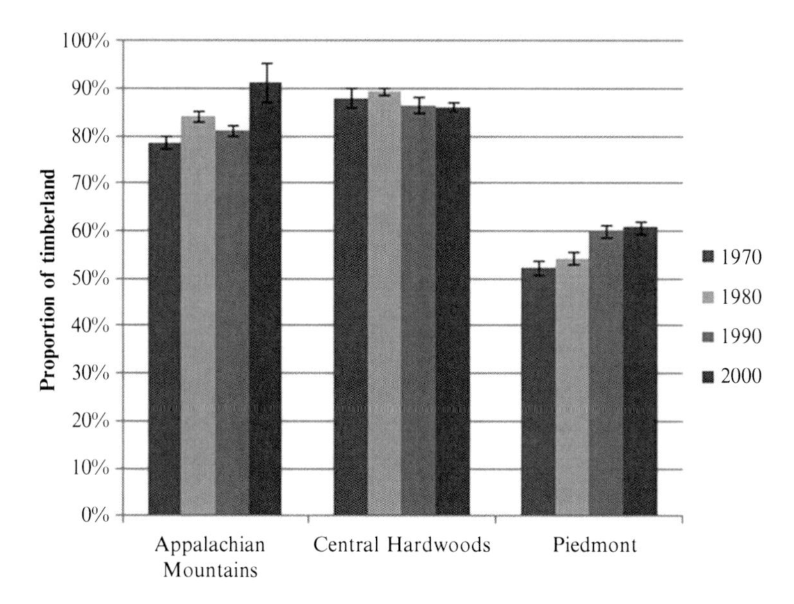

Fig. 9.3 Proportion of timberland in selected hardwood forest types by Bird Conservation Region and time period for all size classes

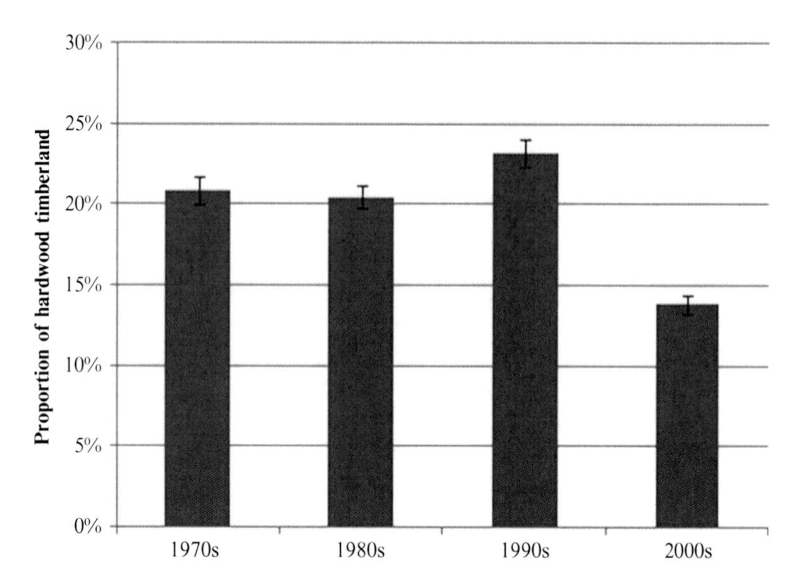

Fig. 9.4 Proportion of timberland that is small-diameter hardwood within three hardwood-dominated Bird Conservation Regions, eastern USA

Table 9.3 Proportion (in percent) of hardwood timberland comprised of small-diameter stands (+/− 1 se) by state and year within three Bird Conservation Regions in the eastern USA

State	1970	1980	1990	2000	P-value
Alabama	38.3 (2.7)^{A}	33.4 (2.4)^{A}	34.8 (2.5)^{A}	24.7 (2.2)^{B}	0.0013
Arkansas	35.0 (4.9)^{A}	32.0 (4.4)^{A}	22.9 (4.0)^{AB}	8.4 (1.2)^{B}	<0.0001
Georgia	16.6 (1.9)^{A}	24.0 (2.0)^{B}	26.4 (2.1)^{C}	18.9 (2.0)^{A}	0.0014
Kentucky	–	18.8 (1.7)^{A}	–	10.1 (0.9)^{B}	<0.0001
Missouri	–	20.9 (1.2)^{A}	–	7.7 (0.8)^{B}	<0.0001
North Carolina	16.6 (2.0)^{A}	17.3 (1.9)^{A}	21.4 (2.3)^{A}	22.7 (2.8)^{A}	0.1643
South Carolina	25.6 (2.8)^{A}	24.0 (2.5)^{A}	31.1 (2.4)^{A}	26.1 (3.2)^{A}	0.3040
Tennessee	17.5 (1.4)^{A}	16.9 (1.3)^{A}	21.0 (1.7)^{AB}	9.3 (0.9)^{C}	<0.0001
Virginia	17.9 (1.7)^{A}	–	15.5 (1.3)^{A}	11.3 (1.2)^{B}	0.0058
West Virginia	–	10.8 (0.8)^{A}	–	10.7 (2.5)^{A}	0.9808

P-values are for ANOVA tests for differences among decades within each state; values are generalized least square means and values in a row with the same letter are not significantly different (P>0.05)

respectively), but declined in the 2000s to 9.1 ± 0.6% of total hardwood timberland area (P < 0.0001). In the Piedmont BCR, small-diameter hardwood forest area increased between the 1970s and the 1990s from 22.3 ± 1.3 to 27.1 ± 1.3%, then decreased in the 2000s to 23.0 ± 1.4% (P = 0.0418).

Within the BCRs of interest, state-level changes in the proportion of hardwood timberland that consisted of small-diameter stands varied by state and by year, and were not consistent across the region, though most states did show overall declines from the 1970s to the 2000s (Table 9.3). Small-diameter area as a proportion of total hardwood timberland decreased in Alabama between the 1990s and 2000s, after three decades of remaining stable ($P=0.0013$). In Arkansas, the small-diameter area declined, but the decline occurred gradually across all four decades ($P<0.0001$). Unlike Arkansas, the small-diameter proportion of hardwood area in Georgia increased from the 1970s to the 1990s, but then decreased in the 2000s to levels that were similar to those noted in the 1980s ($P=0.0014$). Observations for both Kentucky and Missouri only existed for two time periods, the 1980s and the 2000s; the proportion of hardwood area in small-diameter timberland declined between those decades in both states ($P<0.0001$ and $P<0.0001$, respectively). In Tennessee, small-diameter timberland increased in proportion from the 1970s to the 1990s, then declined significantly by the 2000s ($P<0.0001$). Virginia, like Arkansas, experienced a steady decline in the proportion of hardwood timberland in small-diameter stands ($P=0.0058$). Finally, North Carolina, South Carolina, and West Virginia experienced no changes in small-diameter area proportions among decades ($P=0.1643$, 0.3040, and 0.9808, respectively).

The Forest Resources of the United States, 2007 report (Smith et al. 2009) allows for comparisons from 1953 to 2007, but does not discriminate between hardwood and softwood forest types. Although total timberland area increased in the Northeast, proportionally, the area of small-diameter stands has declined since 1977. In the North Central Region, despite increases in total timberland since 1977, declines in small-diameter stands have occurred while large-diameter stands (≥28 cm diameter at breast height [dbh] for hardwoods and ≥23 cm dbh for softwoods) increased proportionally (Fig. 9.5). In contrast, Southeast and South Central Regions have maintained

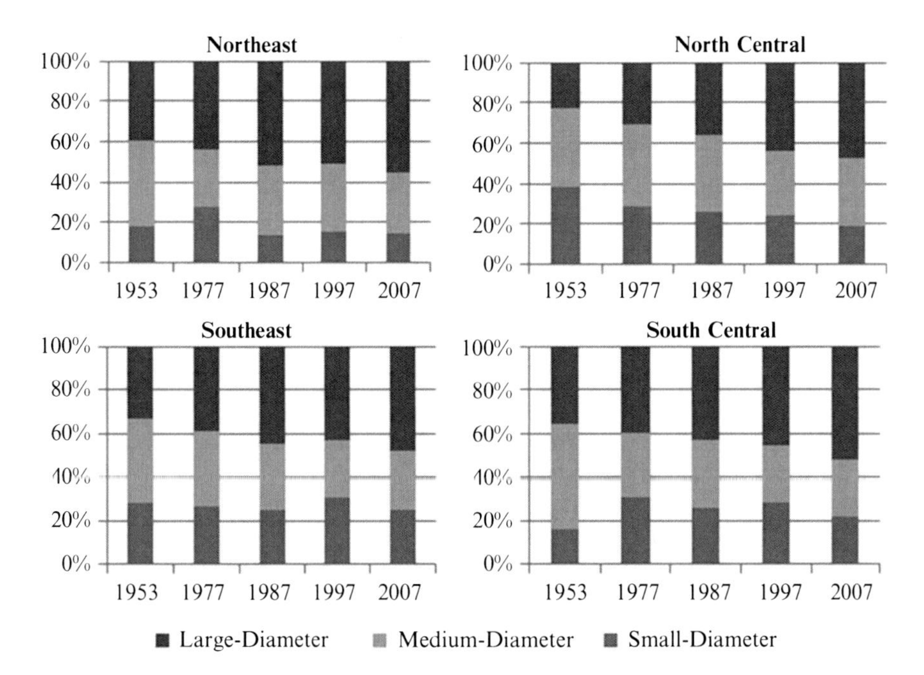

Fig. 9.5 Timberland area in four sub-regions of the USA by stand-size class (data from the Forest Resources of the United States, 2007 (Smith et al. 2009))

relatively constant proportions of small-diameter timberland on an expanding timberland base since the 1950s, with decadal fluctuations in the South Central Region, particularly. While the proportion of large-diameter area has increased steadily in both southern regions, the amount of medium-diameter rather than small-diameter area has decreased.

Our results suggested that hardwood forest area increased from the 1970s–2000s in the Appalachian Mountains and Piedmont BCRs and remained stable in the Central Hardwoods BCR. Because of the stability of the total timber resource, and the relative stability of the overall hardwood resource, we were able to focus on the proportion of that resource that was small-diameter, or early successional, habitat. Declines in small-diameter stands as a proportion of the overall hardwood resource were most notable in the Central Hardwoods and Appalachian Mountains BCRs where declines resulted in small-diameter stands comprising less than 12% of hardwood timberland by the 2000s. In contrast, while we noted proportional declines from the 1990s to the 2000s in the Piedmont BCR, there was no net change from the 1970s, and small-diameter stands still comprised between 34% and 36% of total hardwood timberland. In comparison to our study, Oswalt and Turner (2009) reported that the area of timberland in the Appalachian Hardwood Region (having only slightly different boundaries than our Appalachian Mountains BCR) remained stable during the 1980s–2000s, but acreage in the small-diameter stand size decreased while the larger diameter size classes increased. In addition, they note that total diameter distributions of hardwood trees shifted to larger diameter classes during the same period (Oswalt and Turner 2009).

Within the area of interest at the state level, overall declines in the PTSD from the 1970s to the 2000s were noted in Alabama, Arkansas, Kentucky, Missouri, Tennessee, and Virginia. In contrast, Georgia, Mississippi, and South Carolina experienced increases through the 1990s followed by declines to pre-1990s levels, while North Carolina experienced overall increases and West Virginia experienced no change.

The USDA Forest Service national report (Smith et al. 2009) showed declines in small-diameter timberland acreage across all forest types, not just hardwoods, between the 1990s and 2000s in the Southeast, South Central, and North Central Regions while the area of large-diameter timberland acreage has increased across those regions. Small-diameter area in the southern regions in that report was likely influenced by pine plantation dynamics (Smith et al. 2009). The most notable decline shown in the report was in the North Central Region, where the area of timberland comprised of small-diameter stands has been steadily declining since the 1950s (Smith et al. 2009).

The FIA program has undergone many changes since the 1970s, including switching from measuring plots using a variable-radius prism plot design to a fixed-radius, annual remeasurement plot design, changing plot remeasurement cycles, fluctuating plot lists, and changes in definitions and estimation methods (Bechtold and Patterson 2005). These changes have accompanied the transition of FIA from a series of regional programs to a nationally consistent program that is comparable from state to state across regional boundary lines. Therefore, some changes noted in our analysis may reflect changing FIA methodologies.

9.3.2 Bird Trends

Mean annual indices of relative abundance (individuals per BBS route per year) declined for eight of the nine focal species that occurred in the Central Hardwoods BCR over the three time periods (Fig. 9.6). In the Appalachian Mountains BCR, the pattern of change is clearly stronger than in the other BCRs, as declines are more pronounced for almost all the species (Fig. 9.7). In the Piedmont BCR, the species declined more frequently during 1966–1979 than in 1980–2007 or the overall period (1966–2007) (Fig. 9.8).

Of the eleven focal species, the Eastern Bluebird *(Sialia sialis)* was the only species that increased (0.4–2.4%/year) in all three BCRs and survey-wide (2.2%/year) (Fig. 9.9). Seven of eleven species declined across all of the BCRs in which they occurred (Fig. 9.9). Golden-winged Warblers *(Vermivora chrysoptera)* in the Appalachian Mountains BCR appeared to be undergoing the greatest population decline (−8.9%/year) of any of the 11 focal species (Fig. 9.7). Population trends for 1966–2007 in the three BCRs indicate that there were seven species-time period combinations in which focal species were increasing and 22 combinations in which they were decreasing (Fig. 9.9).

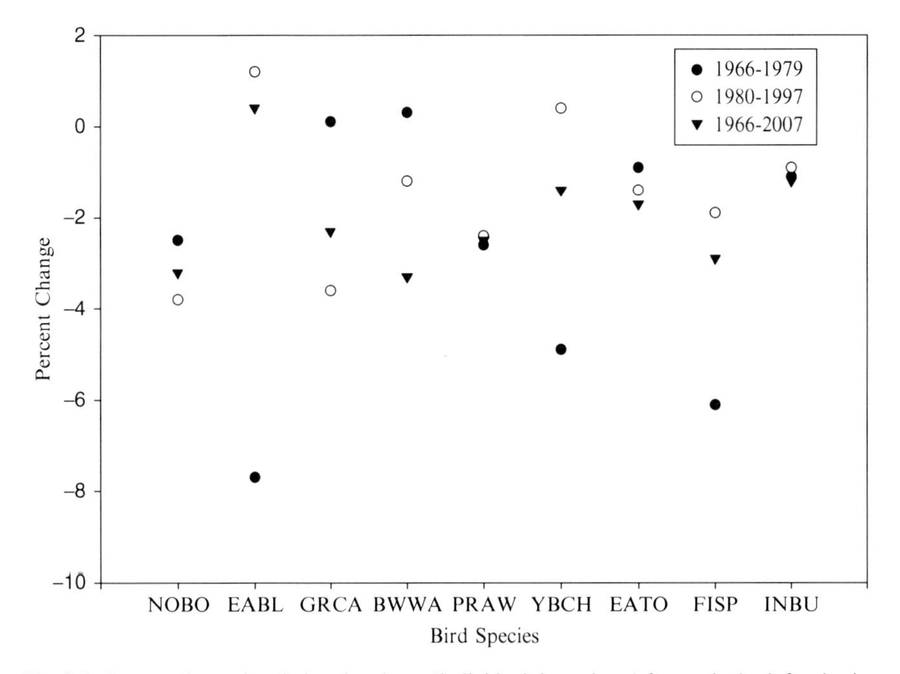

Fig. 9.6 Percent change in relative abundance (individuals/route/year) for scrub-shrub focal avian species in the Central Hardwoods Bird Conservation Region (1966–1979, 1980–1997, 1966–2007) based on North American Breeding Bird Survey data analyses (Sauer et al. 2008). Bird species abbreviations: *NOBO* Northern Bobwhite, *EABL* Eastern Bluebird, *GRCA* Gray Catbird, *BWWA* Blue-winged Warbler, *PRAW* Prairie Warbler, *YBCH* Yellow-breasted Chat, *EATO* Eastern Towhee, *FISP* Field Sparrow, *INBU* Indigo Bunting

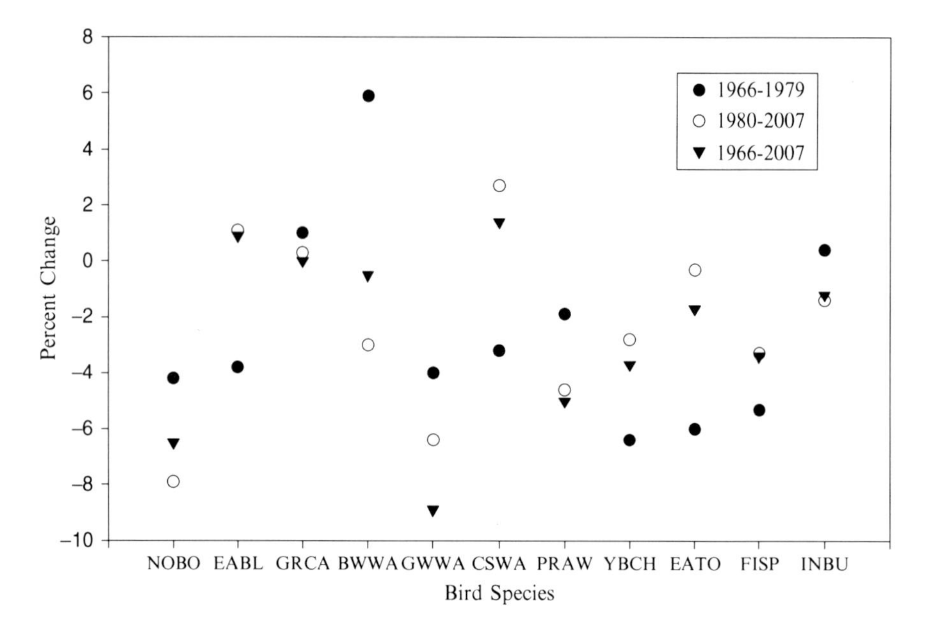

Fig. 9.7 Percent change in relative abundance (individuals/route/year) for scrub-shrub focal avian species in the Appalachian Mountains Bird Conservation Area (1966–1979, 1980–1997, 1966–2007) based on North American Breeding Bird Survey data analyses (Sauer et al. 2008). Bird species abbreviations: *NOBO* Northern Bobwhite, *EABL* Eastern Bluebird, *GRCA* Gray Catbird, *BWWA* Blue-winged Warbler, *GWWA* Golden-winged Warbler, *CSWA* Chestnut-sided Warbler, *PRAW* Prairie Warbler, *YBCH* Yellow-breasted Chat, *EATO* Eastern Towhee, *FISP* Field Sparrow, *INBU* Indigo Bunting

Considering all scrub-shrub breeding bird species, the Central Hardwoods and Appalachian Mountains BCRs experienced the greatest number of significantly declining species, 14 (64%) and 15 (54%) respectively (Table 9.4). These estimated losses ranged from a low of −0.32%/year in the Central Hardwood BCR for the Northern Cardinal *(Cardinalis cardinalis)* to a high of −17.28%/year for the Bewick's Wren *(Thryomanes bewickii)* in the Appalachian Mountains BCR (Table 9.4). In contrast, 23% (five species) and 12% (four species) were estimated as having long-term increases in population trend for the Central Hardwoods and Appalachian Mountains BCRs, respectively (Table 9.4). Fewer species were undergoing significant declines in the Piedmont BCR (seven species), although species with declining trends still outnumbered those with apparent significant increasing trends (Table 9.4).

In all ten states, there have been significant population declines in the Northern Bobwhite *(Colinus virginianus)*, ranging from −1.97%/year in Missouri to −8.86%/year in West Virginia (Table 9.5). Prairie Warblers *(Dendroica discolor)* experienced the highest rate of loss (−22.66%/year) of any species in these states (Table 9.5). Species with significant population declines appeared to be declining in all states in which they were observed (Table 9.5). There were at least five species each in Arkansas and Georgia and 12 species each in Kentucky and Tennessee that

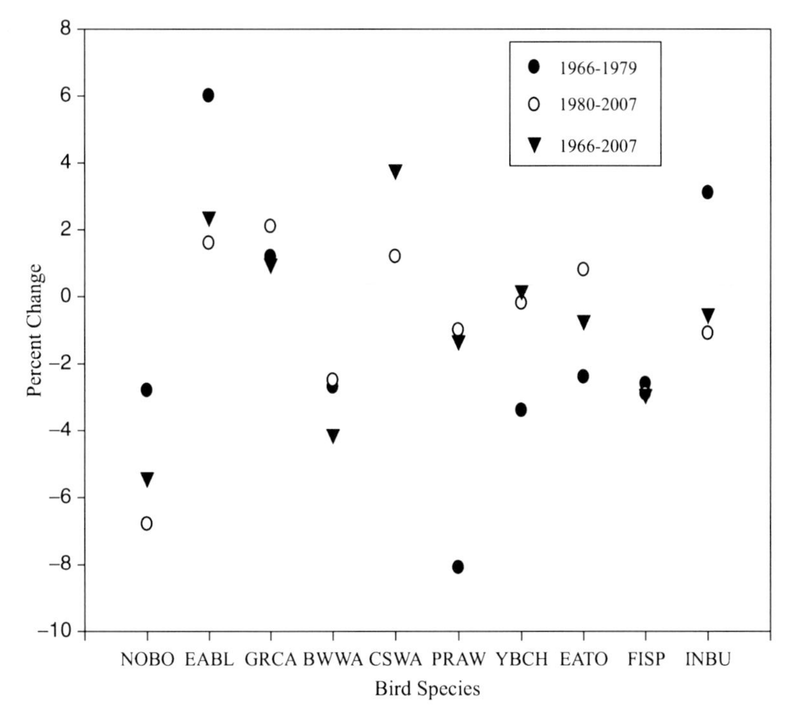

Fig. 9.8 Percent change in relative abundance (individuals/route/year) for scrub-shrub avian focal species in the Piedmont Bird Conservation Area (1966–1979, 1980–1997, 1966–2007) based on North American Breeding Bird Survey data analyses (Sauer et al. 2008). Bird species abbreviations: *NOBO* Northern Bobwhite, *EABL* Eastern Bluebird, *GRCA* Gray Catbird, *BWWA* Blue-winged Warbler, *CSWS* Chestnut-sided Warbler, *PRAW* Prairie Warbler, *YBCH* Yellow-breasted Chat, *EATO* Eastern Towhee, *FISP* Field Sparrow, *INBU* Indigo Bunting

apparently experienced significant long-term population losses (Table 9.5). The proportion of species with significant population declines ranged from a low of 14% in Mississippi to a high of 63% for Tennessee (Table 9.5).

Of the ten states, only Alabama had no species that were apparently undergoing a population increase (Table 9.5). Georgia and Kentucky each had five species that were increasing (Table 9.5). Approximately 28% of the species in Georgia were increasing, the highest proportion of any of these states (Table 9.5). Population trend increases were found for the Carolina Wren *(Thryothorus ludovicianus)* (1.21–3.73%/year), House Wren *(Troglodytes aedon)* (2.57–9.84%/year), and American Goldfinch *(Carduelis tristis)* (1.61–3.29%/year) (Table 9.5). More species were experiencing apparent significant declines in their long-term populations than were increasing and the rates of loss were more pronounced than were the gains (Table 9.5).

Based on our review of regional and state-level BBS data on scrub-shrub birds, it is clear that this group of species has consistently declined across the region over the past 40+years that surveys have been conducted. The relative rates of decline vary

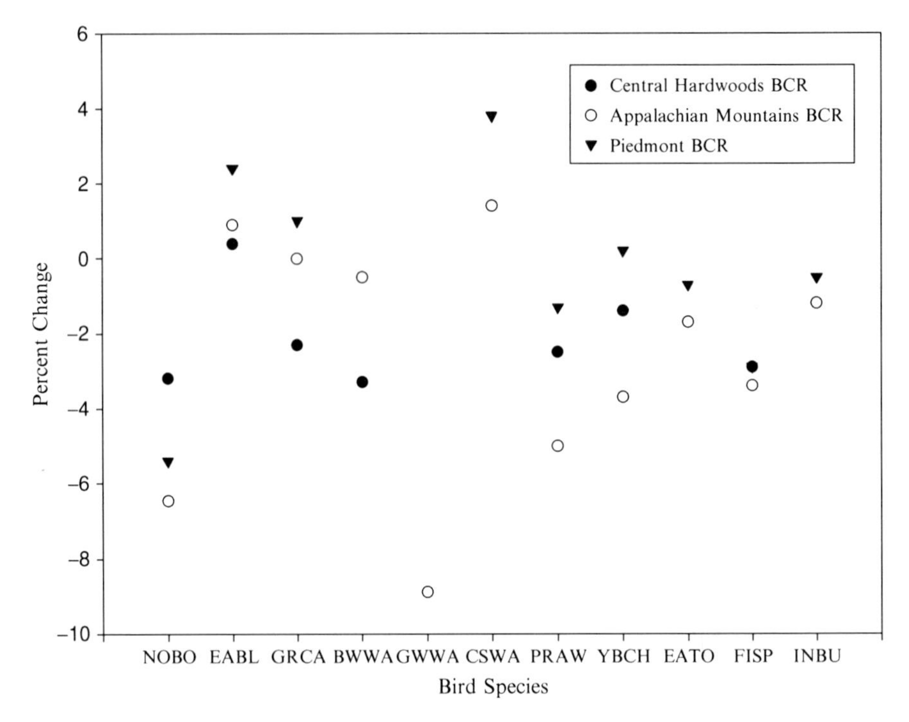

Fig. 9.9 Percent change in relative abundance (individuals/route/year) for scrub-shrub avian focal species by Bird Conservation Region, 1966–2007 based on North American Breeding Bird Survey data analyses (Sauer et al. 2008). Bird species abbreviations: *NOBO* Northern Bobwhite, *EABL* Eastern Bluebird, *GRCA* Gray Catbird, *BWWA* Blue-winged Warbler, *GWWA* Golden-winged Warbler, *CSWS* Chestnut-sided Warbler, *PRAW* Prairie Warbler, *YBCH* Yellow-breasted Chat, *EATO* Eastern Towhee, *FISP* Field Sparrow, *INBU* Indigo Bunting

by species, time period, region, and state. Eastern Bluebirds, for example, generally have been increasing. Eastern Bluebirds will use the early stages of forest succession but also occur in a variety of field habitats and have benefitted from the extensive use of nest boxes in rural areas across the region (Gowaty and Plissner 1998).

9.3.3 Relationship Between Bird Trends and Small–Diameter Forest Trends

Several scrub-shrub species are declining precipitously and have already attracted considerable conservation attention. Based on our regression analysis, the apparent reasons for these declines go beyond the decline in availability of small-diameter hardwood forest habitats as defined in this study. Golden-winged Warblers, for example, are declining along BBS routes at an incredible rate of almost 9% per year in the Appalachian Mountains BCR, resulting in loss of over 98% of the 1966 population. The decline of

Table 9.4 Significant (P<0.10) population trends (% change/year) for scrub-shrub breeding bird species by Bird Conservation Region (BCR) based on North American Breeding Bird Survey data analyses (Sauer et al. 2008) for 1966–2007

	BCR trend (% change/year)		
	Central Hardwoods	Appalachian mountains	Piedmont
Decreasing species			
Northern Bobwhite *(Colinus virginianus)*	−3.15	−6.47	−4.92
Bewick's Wren *(Thryomanes bewickii)*	−4.71	−17.28	
House Wren *(Troglodytes aedon)*		−1.05	
Gray Catbird *(Dumetella carolinensis)*	−2.28		
Brown Thrasher *(Toxostoma rufum)*	−1.48	−1.19	
Blue-winged Warbler *(Vermivora pinus)*	−2.80		−3.13
Golden-winged Warbler *(V. chrysoptera)*		−8.73	
Nashville Warbler *(V. ruficapilla)*		−5.49	
Yellow Warbler *(Dendroica petechia)*	−2.44		
Prairie Warbler *(D. discolor)*	−2.43	−4.97	−1.22
Common Yellowthroat *(Geothlypis trichas)*	−1.04	−0.52	
Yellow-breasted Chat *(Icteria virens)*	−1.38	−3.70	
Eastern Towhee *(Pipilo erythrophthalmus)*	−1.71	−1.68	
Field Sparrow *(Spizella pusilla)*	−2.83	−3.44	−2.77
Lark Sparrow *(Chondestes grammacus)*	−3.10		
Song Sparrow *(Melospiza melodia)*		−0.58	
Northern Cardinal *(Cardinalis cardinalis)*	−0.32		
Blue Grosbeak *(Guiraca caerulea)*			−0.67
Indigo Bunting *(Passerina cyanea)*	−1.25	−1.19	−0.52
American Goldfinch *(Carduelis tristis)*		−1.21	
Species with significant negative trends (%)	64	54	32
Increasing species			
Willow/Alder Flycatcher *(Empidonax spp.)*		1.28	2.41
White-eyed Vireo *(Vireo griseus)*			1.21
Carolina Wren *(Thryothorus ludovicianus)*	2.47	2.33	1.29
House Wren *(T. aedon)*	3.71		
Eastern Bluebird *(Sialia sialis)*	2.4	0.9	2.4
Chestnut-sided Warbler *(D. pensylvanica)*		1.32	
Song Sparrow *(M. melodia)*	0.79		
Northern Cardinal *(C. cardinalis)*			0.50
Blue Grosbeak *(G. caerulea)*	2.32		
American Goldfinch *(Carduelis tristis)*	0.70		2.00
Species with significant positive trends (%)	23	12	26

this species has led to the formation of the Golden-winged Warbler Working Group that is focused on developing and implementing conservation strategies for this and other scrub-shrub species (Buehler et al. 2007). Although Golden-winged Warblers use small-diameter upland hardwood forests, their habitat requirements are more specialized in that they require herbaceous components interspersed with saplings, shrubs, and mature trees (Klaus and Buehler 2001). These conditions are seldom found in regenerating

Table 9.5 Significant ($P < 0.10$) population trends for scrub-shrub breeding bird species by southeastern state based on data from the North American Breeding Bird Survey (Sauer et al. 2008) for 1966–2007

	% Change per year									
	AL	AR	GA	KY	MO	NC	SC	TN	VA	WV
Decreasing species										
Northern Bobwhite (*Colinus virginianus*)	−4.90	−4.78	−4.35	−2.59	−1.97	−4.41	−4.68	−3.97	−3.78	−8.86
Bewick's Wren (*Thryomanes bewickii*)				−6.08				−15.86		
Gray Catbird (*Dumetella carolinensis*)				−2.05		−1.44	−3.06	−4.59	−2.38	
Brown Thrasher (*Toxostoma rufum*)		−5.66		−1.17	−2.23			−1.37		−1.86
Blue-winged Warbler (*Vermivora pinus*)	−5.97			−5.07						−2.64
Golden-winged Warbler (*V. chrysoptera*)										−9.18
Yellow Warbler (*Dendroica petechia*)	−5.94			−2.93		−3.10		−4.13	−3.75	
Prairie Warbler (*D. discolor*)	−1.72	−22.66	−2.36	−2.46	−4.23			−2.59		−5.91
Common Yellowthroat (*Geothlypis trichas*)		−7.76		−0.85		−1.36	−2.54	−0.87		−2.73
Yellow-breasted Chat (*Icteria virens*)				−1.48				−1.86		−4.18
Eastern Towhee (*Pipilo erythrophthalmus*)	−0.85		−1.41		−2.39		−1.87	−1.64	−1.65	−1.29
Field Sparrow (*Spizella pusilla*)	−4.08	−22.70	−2.01	−2.98	−2.13	−1.87		−1.35	−3.08	−3.13
Lark Sparrow (*Chondestes grammacus*)					−2.43					

Northern Cardinal (*Cardinalis cardinalis*)	−0.64							−0.70		
Indigo Bunting (*Passerina cyanea*)	−0.68			−1.57			−2.45	−0.90	−0.67	−2.43
American Goldfinch (*Carduelis tristis*)										−3.84
Proportion of species with significant negative trends	0.44	0.31	0.22	0.58	0.27	0.29	0.33	0.63	0.35	0.55
No. of species with significant negative trends	9	5	5	12	6	6	6	12	6	11
Increasing species										
White-eyed Vireo (*Vireo griseus*)	0.75									
Carolina Wren (*T. ludovicianus*)			1.21	3.73	3.07			1.89	1.56	1.67
House Wren (*Troglodytes aedon*)			9.84	5.06		2.57		7.00		
Chestnut-sided Warbler (*D. pensylvanica*)										2.02
Yellow-breasted Chat (*I. virens*)			1.65							
Song Sparrow (*Melospiza melodia*)			3.83	1.57	1.95			1.51		
Blue Grosbeak (*G. caerulea*)				4.15				2.31		
American Goldfinch (*Carduelis tristis*)	1.94		3.29	1.61		1.84	3.08			
Proportion of species with significant positive trends	0.11	0.00	0.28	0.26	0.09	0.12	0.07	0.21	0.06	0.10
No. of species with significant positive trends	2	0	5	5	2	2	1	4	1	2

forests unless the forests are located in northern regions where tree growth is slow (e.g., Wisconsin), or if management action is taken to slow tree growth and promote herbaceous plant growth, such as with herbicides, grazing, or prescribed burning. Although the regression analysis was suggestive of a relationship with availability of small-diameter hardwood forests, Golden-winged Warbler population declines far exceed the rates of decline in small-diameter forests in the Appalachian Mountains BCR over the past 20 years. The decline in small-diameter forests in concert with the decline of other early successional habitats, however, may be a contributing factor in the decline of this species. Golden-winged Warblers are Nearctic-Neotropical migrants that winter in Central and South America. Extensive deforestation of their wintering habitat is also likely contributing to their decline (Buehler et al. 2007).

Northern Bobwhites also have declined sharply across all three BCRs, averaging 3–6% per year depending on region (Table 9.4). The decline of Northern Bobwhites has attracted considerable conservation attention, leading to formation of the Southeast Quail Study Group and development of the Northern Bobwhite Conservation Initiative (Dimmick et al. 2002). Bobwhites use a diverse configuration of habitats during their annual cycle, using grassland habitats for nesting and brooding but often using small-diameter forests for winter cover, especially in the northern parts of their range (Brennan 1999). Based on the regression results, population declines in this species appear to be more strongly related to other components of their habitat than small-diameter upland forest availability.

There were no consistent relationships between percent annual change in small-diameter upland forest and change in avian relative abundance for any of the 11 species analyzed (Table 9.6). Chestnut-sided Warbler *(Dendroica pensylvanica)* (r^2=0.385),

Table 9.6 Regression coefficients, F values and P values for regression analyses relating annual change in relative abundance of 11 scrub-shrub bird species by state to annual change in amount of small-diameter upland hardwood forest by state across three Bird Conservation Regions in the eastern USA

Species	n	β	95%	CI	r^2	F	P-value
Blue-winged Warbler *(Vermivora pinus)*	7	−1.874	−5.477	1.730	0.263	1.786	0.239
Chestnut-sided Warbler *(D. pensylvanica)*	5	−1.990	−6.615	2.635	0.385	1.875	0.264
Eastern Bluebird *(Sialia sialis)*	10	0.284	−0.376	0.945	0.110	0.986	0.350
Eastern Towhee *(Pipilo erythrophthalmus)*	10	−0.020	−0.518	0.477	0.001	0.009	0.927
Field Sparrow *(Spizella pusilla)*	10	0.042	−0.774	0.857	0.002	0.014	0.909
Gray Catbird *(Dumetella carolinensis)*	10	−0.157	−1.005	0.690	0.022	0.183	0.680
Golden-winged Warbler *(V. chrysoptera)*	5	−175.574	−773.586	422.438	0.225	0.873	0.419
Indigo Bunting *(Passerina cyanea)*	10	−0.269	−0.712	0.174	0.197	1.964	0.199
Northern Bobwhite *(Colinus virginianus)*	10	0.294	−1.782	2.369	0.013	0.107	0.752
Prairie Warbler *(D. discolor)*	10	0.471	−0.208	1.150	0.243	2.563	0.148
Yellow-breasted Chat *(Icteria virens)*	10	0.195	−0.683	1.074	0.032	0.263	0.622

Blue-winged Warbler *(Vermivora pinus)* ($r^2=0.263$), Prairie Warbler ($r^2=0.243$), and Golden-winged Warbler ($r^2=0.225$) had the strongest relationships with small-diameter forest availability but none of these regressions met traditional alpha decision criteria for significance (i.e., $\alpha<0.10$ or 0.05). The other species analyzed showed no apparent relationship with the change in small-diameter forest availability (Table 9.6).

In general, the strongest relationships (r^2s) between birds and small-diameter hardwood forests occurred for scrub-shrub species that are more associated with forested habitats than with field habitats (Blue-winged Warbler, Chestnut-sided Warbler, Prairie Warbler, and Golden-winged Warbler). These species require varying amounts of woody plants (saplings and shrubs) in their habitat that can be found in abundance in regenerating forests (Richardson and Brauning 1995; Nolan et al. 1999; Gill et al. 2001; Klaus and Buehler 2001). The lack of a strong relationship between population declines in these species with small-diameter forest availability suggests that other factors are also linked to the population declines. All four species mentioned above are Nearctic-Neotropical migrants, therefore habitat losses on their wintering grounds or along their migration routes may also be contributing to their population declines.

Declines have also varied by BCR. In general, the Appalachian Mountains BCR appears to be experiencing the greatest declines in small-diameter forested habitats and scrub-shrub birds, the Central Hardwoods BCR is intermediate and the Piedmont BCR is experiencing the least declines. Appalachian Mountains and Central Hardwoods Joint Ventures are underway to address the declines in priority bird species and their habitats. The boundaries of these joint ventures coincide with those of the respective BCRs. The prevalence of pine plantation management in the Piedmont region may explain the improved status of scrub-shrub species that use small-diameter pine forests compared to their status in other regions where pine plantations are less common.

9.4 Conclusion

We demonstrated that the availability of small-diameter upland hardwood forest habitat has changed across the eastern USA over the past four decades, and has declined significantly over the past decade, especially in the Appalachian Mountains BCR. Scrub-shrub birds as a group are also declining significantly across the region over the past four decades, with some species declining precipitously. The decline in small-diameter forested habitats is undoubtedly contributing to the decline for some scrub-shrub species. The FIA database is the only regional database that tracks this forest resource, although its usefulness for tracking change in the habitat availability for specific scrub-shrub birds appears to be somewhat limited. The loss of habitat alone (as measured by FIA data defined by this study) is not solely related to the population trends. Some of the scrub-shrub birds examined are more closely tied to old field habitats. There are no databases that track the availability of this habitat type. In addition, some of the scrub-shrub species are Nearctic-Neotropical migrants that may be experiencing habitat loss along their migration routes or on their wintering grounds.

Acknowledgments We thank the hundreds of observers that contributed to the collection of both FIA and BBS data over the past four decades. Their dedication has made this analysis possible. We also thank the US Forest Service Southern Research Station and the University of Tennessee for support.

Literature Cited

Annand EM, Thompson FR (1997) Forest bird response to regeneration practices in central hardwood forests. J Wildl Manage 61:159–171

Askins RA (2001) Sustaining biological diversity in early successional communities: the challenge of managing unpopular habitats. Wildl Soc Bull 29:407–412

Bechtold WA, Patterson PL (2005) The enhanced forest inventory and analysis program – national sampling design and estimation procedures. Gen Tech Rep SRS-80, USDA Forest Service Southern Research Station, Asheville

Bourque J, Villard MA (2001) Effects of selection cutting and landscape-scale harvesting on the reproductive success of two neotropical migrant bird species. Conserv Biol 15:184–195

Brawn JD, Robinson SK, Thompson FR (2001) The role of disturbance in the ecology and conservation of birds. Annu Rev Ecol Syst 32:251–276

Brennan LA (1999) Northern Bobwhite (*Colinus virginianus*). In: Poole A (ed) Birds of North America online. Cornell Lab of Ornith, Ithaca. Retrieved from the Birds of North America Online: http://bna.birds.cornell.edu.proxy.lib.utk.edu:90/bna/species/397 doi:10.2173/bna.397

Buehler DA, Roth AM, Vallender R, Will TC, Confer JL, Canterbury RA, Swarthout SB, Rosenberg KV, Bulluck LP (2007) Status and conservation priorities of golden-winged Warbler (*Vermivora chrysoptera*) in North America. Auk 124:1439–1445

Bulluck LP, Buehler DA (2006) Avian use of early successional habitats: are regenerating forests, utility right-of-ways and reclaimed surface mines the same? For Ecol Manage 236:76–84

Campbell SP, Witham JW, Hunter ML (2007) Long-term effects of group-selection timber harvesting on abundance of forest birds. Conserv Biol 21:1218–1229

Confer JL, Pascoe SM (2003) Avian communities on utility rights-of-ways and other managed shrublands in the northeastern United States. For Ecol Manage 185:193–205

Dettmers R (2003) Status and conservation of shrubland birds in the northeastern US. For Ecol Manage 185:81–93

Diaz IA, Armesto JJ, Reid S, Sieving KE, Willson MF (2005) Linking forest structure and composition: avian diversity in successional forests of Chiloe Island, Chile. Biol Conserv 123:91–101

Dimmick RW, Gudlin MJ, McKenzie DF (2002) The Northern bobwhite conservation initiative. Misc Publ Southeast Assoc Fish and Wildlife Agencies, Laurens

Fei SL, Steiner KC (2007) Evidence for increasing red maple abundance in the eastern United States. For Sci 53:473–477

Geissler P, Sauer JR (1990) Topics in route-regression analysis. In: Sauer JR, Droege S (eds) Survey designs and statistical methods for the estimation of avian population trends. Biol Rep 90(1), USDI Fish and Wildlife Service, Washington, DC, pp 54–57

Gill FB, Canterbury RA, Confer JL (2001) Blue-winged, Warbler (*Vermivora pinus*). In: Poole A (ed) Birds of North America online. Cornell Lab of Ornith, Ithaca. Retrieved from the Birds of North America Online: http://bna.birds.cornell.edu.proxy.lib.utk.edu:90/bna/species/584, doi:10.2173/bna.584

Gowaty PA, Plissner JH (1998) Eastern Bluebird (*Sialia sialis*). In: Poole A (ed) Birds of North America online. Cornell Lab of Ornithology, Ithaca. Retrieved from the Birds of North America Online: http://bna.birds.cornell.edu.proxy.lib.utk.edu:90/bna/species/381, doi:10.2173/bna.381

Gram WK, Porneluzi PA, Clawson RL, Faaborg J, Richter SC (2003) Effects of experimental forest management on density and nesting success of bird species in Missouri Ozark Forests. Conserv Biol 17:1324–1337

Hunter WC, Buehler DA, Canterbury RA, Confer JL, Hamel PB (2001) Conservation of disturbance-dependent birds in eastern North America. Wildl Soc Bull 29:440–455

James FC, McCullogh CE, Wiedenfeld DA (1996) New approaches to the analysis of population trends in land birds. Ecology 77:13–27

King DI, Byers BE (2002) An evaluation of powerline rights-of-way as habitat for early-successional shrubland birds. Wildl Soc Bull 30:868–874

King DI, Griffin CR, DeGraaf RM (1998) Nest predator distribution among clearcut forest, forest edge and forest interior in an extensively forested landscape. For Ecol Manage 104:151–156

King DI, Degraaf RM, Griffin CR (2001) Productivity of early successional shrubland birds in clearcuts and groupcuts in an eastern deciduous forest. J Wildl Manage 65:345–350

Klaus NA, Buehler DA (2001) Golden-winged Warbler breeding habitat characteristics and nest success in clearcuts in the southern Appalachian Mountains. Wilson Bull 113:297–301

Krementz DG, Christie JS (2000) Clearcut stand size and scrub-successional bird assemblages. Auk 117:913–924

Lacki MJ, Fitzgerald JL, Hummer JW (2004) Changes in avian species composition following surface mining and reclamation along a riparian forest corridor in southern Indiana. Wetlands Ecol Manage 12:447–457

Link W, Sauer JR (1994) Estimating equations estimates of trends. Bird Popul 2:23–32

Litvaitis JA (ed) (2003) Early-successional forests and shrubland habitats in the northeastern United States: critical habitats dependent on disturbance. For Ecol Manage 185:1–215

Marshall MR, DeCecco JA, Williams AB, Gale GA, Cooper RJ (2003) Use of regenerating clearcuts by late-successional bird species and their young during the post-fledging period. For Ecol Manage 183:127–135

Moran EF, Brondizio ES, Tucker JM, da Silva-Forsberg MC, McCracken S, Falesi I (2000) Effects of soil fertility and land-use on forest succession in Amazonia. For Ecol Manage 139:93–108

Nolan Jr V, Ketterson ED, Buerkle CA (1999) Prairie Warbler (*Dendroica discolor*). In: Poole A (ed) Birds of North America online. Cornell Lab of Ornithology, Ithaca. Retrieved from the Birds of North America Online: http://bna.birds.cornell.edu.proxy.lib.utk.edu:90/bna/species/455, doi:10.2173/bna.455

Oswalt CM, Turner JA (2009) Status of hardwood forest resources in the Appalachian region including estimates of growth and removals. Res Bull SRS-142, USDA Forest Service Southern Research Station, Asheville

Pagen RW, Thompson FR, Burhans DE (2000) Breeding and post-breeding habitat use by forest migrant songbirds in the Missouri Ozarks. Condor 102:738–747

Peterjohn BG, Sauer JR, Robbins C (1995) The North American breeding bird survey and population trends of neotropical migratory birds. In: Finch DM, Martin TE (eds) Ecology and management of neotropical migrant birds: a synthesis and review of critical issues. Oxford University Press, Oxford, pp 3–39

Peterjohn BG, Sauer JR, Link W (1997) The 1994 and 1995 summary of the North American breeding bird survey. Bird Popul 3:48–66

Raphael MG, Morrison ML, Yoder-Williams MP (1987) Breeding bird populations during twenty-five years of postfire succession in the Sierra Nevada. Condor 89:614–626

Richardson M, Brauning DW (1995) Chestnut-sided Warbler (*Dendroica pensylvanica*). In: Poole A (ed) Birds of North America online. Cornell Lab of Ornithology, Ithaca. Retrieved from the Birds of North America Online: http://bna.birds.cornell.edu.proxy.lib.utk.edu:90/bna/species/190, doi:10.2173/bna.190

Rodewald AD, Vitz AC (2005) Edge- and area-sensitivity of shrubland birds. J Wildl Manage 69:681–688

Rodewald AD, Yahner RH (2001a) Influence of landscape composition on avian community structure and associated mechanisms. Ecology 82:3493–3504

Rodewald AD, Yahner RH (2001b) Avian nesting success in forested landscapes: influence of landscape composition, stand and nest-patch microhabitat, and biotic interactions. Auk 118:1018–1028

Sallabanks RE, Arnett B, Marzluff JM (2000) An evaluation of research on the effects of timber harvest on bird populations. Wildl Soc Bull 28:1144–1155

Sauer JR, Droege S (1990) Survey designs and statistical methods for the estimation of avian population trends. Biol Rep 90, USDI Fish and Wildlife Service, Washington, DC

Sauer JR, Hines JE, Fallon F (2008) The North American breeding bird survey, results and analysis 1966–2007. Version 5.15.2008, USDI Geological Survey, Laurel

Smith WB, Miles PD, Perry CH, Pugh SA (2009) Forest resources of the United States. Gen Tech Rep WO-78, USDA Forest Service, Washington, DC

Thomas L, Martin K (1996) The importance of analysis method for breeding bird survey population trend estimates. Conserv Biol 10:479–490

Thompson FR, Dijak WD, Kulowiec TG, Hamilton DA (1992) Breeding bird populations in Missouri Ozark forests with and without clearcutting. J Wildl Manage 56:23–30

Thompson FR III, Degraaf RM, Trani MK (eds) (2001) Conservation of woody, early successional habitats and wildlife in the eastern United States. Wildl Soc Bull 29:407–494

Tingley MW, Orwig DA, Field R, Motzkin G (2002) Avian response to removal of a forest dominant: consequences of hemlock woolly adelgid infestations. J Biogeogr 29:1505–1516

USDA Forest Service. 2009. The forest inventory and analysis database: database description and users manual version 4.0 for phase 2. Draft Rev 2. Forest Inventory and Analysis Program, USDA Forest Service. Available online at: http://www.fia.fs.fed.us/tools-data/docs/default.asp

Vitz AC, Rodewald AD (2006) Can regenerating clearcuts benefit mature-forest songbirds? An examination of post-breeding ecology. Biol Conserv 127:477–486

Chapter 10
Bats and Gaps: The Role of Early Successional Patches in the Roosting and Foraging Ecology of Bats

Susan C. Loeb and Joy M. O'Keefe

Abstract Early successional habitats are important foraging and commuting sites for the 14 species of bats that inhabit the Central Hardwood Region, especially larger open-adapted species such as hoary bats (*Lasiurus cinereus*), red bats (*L. borealis*), silver-haired bats (*Lasionycteris noctivagans*), and big brown bats (*Eptesicus fuscus*). Forest gaps, small openings, and the edges between early successional patches and mature forest are especially important habitats because they are used by both open-adapted and clutter-adapted species. Several bat species select roosts in close proximity to early successional patches, perhaps to minimize foraging and commuting costs. Future research on effects of early successional patch size, shape, vegetation structure, and connectivity on bat use, and the distribution of early successional habitats in relation to mature forest, roosting sites, and water sources will assist managers in providing the optimal types and distribution of early successional patches on the landscape.

10.1 Introduction

Fourteen species of bats inhabit the forests of the Central Hardwood Region, USA (Whitaker and Hamilton 1998, Fig. 1.1, Table 10.1). These bats rely on forest ecosystems for resources essential for survival and reproduction, including day and

S.C. Loeb (✉)
Research Ecologist with the Upland Hardwood Ecology and Management Research Work Unit, USDA Forest Service, Southern Research Station, Department of Forestry and Natural Resources, Clemson University, Clemson, SC 29634, USA
e-mail: sloeb@fs.fed.us

J.M. O'Keefe
Department of Biology, Indiana State University, 600 Chestnut Street, Terre Haute, IN 47809, USA
e-mail: Joy.O'Keefe@indstate.edu

C.H. Greenberg et al. (eds.), *Sustaining Young Forest Communities*, Managing Forest Ecosystems 21, DOI 10.1007/978-94-007-1620-9_10,

Table 10.1 Bats of the Central Hardwood Region and their primary roosting and foraging habitats

Common name	Scientific name	Summer roosting habitat	Winter roosting habitat	Major foraging areas
Rafinesque's big-eared bat	*Corynorhinus rafinesquii*	Caves, mines, or large hollow trees	Caves, mines, or large hollow trees	Oak-hickory forests
Ozark big-eared bat	*C. townsendii ingens*	Caves	Caves	Upland hardwood forests, edges
Virginia big-eared bat	*C. townsendii virginianus*	Caves	Caves	Cliffs and mixed mesophytic forests
Big brown bat	*Eptesicus fuscus*	Live and dead hardwood and pine trees; buildings	Caves, mines, rock outcrops, buildings	Forests, edges, and openings
Silver-haired bat	*Lasionycteris noctivagans*	Live and dead trees	Live trees and rock outcrops	Open areas
Red bat	*L. borealis*	Live hardwoods	Live hardwoods and leaf litter	Forests, forest edges, openings
Hoary bat	*L. cinereus*	Live hardwoods and pines	Live hardwoods and leaf litter[a]	Forest edges and openings; above canopy
Southeastern bat	*Myotis austroriparius*	Large hollow trees	Caves	Riparian areas, bottomland hardwoods
Gray bat	*M. grisescens*	Caves	Caves	Riparian areas
Small-footed bat	*M. leibii*	Talus slopes, cliff faces, bridges	Caves, mines, talus slopes, artificial structures	Closed canopy forest
Little brown bat	*M. lucifugus*	Buildings and snags	Caves and mines	Riparian areas, open canopy forests
Northern long-eared bat	*M. septentrionalis*	Hardwood and pine snags	Caves and mines	Closed canopy forests
Indiana bat	*M. sodalis*	Hardwood and pine snags	Caves and mines	Forest interior, edges, and open areas
Evening bat	*Nycticeius humeralis*	Live and dead trees	Live and dead trees	Riparian areas
Tri-colored bat	*Perimyotis subflavus*	Live hardwoods	Caves, mines, artificial structures	Forest interior, edges, and open areas

[a] No studies on winter roost habits – assume similar to red bats based on anecdotal information

night roosts (Barclay and Kurta 2007; Carter and Menzel 2007; Ormsbee et al. 2007), foraging sites (Lacki et al. 2007), and drinking water (Hayes and Loeb 2007). In turn, bats play an important role in forest ecosystem function (Marcot 1996). Eastern bats consume large amounts of insects per day (e.g., Kurta et al. 1990); many of these are important agricultural and forest pests (Jones et al. 2009). Bats also redistribute nutrients across the landscape and can create nutrient hotspots below their tree roosts (Duchamp et al. 2010). Because tree roosts are often near gaps or openings (Kalcounis-Rueppell et al. 2005), nutrient hotspots may be important in forest regeneration. Thus, understanding how management of forested ecosystems, including early successional habitats, affects bats is critical for the conservation and management of these species and for maintaining forest ecosystem health.

Many of the bats that inhabit the Central Hardwood Region are of conservation concern. The Indiana bat (*Myotis sodalis*), gray bat (*M. grisescens*), Ozark big-eared bat (*Corynorhinus townsendii ingens*), and Virginia big-eared bat (*C. t. virginianus*) are federally listed endangered species (USDI Fish and Wildlife Service 1984; 2007) and Rafinesque's big-eared bat, eastern small-footed bat, and southeastern bat are considered species of special concern (Harvey et al. 1999). Factors leading to the decline of these species include habitat loss and fragmentation, and disturbance and destruction of winter hibernation sites and summer maternity sites. Recently, these species and many other bats in the eastern USA have been facing additional threats. Since winter 2006–2007, five cave associated bats [the little brown bat (*M. lucifugus*), northern long-eared bat (*M. septentrionalis*), Indiana bat, small-footed bat (*M. leibii*), and tri-colored bat (*Perimyotis subflavus*)] have experienced severe population declines in the northeastern USA due to white-nose syndrome (Blehert et al. 2009). The big brown bat (*Eptesicus fuscus*) has also been affected by white-nose syndrome and the gray bat, Virginia big-eared bat, Ozark big-eared bat, Rafinesque's big-eared bat, and southeastern bat (*M. austroriparius*) may be at risk as this disease moves south and west (Szymanski et al. 2009). While non-cave hibernating bats such as the hoary bat (*Lasiurus cinereus*), red bat (*L. borealis*), and silver-haired bat (*Lasionycteris noctivagans*) have not been affected by white-nose syndrome to date, they are being impacted by wind energy development, particularly in the Appalachians (Arnett et al. 2008). For example, based on current mortality rates and projected number of turbines, estimated mortalities in the year 2020 in the mid-Atlantic region range from 33,000 to 110,000 individuals (Kunz et al. 2007). Because bats have low reproductive rates (Barclay and Harder 2003), population recovery will likely be slow (Racey and Entwistle 2003).

Although all bats in the eastern USA are insectivorous, they vary considerably in body size, wing morphology, and life history characteristics (Whitaker and Hamilton 1998). For example, body size ranges from 4 to 6 g for the small-footed bat and tri-colored bat to more than 25 g for the hoary bat. Some species are year round residents in an area (e.g., Rafinesque's big-eared bats) while others are short-distance migrants (e.g., tri-colored bats), regional migrants (e.g., Indiana bats, little brown bats), or long-distance migrants (hoary bat, red bat, and silver-haired bat). Further, some species rely on trees as their primary roosts while others roost in caves, mines, or artificial structures (Table 10.1). Thus, an examination of the responses of bats to early successional communities or vegetation structure requires consideration of this variation in morphology and life history strategies.

Given concerns over the decline of early successional habitats (Greenberg et al., Chap. 1), combined with considerable concern over the viability of bats in the Central Hardwood Region, it is important to assess how forest management and the creation of early successional habitats may affect bats. Thus, our objective was to evaluate the use and importance of early successional habitats to bats in the Central Hardwood Region. Although there has been a substantial increase in research on bats in forest ecosystems in the past 15 years (Brigham 2007), there are still limited data on bats in the Central Hardwood Region. Thus, we drew upon research in other forest ecosystems to make inferences about the possible importance of early successional habitats for bats in the Central Hardwood Region. We primarily considered early successional habitats such as recently clearcut forests, wildlife openings, meadows, forest gaps, and grassy forest roads. We first considered use of early successional patches for foraging and commuting and then in roost site selection; we concluded with sections on managing early successional areas for bats in the Central Hardwood Region. We also recommend future research directions, including the importance of considering early successional habitats within the context of the larger landscape matrix.

10.2 Use of Early Successional Patches for Foraging and Commuting

Understanding and predicting bat foraging and commuting habitat is greatly aided by considering echolocation call structure and ecomorphology. Using these characteristics, bats can generally be placed along a continuum of open to closed canopy specialists (Norberg and Rayner 1987; Fenton 1990). Species adapted to open environments are fast, agile flyers and have high aspect ratios (long, narrow wings), high wing loading (large body relative to wing area), and pointed wing tips. Their echolocation calls are usually high intensity, low frequency, and narrowband (Fig. 10.1a). In contrast, species with low aspect ratios, low wing loading, and rounded wingtips are slow but maneuverable flyers. Their echolocation calls are usually high frequency, lower intensity broadband calls (Fig. 10.1b). These species are better adapted to foraging in cluttered environments (i.e., those with many physical and acoustical obstructions such as branches and leaves). Species that are intermediate in these characteristics are often found using edges. Bats of the Central Hardwood Region can be generally placed along this continuum (Table 10.2). For example, hoary bats and silver-haired bats are considered open-space species, whereas Rafinesque's big-eared bats and northern long-eared bats are more likely to be found in cluttered spaces.

Most studies of bat foraging activity and habitat use have used acoustic detectors, although some studies have used radio-telemetry (Fig. 10.2). While the use of acoustic detectors over the past two decades has advanced our understanding of habitat use by bats (Brigham 2007; Lacki et al. 2007), there are many pitfalls and

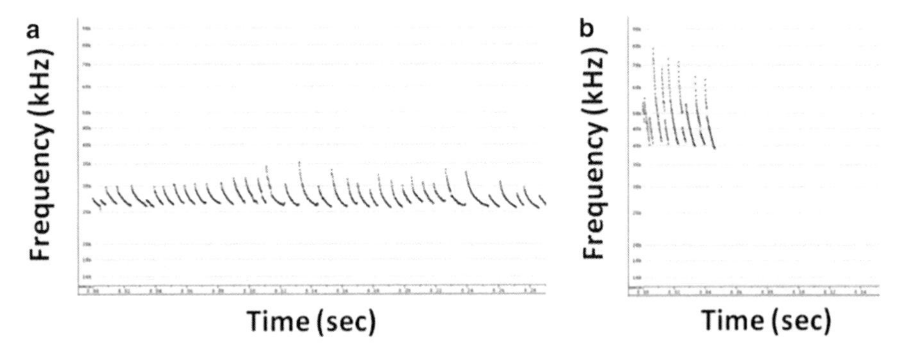

Fig. 10.1 The search phase echolocation calls of (**a**) a big brown bat, *Eptesicus fuscus*, and (**b**) a northern long-eared bat, *Myotis septentrionalis*

Table 10.2 Wing morphology indices from Norberg and Rayner (1987) and echolocation call structures of bats in the Central Hardwood Region

Species	Wing aspect ratio	Wing loading	Wingtip shape index	Echolocation call
Hoary bat *(Lasiurus cinereus)*	8.1	16.5	1.60	Narrow band, high intensity
Red bat *(L. borealis)*,	6.7	14.0	1.26	Broad band, medium intensity
Big brown bat *(Eptesicus fuscus)*	6.4	9.4	1.09	Narrow band, high intensity
Evening bat *(Nycticeius humeralis)*	6.8	10.7	1.01	Broad band, medium intensity
Silver-haired bat *(Lasionycteris noctivagans)*	6.6	8.2	1.68	Narrow band, high intensity
Gray bat *(M. grisescens)*,	6.4	8.2	1.79	Broad band, medium intensity
Tri-colored bat *(Perimyotis subflavus)*	6.2	5.6	2.05	Broad band, medium intensity
Small-footed bat *(Myotis leibii)*	6.1	6.7	2.96	Broad band, low intensity
Little brown bat (*M. lucifugus)*	6.0	7.5	3.20	Broad band, medium intensity
Northern long-eared bat *(Myotis septentrionalis)*	5.8	6.8	2.24	Broad band, low intensity
Indiana bat *(Myotis sodalis)*	5.4	6.5	5.56	Broad band, medium intensity
Rafinesque's big-eared bat *(Corynorhinus rafi nesquii)*	5.9	5.9	–	Broad band, very low intensity
Virginia and Ozark big-eared bat *(C. t. virginianus)*	5.9	7.2	2.31	Broad band, very low intensity

High aspect ratios indicate long narrow wings, high wing loadings indicate heavier body size relative to wing size, and higher wingtip shape indices indicate more rounded wingtips. Species are listed from top to bottom along the open-closed canopy continuum based on ecomorphology and echolocation call characteristics. No data are available for the southeastern bat

assumptions of the technique (Hayes 2000; Gannon et al. 2003). For example, it is not possible to distinguish age and sex of bats detected by recorders and thus, we must assume that habitat use is consistent among all sub-populations. Because there is currently no way to tell whether multiple recordings of a species at a site represent one individual detected multiple times or several individuals, the number of calls recorded cannot be used as an index of abundance (Weller 2007). It is also difficult to infer habitat selection or preference based on acoustic data because bat detectors

Fig. 10.2 Techniques used to determine use of different habitat types by bats. (**a**) An Anabat II bat detector system, (**b**) the Anabat II system deployed in the field, (**c**) an Indiana bat with a radio transmitter, and (**d**) tracking a bat with an antenna and receiver

measure resource use by populations and not individuals (Miller et al. 2003). If differential detectability among species and habitat types is not taken into account, areas where detection is best (e.g., open areas) may appear to be favored or species with higher intensity calls may appear to be more active (MacKenzie 2006). Finally, it is not possible to use acoustic detectors to assess foraging and commuting behavior for species with very low intensity echolocation calls, like big-eared bats (*Corynorhinus* spp.) (Fenton 2003). Nonetheless, acoustic detectors yield considerable data on relative activity of bats in early successional habitats and other habitat types, and allow researchers to examine use by multiple species in a variety of habitat types at the same time, thus controlling for factors such as weather, time of night, and stage of the reproductive cycle.

Much of the information on bat use of early successional habitats comes from studies that have examined the effects of timber harvesting on bat activity. In general, these studies have shown that overall activity and foraging activity are greater in recent clearcuts than in adjacent forest (Erickson and West 1996; Krusic et al. 1996; Grindal and Brigham 1998, 1999; Ellis et al. 2002). However, responses to harvesting often vary by species and in accordance with predictions based on ecomorphology and echolocation call structure. For example, open-space species such as silver-haired bats and hoary bats are more active in recently clearcut stands (2–8 years post-harvest) than in mature forests whereas clutter-adapted bats such as *Myotis* spp. are more active in mature forests than in recent clearcuts (Patriquin and Barclay 2003; Owen et al. 2004; Morris et al. 2010). However, some small species such as northern long-eared bats, Indiana bats, and tri-colored bats use open areas as well as mature forests (Ellis et al. 2002; Sparks et al. 2005; O'Keefe 2009).

While relative adaptations to clutter may influence use and avoidance of early successional habitats by large and small bats, predation risk and food availability may also contribute to the observed pattern. Faster flight might enable larger bats to evade predators in open areas while smaller bats might need the protective cover provided by forest canopy (Rydell et al. 1996). Food availability can also play a role although there are few data to support this. In some areas, nocturnal insect abundance is greater in early successional forest than mature forest (Lunde and Harestad 1986), while in other areas it is the same or greater in mature forest than in clearcuts (Kalcounis and Brigham 1995; Grindal and Brigham 1998; Burford et al. 1999; Grindal and Brigham 1999; Dodd et al. 2008; Morris et al. 2010). Some studies have found that bat activity is positively related to insect abundance (Kalcounis and Brigham 1995; Tibbels and Kurta 2003), while others have found no relationship between bat activity and insect abundance (Lunde and Harestad 1986; Grindal and Brigham 1999; Obrist et al. 2011). Morris et al. (2010) concluded that insect availability can be important in determining bat habitat use in an intensively managed forest landscape in the North Carolina Coastal Plain, but it plays a secondary role to stand structural characteristics. However, it should be noted that most studies have only identified potential prey to Order or Family and thus, factors such as prey size or preference have not been investigated. There is a positive relationship between bat body size and insect prey size, although larger bats take both large and small prey items (Aldridge and Rautenbach 1987; Barclay and Brigham 1991). If larger insects are restricted to open areas or are more vulnerable to capture by bats in open areas, then larger bats may select these areas because of the availability of preferred prey items. Conversely, smaller insects may seek the protective cover of forest, thus attracting small bats to late seral stage forests. We are not aware of any studies that have examined the relationship between insects and their habitat associations relative to body size or vulnerability to capture, but these types of studies are necessary to fully understand bat foraging in relation to early and late successional habitats.

Edges between early successional patches and stands in mid to late seral stages appear to be particularly important to medium- and small-sized bat species. For example, in Alberta and British Columbia, overall bat activity is greater along the edges of forests adjacent to recently clearcut patches than in the center (Crampton and Barclay 1996; Grindal and Brigham 1999). In a similar study in Alberta,

Hogberg et al. (2002) found that activity of *Myotis* spp. was also greater along edges of clearcuts than in the center. Hoary bats, red bats, and *Myotis* spp. are positively associated with edge in Ontario, Canada (Furlonger et al. 1987) as are bats of five genera in intensively managed forests in the Coastal Plain of South Carolina (Hein et al. 2009a). Activity of hoary bats, Brazilian free-tailed bats (*Tadarida brasiliensis*), tri-colored bats, big brown bats, evening bats (*Nycticeius humeralis*), and red bats is significantly greater at forest edges than in interior forest in the Coastal Plain of North Carolina (Morris et al. 2010). In Oklahoma, adult female Ozark big-eared bats forage in open areas, edges, and intact forests, but not all habitat types are used equally; edges are selected and interior forests are avoided, particularly during the post-lactation period (Clark et al. 1993). Radio-telemetry studies of Indiana bats and northern long-eared bats also suggest that edges are important for commuting (Murray and Kurta 2004; Henderson and Broders 2008).

Wind, insect availability, predation risk, and navigation may drive bats' preferences for edges (Verboom and Spolestra 1999). Foraging and commuting are more efficient along leeward edges, which are more protected from wind than open areas. Insects also tend to concentrate along leeward edges (Lewis 1970; Whitaker et al. 2000) making them particularly rich foraging grounds. Edges may also serve as navigation aids for short-distance movements (Verboom et al. 1999) and long-distance migration (Furmankiewicz and Kucharska 2009). Although several authors have suggested that edges provide protection from predators (Clark et al. 1993; Walsh and Harris 1996; Verboom and Spolestra 1999), Lesiński et al. (2009) found that tawny owl (*Strix aluco*) predation on bats was greater along forest edges than in forest interiors or open areas. Thus, the protective nature of edges needs further investigation.

Small openings or gaps within interior forests also appear to be important foraging and commuting areas (Hayes and Loeb 2007). For example, bat activity in wildlife openings or logging decks larger than 0.25 ha in red pine (*Pinus resinosa*) stands in Michigan is significantly greater than in adjacent thinned or unthinned stands (Tibbels and Kurta 2003) and bat activity in bottomland hardwood forests in South Carolina is significantly greater in 0.03 ha and 0.50 ha gaps than in 70-year-old forest (Menzel et al. 2002). Several studies in the Appalachian Mountains have also found that small openings and gaps are important foraging and commuting sites (Ford et al. 2005; Loeb and O'Keefe 2006; Schirmacher et al. 2007). In the upper Piedmont and mountains of South Carolina, big brown bats, red bats, tri-colored bats, and northern long-eared bats are more likely to be recorded at sample points with open vegetation regardless of stand age class than at points with medium or dense vegetation, indicating that they are using small gaps within mature forest as well as recently harvested stands and open fields (Loeb and O'Keefe 2006). In the Central Appalachians of West Virginia, occurrence of big brown bats, red bats, hoary bats, and little brown bats is positively related to gap width (Ford et al. 2005), and in the Allegheny Plateau presence of big brown bats, little brown bats, and tri-colored bats is positively associated with forest openings (Schirmacher et al. 2007). Bats may be attracted to small gaps and openings within mature forest for several reasons. Similar to recent clearcuts and other large early successional habitats, structural and acoustic clutter are lower in gaps than in intact forest and insect availability may also be greater in gaps

(Tibbels and Kurta 2003). Further, many bats select roost structures close to gaps (see sect. 10.3) either for thermal benefits from increased solar exposure or access to foraging sites (e.g., Willis and Brigham 2005; Perry et al. 2007a; O'Keefe et al. 2009). Forest roads and trails are also important foraging and commuting habitats. Several studies in Australia have found that bat activity is significantly greater on trails than in mature forest, even when the trails traverse dense regrowth forests (Law and Chidel 2001, 2002; Adams et al. 2009). Similarly, Zimmerman and Glanz (2000) found that bat activity in Maine was positively associated with gravel roads that functioned as edge between forests and open areas. In an intensively managed forest in the Coastal Plain of South Carolina, the odds of big brown bat, tri-colored bat, and Seminole bat (*L. seminolus*) occurrence were over five times greater if a road was present (Hein et al. 2009a). Forest trails and roads may be important because they combine the features of small gaps (i.e., reduced clutter) and edges (i.e., navigational aids and potential cover from predators).

To illustrate relative use of various early successional areas and mature forest by bats, we present data collected on the Nantahala National Forest in southwestern North Carolina. We used AnabatII bat detectors and zero-crossings interface modules (Titley Electronics, Australia) to passively record bat activity for two full nights each in one 80 year-old yellow-poplar-oak (*Liriodendron tulipifera-Quercus* spp.) forest, three gated roads 6–10 m wide, and three wildlife openings ≤ 3 ha. Gated forest roads and wildlife openings were planted in grasses and clover and maintained by annual or biennial mowing. Bat activity was far greater on gated roads and in wildlife openings than in the forested site (Fig. 10.3a). However, as in many other studies, use of roads and openings varied by species or phonic group. Hoary bats, which are open-space bats (Table 10.2), used wildlife openings to a much greater extent than roads (Fig. 10.3b). Clutter-adapted northern long-eared bats, which have echolocation calls and body morphology designed for gleaning insects from vegetation (Faure et al. 1993), rarely used openings but often used roads (Fig. 10.3b). Big brown bats, red bats, and tri-colored bats appeared to be more flexible in their use of early successional habitats than open or closed canopy specialists.

Many studies have examined the relationship between seral stage and bat activity at the stand level, but findings from recent landscape scale studies suggest that the importance of early successional patches may vary with scale. In the Ozark Highlands of Missouri, occupancy by big brown bats is positively associated with non-forested areas such as pastures or grasslands at the site level, but negatively associated with the percent of non-forested area in the surrounding landscape, whereas red bats are positively associated with non-forested habitat at both site and landscape scales (Amelon 2007). Site level characteristics have insignificant effects on little brown bat occupancy but the amount of non-forested habitats and pine forest in the surrounding landscape has a positive effect on their presence (Amelon 2007). Site occupancy of Indiana bats in Missouri is also positively related to the proportion of non-forested habitats in the surrounding landscape (Yates and Muzika 2006).

All early successional habitats are not equivalent and several factors may affect their relative use by bats including size, vegetation type, and position on the landscape. For example, big brown and hoary bat activity is greater in open fields with

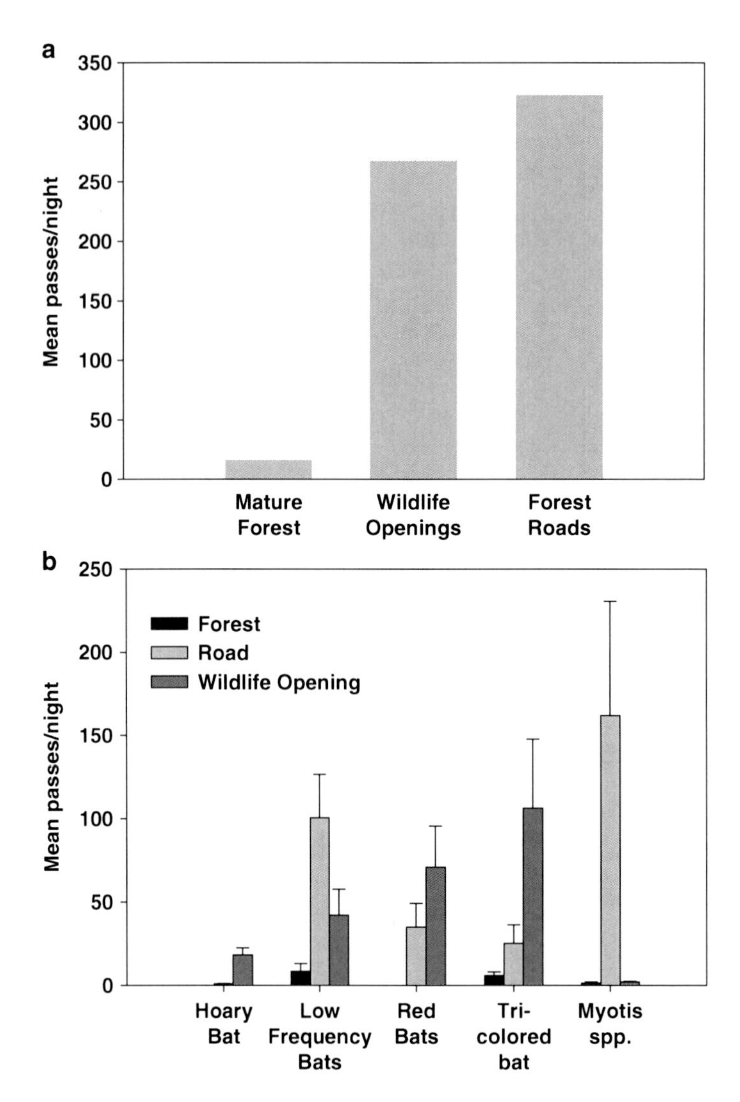

Fig. 10.3 Bat activity in intact forest, gated roads, and wildlife openings on the Nantahala National Forest, North Carolina in summer 2004. (**a**) Mean number of bat passes recorded per night for all species. (**b**) Mean number of passes per night by species or species group. Low frequency bats consisted of big brown bats and silver-haired bats and *Myotis* spp. consisted of northern long-eared bats, little brown bats, and small-footed bats

herbaceous vegetation maintained by mowing than in clearcut areas with regenerating saplings (Brooks 2009), and ponds in 0.4 ha fields are used more than ponds in 6.2–18.2 ha clearcuts <10 years old (Huie 2002). Differential use of open spaces may be due to differences in patch characteristics. For example, vegetation type and management history can affect insect abundance and diversity (Swengel 2001), and there is some indication that patch size can affect bat activity. Grindal and Brigham (1998) found that bat activity in British Colombia tends to decrease with increasing clearcut size (from 0.5 to 1.5 ha); in the Coastal Plain of South Carolina, foraging activity of hoary bats is greater in small gaps (0.03 ha) but activity of other species does not differ between large and small gaps (Menzel et al. 2002). Patch size effects may be related to the reluctance of some bats to cross large open areas. Swystun et al. (2001) found that edges of isolated forest patches were not used as often as edges of forest patches close to mature forest, and the close adherence of northern long-eared bats and Indiana bats to hedgerows while commuting suggests that some bats may avoid large open spaces (Murray and Kurta 2004; Henderson and Broders 2008). Thus, position on the landscape and connectivity may be important factors affecting bat activity in early successional patches. However, we are unaware of any studies that relate use of early successional patches to their distribution on the landscape.

Unlike early successional habitats, forests in mid-seral stages (e.g., closed canopy sapling-pole stage) are rarely used by bats (Crampton and Barclay 1996; Erickson and West 1996; Krusic et al. 1996; Law and Chidel 2001; Ellis et al. 2002; Loeb and O'Keefe 2006; Adams et al. 2009). Thus, early successional habitats that are allowed to regenerate to closed canopy forest will likely become unsuitable bat foraging and commuting habitat for a given period. However, the horizontal edges created by the canopies of mid-successional forests may be attractive to bats and thinning may be used in mid-successional closed canopy forest to minimize the time these stands are unsuitable (Guldin et al. 2007) and create more bat-suitable stand structures (Humes et al. 1999; Loeb and Waldrop 2008).

10.3 Early Successional Habitats and Roost Site Selection

Bat roosts have many critical functions including serving as sites for social interactions and rearing young, and providing protection from the elements and predators (Kunz and Lumsden 2003). Bats that roost in trees during the non-hibernation period can be divided into those that roost in crevices or cavities (including those that roost between the bark and bole) and those that roost in foliage (Kunz and Lumsden 2003). Both groups are represented in the Central Hardwood Region (Table 10.1).

Over the past 15–20 years many studies have investigated use and selection of day roosts by tree-roosting bats during the summer reproductive period; however, less work has been conducted on foliage-roosting species (Barclay and Kurta 2007; Carter and Menzel 2007). Most studies have focused on tree and stand level features and less attention has been given to relationships between larger-scale features and roost site selection (Kalcounis-Rueppell et al. 2005). In general, tree-roosting bats select roosts within mature forests (Barclay and Brigham 1996) and avoid using recently harvested stands and other open areas even when roost trees may be available (e.g., Arnett and Hayes 2009). However, there are some exceptions. In some areas of the Pacific Northwest, western long-eared bats (*M. evotis*) roost in stumps in 8–9 year old clearcuts (Vonhof and Barclay 1997) or rock crevices in open areas close to forest edges (Rancourt et al. 2005). In the Southern Appalachians, red bats occasionally roost in < 5 year-old hardwood stands (O'Keefe et al. 2009) and in the Piedmont of South Carolina they have been found roosting in pastures and yards (Leput 2004). Though maternity colonies of northern long-eared bats usually roost in large dead trees in mature stands, a few colonies in the Southern Appalachians were found roosting in leave trees within 2–4 year old shelterwood cuts (O'Keefe 2009). Small canopy gaps in mature forest are also important roost sites for bats. For example, Indiana bats in Missouri and Illinois primarily roost within or on the edges of forest openings (Callahan et al. 1997; Carter and Feldhamer 2005) and silver-haired bats in northeastern Washington roost exclusively in forest gaps (Campbell et al. 1996). In general, cavity or crevice-roosting bats tend to select roosts in areas with lower canopy cover than the surrounding forest (Kalcounis-Rueppell et al. 2005), presumably for increased solar radiation and, thus, reduced thermoregulatory costs (Barclay and Kurta 2007).

Although early successional patches other than small gaps are not important for roosting per se, their distribution on the landscape may be important for roost site selection. Because early successional habitats are important foraging and commuting sites for many bat species, and because flight is an expensive mode of locomotion (Altringham 1996), bats may try to minimize their commuting costs by selecting roosts that are close to early successional habitats. However, it should be noted that some authors have suggested that commuting costs are trivial compared to other energetic demands (Kurta et al. 2002; Lumsden et al. 2002).

Despite considerable variation within and among species, there is evidence that some bats select roosts based on proximity to early successional patches. For example, red bats which are often considered to be an open-adapted species (Table 10.2), use edges, forests, and open areas for foraging (Elmore et al. 2005; Ford et al. 2006; Amelon 2007; Morris et al. 2010). Thus, it is not surprising that they have been found to roost closer than expected to edges, particularly forest roads in a variety of forest types (Mager and Nelson 2001; Leput 2004; Perry et al. 2008; O'Keefe et al. 2009). However, in some forest landscapes red bats avoid roosting near edges and open areas. For example, in an intensively managed pine landscape in Mississippi, red bats roost farther from 0 to 8 year-old clearcut edges than expected (Elmore et al. 2004) and in mature second-growth mixed mesophytic forests in Kentucky, they roost an average of 277 m from the forest edge (Hutchinson and Lacki 2000). Evening bats fall closer to the center of the clutter continuum (Table 10.2) but, like red bats, vary in their response to early successional habitats in the surrounding landscape with regards to roosting. In Arkansas and coastal Georgia, evening bats select roosts in landscapes with more open areas such as fields, wildlife openings, clearcuts < 8 years of age, and group selection cuts (Miles et al. 2006; Perry et al. 2008), but in coastal South Carolina evening bats roost farther from openings (fields, wildlife openings, and clearcuts < 5 years old) than expected (Hein et al. 2009b).

Variation in roost site selection relative to early successional habitats is also seen among some of the more clutter-adapted species, with responses varying with the amount and configuration of early successional patches on the landscape. In a landscape dominated by mature hardwood forests in the Southern Appalachians, adult male tri-colored bats roost closer than expected to small nonlinear openings and recent (≤ 5 years) 2-aged shelterwood harvests than expected (O'Keefe et al. 2009) and in Indiana woodlands, adult female tri-colored bats roost close (52 m) to the forest edge, though non-reproductive females roost significantly closer to edges than reproductive females (Veilleux et al. 2004). In contrast, tri-colored bats in a managed forest in Arkansas roost farther than expected from roads, but are more likely to use areas with small openings (i.e., group selection harvests, Perry et al. 2008) which corresponds with our finding that tri-colored bat activity was greater in nonlinear openings than along roads (see Fig. 10.3). Northern long-eared bats are considered to be clutter specialists (Faure et al. 1993), but proximity to early successional habitats may affect landscape-scale roost site selection for this species. In Arkansas, northern long-eared bats select roosts in areas with more group selection harvests within 250 m, though this same population tends to roost farther from roads than expected (Perry et al. 2008). In Kentucky, lactating female northern long-eared bats select roosts that are close to roads, but pregnant and post-lactating females do not (Lacki and Schwierjohann 2001).

Several factors may be driving the considerable variation in roost site selection relative to early successional habitats. Studies of eastern red bats have been conducted in forests undergoing a wide range of forest management activities (Elmore et al. 2004; Leput 2004; Perry et al. 2007b; O'Keefe et al. 2009), in large contiguous tracts of non-managed forest (Hutchinson and Lacki 2000), and in urban areas (Mager and Nelson 2001; Limpert et al. 2007). Thus, variation in the amount and distribution of early successional habitats on the landscape may account for the differences among the various studies. Even among forested landscapes, the distribution of early successional patches on the landscape can be quite variable due to both geography and land use history (e.g., forest management practices, agriculture, and urbanization). Historically, large scale disturbances such as hurricanes and fires have been more important in the creation of early successional habitats in the Coastal Plain and grassland biomes in the eastern USA, whereas small scale disturbances are more important in creating early successional habitats in upland hardwood forests of the Central Hardwoods Region (Runkle 1990; White et al., Chap. 3). Further, in the pine forests of the Coastal Plain final harvests tend to be clearcuts, whereas there is greater reliance on partial cuts and natural regeneration in upland hardwood systems (Wear and Greis 2002). In addition, small gaps are very difficult to map and may often be ignored in coarser, landscape scale studies. Thus, influences of early successional habitats on roost site selection may be underrepresented in landscapes where small gaps represent the primary form of early successional habitats. In contrast, where large scale forest disturbances are more common such as the Coastal Plain, early successional habitats are more available and are easier to map and include in roost-related landscape studies.

Sex and reproductive condition of bats may also contribute to variation in roost selection relative to early successional habitats. During the reproductive period, females have high energy demands related to gestation and lactation (Kurta et al. 1989, 1990). Because daily torpor can delay gestation and growth of young (Racey and Swift 1981), pregnant and lactating females use torpor less often than adult males and non-reproductive females (Hamilton and Barclay 1994). Thus, reproductive females often select warmer roosts that allow them to maintain higher body temperatures while minimizing energetic costs (Lausen and Barclay 2006; Willis 2006), regardless of distance to foraging sites. In contrast, thermal considerations may be less important for males and non-reproductive females and distance to foraging areas may be a more important factor in their choice of roost sites. For example, O'Keefe et al. (2009) found that tree and stand characteristics are not important factors determining roost selection by male red bats and tri-colored bats, but males of both species roost closer to openings than expected. In contrast, studies of female red bats have found that tree and stand characteristics are important factors influencing roost site selection (Hutchinson and Lacki 2000; Veilleux et al. 2003; Elmore et al. 2005; Perry and Thill 2007; Perry et al. 2007a). Similarly, in southeastern Australia, pregnant and lactating female lesser long-eared bats (*Nyctophilus geoffroyi*) roost in large diameter snags in floodplain forests despite the fact that these snags are 4–10 km from foraging sites in a farmland mosaic whereas males, which do not rely on large-diameter snags, roost within 2 km of foraging sites in the same farmland mosaic (Lumsden et al. 2002).

Roost permanency, abundance and distribution of suitable roosts, predation risk, and social factors are also important in roost site selection by bats (Kunz and Lumsden 2003; Miller et al. 2003; Kalcounis-Rueppell et al. 2005) and may determine whether bats roost in or near early successional habitats. Some bats may avoid roosting in or near early successional habitats because trees in these sites are more exposed to predators (Russo et al. 2007) or to wind and rain (Callahan et al. 1997). Further, because bats commonly move among several roosts within a relatively small area (Barclay and Kurta 2007; Carter and Menzel 2007), the abundance of suitable roosts within a site may also be a more important factor governing roost selection.

10.4 Management of Early Successional Habitats for Bats in the Central Hardwood Region

It is generally agreed that roosts are the most limiting factor for bats in forested landscapes and conservation strategies should focus on protecting existing roosts and ensuring the availability of future roosts (Hayes 2003; Duchamp et al. 2007). Because most tree roosting bats use large trees or snags in mature forests (Kalcounis-Rueppell et al. 2005), creation and maintenance of early successional patches would seem counter to wise conservation strategies for bats. However, as we have illustrated, early successional habitats are important foraging and commuting sites for many species and the distribution of early successional patches may influence roost site selection. Nonetheless, the importance of roosts to survival and reproductive success of bats and reliance of so many bat species on large trees and snags must be kept in mind when considering management of early successional habitats in the Central Hardwood Region.

Early successional habitats, particularly in the form of gaps, small openings, and forest roads are important foraging sites for many species of bats and many species roost in, or at the edge, of gaps. Thus, creating many small gaps within mature forest may benefit bats as well as birds (Blake and Hoppes 1986) and reptiles (Greenberg 2001; Moorman et al., Chap. 11). Conserving or creating potential roost trees in or at the edge of small gaps will provide valuable roosting habitat, particularly for maternity colonies. Although retaining relict, cull, and dead trees is often counter to many silvicultural goals and may be perceived as a safety hazard, there are various ways to manage stands in which these structures are retained while minimizing the impact on subsequent regeneration (Guldin et al. 2007) and safety.

Because many bats are reluctant to cross large expanses of open area (e.g., Murray and Kurta 2004; Henderson and Broders 2008) and edges are often the preferred foraging site within clearcuts (e.g., Crampton and Barclay 1996; Grindal and Brigham 1999; Hogberg et al. 2002), most bats will probably benefit from smaller-sized cuts (<10 ha). Small cuts will increase the amount of edge relative to the amount of open area and minimize the distances bats will have to fly to traverse open spaces. Smaller cuts may also benefit other wildlife, particularly amphibians (Moorman et al., Chap. 11). However, as cut areas regenerate to dense closed-canopy second growth forest, the interior portion of these forests will cease to serve as bat foraging habitat, at least until the stands are thinned or mature into more suitable forest structures. Further, we have often observed a strong increase in bat activity one year after silvicultural treatments such as thinning and 2-age shelterwood cuts, with declines in activity in subsequent years (Loeb and Waldrop 2008; O'Keefe 2009). Thus, the benefits of regeneration cuts for bats may be short-term, at least at the stand scale. But, if regeneration cuts are conducted at a sustained rate over time across the landscape, harvesting may provide long-term benefits at larger spatial scales.

Bats in fragmented habitats often fly along tree lines and hedgerows (Murray and Kurta 2004; Henderson and Broders 2008), so leaving strips of mature trees within shelterwood cuts or using group selection harvests may facilitate bat movements in cuts, make cuts more desirable foraging sites, and provide current or future roost sites near foraging habitats. However, creation of many small cuts or group selection harvest may have some indirect negative impacts on bats and other organisms due to a higher rate of soil erosion resulting from multiple re-entries (Hood et al. 2002). Leaving too many trees may also interfere with regeneration of shade-intolerant trees (Guldin et al. 2007). In riparian areas, leaving trees in streamside management zones may be sufficient for providing adequate cover and roosting habitat (Hayes and Loeb 2007; O'Keefe 2009).

Although heavily traveled roads can impede movement of some bat species (Kerth and Melber 2009) and are sources of mortality for other species (Lesiński 2007, 2008; Russell et al. 2009), our data and those of several others (Zimmerman and Glanz 2000; Law and Chidel 2001, 2002; Adams et al. 2009; Hein et al. 2009a) demonstrate that gated or lightly traveled forest roads are often important foraging and commuting sites for bats. Forest roads may be particularly important in densely forested habitats by serving as commuting routes between roosting and foraging sites. However, roads may have negative impacts on forest ecosystems (e.g., sedimentation), but impacts can be minimized through best management practices such as road closure, revegetation, and sediment control (Swift and Burns 1999; Grace and Clinton 2007).

Because bats are so mobile and use a variety of habitat types for foraging, roosting, and commuting, habitat management for bats requires a landscape approach and consideration of both spatial and temporal factors (Duchamp et al. 2007). Unfortunately, there is considerable variation among species in the Central Hardwoods Region in roost structure use (Table 10.1) and movement patterns (Lacki et al. 2007), which may make planning at the landscape scale more difficult. Thus, providing a diversity of habitat types of varying age classes and forest structures across the landscape (especially mature forest structures for roosting) while ensuring good connectivity among them, may be the best strategy for maintaining viable bat populations.

10.5 Future Research

There are still many questions about the use of early successional habitats by bats and the best ways to manage these habitats to meet the needs of bats. One of the first areas of research concerns the applicability of existing knowledge to times outside the summer reproductive period. Most research has been conducted during summer, a critical period for reproduction and growth of the young. Spring and fall also are critical due to energetic demands of migration and hibernation, but little is known about bat roosting and foraging habitat needs and how management activities may affect them during these times (Cryan and Veilleux 2007). Bats lose approximately 15%–30% of their body weight during hibernation (Hall 1962; Thomas et al. 1990; Johnson et al. 1998) and must regain some of that weight upon emergence in spring. Replenishing fat reserves is particularly important for females who must migrate to summer maternity sites and prepare for the reproductive period. During late summer and fall, bats must put on sufficient fat to migrate to hibernation sites and meet the physiological demands of winter. Foraging resources may be particularly critical during migration periods, yet we know little about the foraging habitats used by bats during these periods.

Species that do not hibernate in caves face increased energy demands during late summer and early fall because they often migrate up to 1,000 km from their summer range to more southerly areas where they roost in trees, bushes, and leaf litter (Cryan and Veilleux 2007). However, many individuals either remain within the Central Hardwood Region or migrate from more northerly climes to the Central Hardwood Region. While some research has been conducted on roost site selection of long distance migrants during winter (Saugey et al. 1998; Boyles and Robbins 2006; Mormann and Robbins 2007), no studies have examined the relationship between winter roost sites and early successional habitat types. Further, many bats in the Central Hardwood Region are active during warm nights in winter and at least some species (red bats, silver-haired bats, big brown bats, hoary bats) feed during arousals (Boyles et al. 2006; Dunbar et al. 2007). Thus, it is important to determine whether early successional habitats are important for foraging bats in the Central Hardwood Region during winter.

As touched on in previous sections, we know little about how the spatial configuration and structure of early successional patches affects use by bats. For example, far more research is needed on effects of size and shape of early successional patches before we can develop effective management plans that meet the needs of bats. Other landscape-scale factors that may affect bats' use of early successional patches include distance to other early successional patches, distance to water sources, distance to flight corridors such as roads or trails, and the age structure of the surrounding forest matrix. We also need a better understanding of how these factors affect insect use of early successional habitats, particularly the insect taxa and size classes that are preferred food items of bats. Future research should also compare use of recently harvested areas and other types of early successional habitats of similar size and shape to tease apart the effects of vegetation and other patch characteristics (e.g., size and shape). This will require that researchers clearly describe patch vegetation and structural characteristics when reporting the results of their studies.

Because gaps appear to be an important habitat component for all eastern bat species and were historically the most prevalent form of early successional habitats in the Central Hardwood Region (Runkle 1990), we need a better understanding of how bats use gaps and the relationship between roost site selection and gap dynamics. Use of new technologies such as LIDAR (Light Detection and Ranging), may provide more accurate data on small gap distribution in mature forest (e.g., Vepakomma et al. 2008), allowing researchers to better test hypotheses about the relationship between habitat selection and forest gaps.

10.6 Conclusions

Early successional habitats are just one of many habitat types used by bats in the Central Hardwood Region and other forested ecosystems. They may be important to bats for foraging but many questions remain about characteristics of early successional patches that contribute to their relative use. Because most bats roost in mature forest, the creation and management of early successional patches on the landscape must be balanced by the maintenance of sufficient mature forest with large trees and snags for roosting. Given the numerous other threats facing bats at the present time (habitat loss to urbanization, climate change, white-nose syndrome, wind energy development) and the precipitous declines in many populations, managing existing forests so that they provide all the resource needs of bats (roosts, foraging and commuting habitats, and clean water) is critical for the conservation of these important species.

Literature Cited

Adams MD, Law BS, French KO (2009) Vegetation structure influences the vertical stratification of open- and edge-space aerial-foraging bats in harvested forests. For Ecol Manage 258:2090–2100

Aldridge HDJN, Rautenbach IL (1987) Morphology, echolocation and resource partitioning in insectivorous bats. J Anim Ecol 56:763–778

Altringham JD (1996) Bats: biology and behavior. Oxford University Press Inc., New York

Amelon SK (2007) Multi-scale factors influencing detection, site occupancy and resource use by foraging bats in the Ozark Highlands of Missouri. Dissertation, University of Missouri, Columbia

Arnett EB, Hayes JP (2009) Use of conifer snags as roosts by female bats in western Oregon. J Wildl Manage 73:214–225

Arnett EB, Brown WK, Erickson WP, Fiedler JK, Hamilton BL, Henry TH, Jain A, Johnson GD, Kerns J, Koford RR, Nicholson CP, O'Connell TJ, Piorkowski MD, Tankersley RD Jr (2008) Patterns of bat fatalities at wind energy facilities in North America. J Wildl Manage 72:61–78

Barclay RMR, Brigham RM (1991) Prey detection, dietary niche breadth, and body size in bats: why are aerial insectivorous bats so small? Am Nat 137:693–703

Barclay RMR, Brigham RM (1996) Bats and forests symposium, Canada Research Branch, BC Ministry of Forestry, Victoria, p 292

Barclay RMR, Harder LD (2003) Life histories of bats: life in the slow lane. In: Kunz TH, Fenton MB (eds) Bat ecology. The University of Chicago Press, Chicago, pp 209–253

Barclay RMR, Kurta A (2007) Ecology and behavior of bats roosting in tree cavities and under bark. In: Lacki MJ, Hayes JP, Kurta A (eds) Bats in forests: conservation and management. Johns Hopkins University Press, Baltimore, pp 17–59

Blake JG, Hoppes WG (1986) Influence of resource abundance on use of tree-fall gaps by birds in an isolated woodlot. Auk 103:328–340

Blehert DS, Hicks AC, Behr M, Meteyer CU, Berlowski-Zier BM, Buckles EL, Coleman JTH, Darling SR, Gargas A, Niver R, Okoniewski JC, Rudd RJ, Stone WB (2009) Bat white-nose syndrome: an emerging fungal pathogen? Science 323:227

Boyles JG, Robbins LW (2006) Characteristics of summer and winter roost trees used by evening bats (*Nycticeius humeralis*) in southwestern Missouri. Am Midl Nat 155:210–220

Boyles JG, Dunbar MB, Whitaker JO Jr (2006) Activity following arousal in winter in North America. Mamm Rev 36:267–280

Brigham RM (2007) Bats in forests: what we know and what we need to learn. In: Lacki MJ, Hayes JP, Kurta A (eds) Bats in forests: conservation and management. Johns Hopkins University Press, Baltimore, pp 1–15

Brooks RT (2009) Habitat-associated and temporal patterns of bat activity in a diverse forest landscape of southern New England, USA. Biodivers Conserv 18:529–545

Burford LS, Lacki MJ, Covell CV Jr (1999) Occurrence of moths among habitats in a mixed mesophytic forest: implications for management of forest bats. For Sci 45:323–332

Callahan EV, Drobney RD, Clawson RL (1997) Selection of summer roosting sites by Indiana bats (*Myotis sodalis*) in Missouri. J Mammal 78:818–825

Campbell LA, Hallett JG, O'Connell MA (1996) Conservation of bats in managed forests: use of roosts by *Lasionycteris noctivagans*. J Mammal 77:976–984

Carter TC, Feldhamer GA (2005) Roost tree use by maternity colonies of Indiana bats and northern long-eared bats in southern Illinois. For Ecol Manage 219:259–268

Carter TC, Menzel JM (2007) Behavior and day-roosting ecology of North American foliage-roosting bats. In: Lacki MJ, Hayes JP, Kurta A (eds) Bats in forests: conservation and management. Johns Hopkins University Press, Baltimore, pp 61–81

Clark BS, Leslie DM Jr, Carter TS (1993) Foraging activity of adult female Ozark big-eared bats (*Plecotus townsendii ingens*) in summer. J Mammal 74:422–427

Crampton LH, Barclay RMR (1996) Habitat selection by bats in fragmented and unfragmented aspen mixedwood stands of different ages. In: Barclay RMR, Brigham RM (eds) Bats and forests symposium, Canada Research Branch, BC Ministry of Forestry, Victoria, pp 238–259

Cryan PM, Veilleux JP (2007) Migration and use of autumn, winter, and spring roosts by tree bats. In: Lacki MJ, Hayes JP, Kurta A (eds) Bats in forests: conservation and management. Johns Hopkins University Press, Baltimore, pp 153–175

Dodd LE, Lacki MJ, Rieske LK (2008) Variation in moth occurrence and implications for foraging habitat of Ozark big-eared bats. For Ecol Manage 255:3866–3872

Duchamp JE, Arnett EB, Larson MA, Swihart RK (2007) Ecological considerations for landscape management of bats. In: Lacki MJ, Hayes JP, Kurta A (eds) Bats in forests: conservation and management. Johns Hopkins University Press, Baltimore, pp 237–261
Duchamp JE, Sparks DW, Swihart RK (2010) Exploring the "nutrient hot spot" hypothesis at trees used by bats. J Mammal 91:48–53
Dunbar MB, Whitaker JO Jr, Robbins LW (2007) Winter feeding by bats in Missouri. Acta Chiropt 9:305–310
Ellis AM, Patton LL, Castleberry SB (2002) Bat activity in upland and riparian habitats in the Georgia Piedmont. Proc Annu Conf Southeast Assoc Fish Wildl Agen 56:210–218
Elmore LW, Miller DA, Vilella FJ (2004) Selection of diurnal roosts by red bats (*Lasiurus borealis*) in an intensively managed pine forest in Mississippi. For Ecol Manage 199:11–20
Elmore LW, Miller DA, Vilella FJ (2005) Foraging area size and habitat use by red bats (*Lasiurus borealis*) in an intensively managed pine landscape in Mississippi. Am Midl Nat 153:405–417
Erickson JL, West SD (1996) Managed forests in the western Cascades: the effects of seral stage on bat habitat use patterns. In: Barclay RMR, Brigham RM (eds) Bats and forests symposium, Canada Research Branch, BC Ministry of Forestry, Victoria, pp 215–227
Faure PA, Fullard JH, Dawson JW (1993) The gleaning attacks of the northern long-eared bat, *Myotis septentrionalis*, are relatively inaudible to moths. J Exp Biol 178:173–189
Fenton MB (1990) The foraging behaviour and ecology of animal-eating bats. Can J Zool 68:411–422
Fenton MB (2003) Science and the conservation of bats: where to next? Wildl Soc Bull 31:6–15
Ford WM, Menzel MA, Rodrigue JL, Menzel JM, Johnson JB (2005) Relating bat species presence to simple habitat measures in a central Appalachian forest. Biol Conserv 126:528–539
Ford WM, Menzel JM, Menzel MA, Edwards JW, Kilgo JC (2006) Presence and absence of bats across habitat scales in the upper coastal plain of South Carolina. J Wildl Manage 70:1200–1209
Furlonger CL, Dewar HJ, Fenton MB (1987) Habitat use by foraging insectivorous bats. Can J Zool 65:284–288
Furmankiewicz J, Kucharska M (2009) Migration of bats along a large river valley in southwestern Poland. J Mammal 90:1310–1317
Gannon WL, Sherwin RE, Haymond S (2003) On the importance of articulating assumptions when conducting acoustic studies of habitat use by bats. Wildl Soc Bull 31:45–61
Grace JM III, Clinton BD (2007) Protecting soil and water in forest road management. Trans Am Soc Agric Biol Eng 50:1579–1584
Greenberg CH (2001) Response of reptile and amphibian communities to canopy gaps created by wind disturbance in the Southern Appalachians. For Ecol Manage 148:135–144
Grindal SD, Brigham RM (1998) Short-term effects of small-scale habitat disturbance on activity by insectivorous bats. J Wildl Manage 62:996–1003
Grindal SD, Brigham RM (1999) Impacts of forest harvesting on habitat use by foraging insectivorous bats at different spatial scales. Ecoscience 6:25–34
Guldin JM, Emmingham WH, Saugey DA, Carter SA (2007) Silvicultural practices and management of habitat for bats. In: Lacki MJ, Hayes JP, Kurta A (eds) Bats in forests: conservation and management. Johns Hopkins University Press, Baltimore, pp 177–205
Hall JS (1962) A life history and taxonomic study of the Indiana bat, *Myotis sodalis*. Reading Publ Mus and Art Gallery, Reading
Hamilton IM, Barclay RMR (1994) Patterns of daily torpor and day-roost selection by male and female big brown bats (*Eptesicus fuscus*). Can J Zool 72:744–749
Harvey MJ, Altenbach JS, Best TL (1999) Bats of the United States. Arkansas Game and Fish Commission, Little Rock
Hayes JP (2000) Assumptions and practical considerations in the design and interpretation of echolocation-monitoring studies. Acta Chiropt 2:225–236
Hayes JP (2003) Habitat ecology and conservation of bats in western coniferous forests. In: Zabel CJ, Anthony RG (eds) Mammal community dynamics: management and conservation in the coniferous forests of western North America. Cambridge University Press, New York, pp 81–119

Hayes JP, Loeb SC (2007) The Influences of forest management on bats in North America. In: Lacki MJ, Hayes JP, Kurta A (eds) Bats in forests: conservation and management. Johns Hopkins University Press, Baltimore, pp 207–235

Hein CD, Castleberry SB, Miller KV (2009a) Site-occupancy of bats in relation to forested corridors. For Ecol Manage 257:1200–1207

Hein CD, Miller KV, Castleberry SB (2009b) Evening bat summer roost-site selection on a managed pine landscape. J Wildl Manage 73:511–517

Henderson LE, Broders HG (2008) Movements and resource selection of the northern long-eared myotis (*Myotis septentrionalis*) in a forest-agriculture landscape. J Mammal 89:952–963

Hogberg LK, Patriquin KJ, Barclay RM (2002) Use by bats of patches of residual trees in logged areas of the boreal forest. Am Midl Nat 148:282–288

Hood SM, Zedaker SM, Aust WM, Smith DW (2002) Universal soil loss equation (USLE)–predicted soil loss for harvesting regimes in Appalachian hardwoods. North J Appl For 19:53–58

Huie KM (2002) Use of constructed woodland ponds by bats in the Daniel Boone National Forest. Thesis, Eastern Kentucky University, Richmond

Humes ML, Hayes JP, Collopy MW (1999) Bat activity in thinned, unthinned, and old-growth forests in western Oregon. J Wildl Manage 63:553–561

Hutchinson JT, Lacki MJ (2000) Selection of day roosts by red bats in mixed mesophytic forests. J Wildl Manage 64:87–94

Johnson SA, Brack V Jr, Rolley RE (1998) Overwinter weight loss of Indiana bats (*Myotis sodalis*) from hibernacula subject to human visitation. Am Midl Nat 139:255–261

Jones G, Jacobs DS, Kunz TH, Willig MR, Racey PA (2009) Carpe noctem: the importance of bats as bioindicators. End Spec Res 8:93–115

Kalcounis MC, Brigham RM (1995) Intraspecific variation in wing loading affects habitat use by little brown bats (*Myotis lucifugus*). Can J Zool 73:89–95

Kalcounis-Rueppell MC, Psyllakis JM, Brigham RM (2005) Tree roost selection by bats: an empirical synthesis using meta-analysis. Wildl Soc Bull 33:1123–1132

Kerth G, Melber M (2009) Species-specific barrier effects of a motorway on the habitat use of two threatened forest-living bat species. Biol Conserv 142:270–279

Krusic RA, Yamasaki M, Neefus CD, Pekins PJ (1996) Bat habitat use in White Mountain National Forest. J Wildl Manage 60:625–631

Kunz TH, Lumsden LF (2003) Ecology of cavity and foliage roosting bats. In: Kunz TH, Fenton MB (eds) Bat ecology. The University of Chicago Press, Chicago, pp 3–89

Kunz TH, Arnett EB, Erickson WP, Hoar AR, Johnson GD, Larkin RP, Strickland MD, Thresher RW, Tuttle MD (2007) Ecological impacts of wind energy development on bats: questions, hypotheses, and research needs. Front Ecol Environ 5:315–324

Kurta A, Bell GP, Nagy KA, Kunz TH (1989) Water balance of free-ranging little brown bats (*Myotis lucifugus*) during pregnancy and lactation. Can J Zool 67:2468–2472

Kurta A, Kunz TH, Nagy KA (1990) Energetics and water flux of free-ranging big brown bats (*Eptesicus fuscus*) during pregnancy and lactation. J Mammal 71:59–65

Kurta A, Murray SW, Miller DH (2002) Roost selection and movements across the summer landscape. In: Kurta A, Kennedy J (eds) The Indiana bat: biology and management of an endangered species. Bat Conservation International, Austin, pp 118–129

Lacki MJ, Schwierjohann JH (2001) Day-roost characteristics of northern bats in mixed mesophytic forest. J Wildl Manage 65:482–488

Lacki MJ, Amelon SK, Baker MD (2007) Foraging ecology of bats in forests. In: Lacki MJ, Hayes JP, Kurta A (eds) Bats in forests: conservation and management. Johns Hopkins University Press, Baltimore, pp 83–127

Lausen CL, Barclay RMR (2006) Benefits of living in a building: big brown bats (*Eptesicus fuscus*) in rocks versus buildings. J Mammal 87:362–370

Law B, Chidel M (2001) Bat activity 22 years after first-round intensive logging of alternate coupes near Eden, New South Wales. Aust For 64:242–247

Law B, Chidel M (2002) Tracks and riparian zones facilitate the use of Australian regrowth forest by insectivorous bats. J Appl Ecol 39:605–617

Leput DW (2004) Eastern red bat (*Lasiurus borealis*) and eastern pipistrelle (*Pipistrellus subflavus*) maternal roost selection: implications for forest management. Thesis, Clemson University, Clemson
Lesiński G (2007) Bat road casualties and factors determining their number. Mammalia 71:138–142
Lesiński G (2008) Linear landscape elements and bat casualties on roads – an example. Ann Zool Fennici 45:277–280
Lesiński G, Gryz J, Kowalski M (2009) Bat predation by tawny owls *Strix aluco* in differently human-transformed habitats. Ital J Zool 76:415–421
Lewis T (1970) Patterns of distribution of insects near a windbreak of tall trees. Ann Appl Biol 65:213–220
Limpert DL, Birch DL, Scott MS, Andre M, Gillam E (2007) Tree selection and landscape analysis of eastern red bat day roosts. J Wildl Manage 71:478–486
Loeb SC, O'Keefe JM (2006) Habitat use by forest bats in South Carolina in relation to local, stand, and landscape characteristics. J Wildl Manage 70:1210–1218
Loeb SC, Waldrop TA (2008) Bat activity in relation to fire and fire surrogate treatments in southern pine stands. For Ecol Manage 255:3185–3192
Lumsden LF, Bennett AF, Silins JE (2002) Location of roosts of the lesser long-eared bat *Nyctophilus geoffroyi* and Gould's wattled bat *Chalinolobus gouldii* in a fragmented landscape in south-eastern Australia. Biol Conserv 106:237–249
Lunde RE, Harestad AS (1986) Activity of little brown bats in coastal forests. Northwest Sci 60:206–209
MacKenzie DI (2006) Modeling the probability of resource use: the effect of, and dealing with, detecting a species imperfectly. J Wildl Manage 70:367–374
Mager KJ, Nelson TA (2001) Roost-site selection by eastern red bats (*Lasiurus borealis*). Am Midl Nat 145:120–126
Marcot BG (1996) An ecosystem context for bat management: a case study of the interior Columbia River Basin, USA. In: Barclay RMR, Brigham RM (eds) Bats and forests symposium, Canada Research Branch, BC Ministry of Forestry, Victoria, pp 19–36
Menzel MA, Carter TC, Menzel JM, Ford WM, Chapman BR (2002) Effects of group selection silviculture in bottomland hardwoods on the spatial activity patterns of bats. For Ecol Manage 162:209–218
Miles AC, Castleberry SB, Miller DA, Conner LM (2006) Multi-scale roost-site selection by evening bats on pine-dominated landscapes in southwest Georgia. J Wildl Manage 70:1191–1199
Miller DA, Arnett EB, Lacki MJ (2003) Habitat management for forest-roosting bats of North America: a critical review of habitat studies. Wildl Soc Bull 31:30–44
Mormann BM, Robbins LW (2007) Winter roosting ecology of eastern red bats in southwest Missouri. J Wildl Manage 71:213–217
Morris AD, Miller DA, Kalcounis-Rueppell MC (2010) Use of forest edges by bats in a managed pine forest landscape. J Wildl Manage 74:26–34
Murray SW, Kurta A (2004) Nocturnal activity of the endangered Indiana bat (*Myotis sodalis*). J Zool 262:197–206
Norberg UM, Rayner JMV (1987) Ecological morphology and flight in bats (Mammalia; Chiroptera): wing adaptations, flight performance, foraging strategy and echolocation. Philos Trans R Soc Lond B 316:335–427
Obrist MK, Rathey E, Bontadina F, Martinoli A, Conedera M, Christe P, Moretti M (2011) Response of bat species to sylvo pastoral abandonment. For Ecol Manage 261:789–798
O'Keefe JM (2009) Roosting and foraging ecology of forest bats in the Southern Appalachian Mountains. Dissertation, Clemson University, Clemson
O'Keefe JM, Loeb SC, Lanham JD, Hill HS Jr (2009) Macrohabitat factors affect day roost selection by eastern red bats and eastern pipistrelles in the Southern Appalachian Mountains, USA. For Ecol Manage 257:1757–1763
Ormsbee PC, Kiser JD, Perlmeter SI (2007) Importance of night roosts to the ecology of bats. In: Lacki MJ, Hayes JP, Kurta A (eds) Bats in forests: conservation and management. Johns Hopkins University Press, Baltimore, pp 130–151

Owen SF, Menzel MA, Edwards JW, Ford WM, Menzel JM, Chapman BR, Wood PB, Miller KV (2004) Bat activity in harvested and intact forest stands in the Allegheny Mountains. North J Appl For 21:154–159
Patriquin KJ, Barclay RMR (2003) Foraging by bats in cleared, thinned and unharvested boreal forest. J Appl Ecol 40:646–657
Perry RW, Thill RE (2007) Tree roosting by male and female eastern pipistrelles in a forested landscape. J Mammal 88:974–981
Perry RW, Thill RE, Carter SA (2007a) Sex-specific roost selection by adult red bats in a diverse forested landscape. For Ecol Manage 253:48–55
Perry RW, Thill RE, Leslie DM (2007b) Selection of roosting habitat by forest bats in a diverse forested landscape. For Ecol Manage 238:156–166
Perry RW, Thill RE, Leslie DM Jr (2008) Scale-dependent effects of landscape structure and composition on diurnal roost selection by forest bats. J Wildl Manage 72:913–925
Racey PA, Entwistle AC (2003) Conservation ecology of bats. In: Kunz TH, Fenton MB (eds) Bat ecology. The University of Chicago Press, Chicago, pp 680–743
Racey PA, Swift SM (1981) Variation in gestation length in a colony of pipistrelle bats (*Pipistrellus pipistrellus*) from year to year. J Reprod Fertil 61:123–129
Rancourt SJ, Rule MI, O'Connell MA (2005) Maternity roost site selection of long-eared myotis, *Myotis evotis*. J Mammal 86:77–84
Runkle JR (1990) Gap dynamics in an Ohio *Acer-Fagus* forest and speculations on the geography of disturbance. Can J For Res 20:632–641
Russell AL, Butchkoski CM, Saidak L, McCracken GF (2009) Road-killed bats, highway design, and the commuting ecology of bats. End Spec Res 8:49–60
Russo D, Cistrone L, Jones G (2007) Emergence time in forest bats: the influence of canopy closure. Acta Oecol 31:119–126
Rydell J, Entwistle AC, Racey PA (1996) Timing of foraging flights of three species of bats in relation to insect activity and predation risk. Oikos 76:243–252
Saugey DA, Crump BG, Vaughn RL, Heidt GA (1998) Notes on the natural history of *Lasiurus borealis* in Arkansas. J Ark Acad Sci 52:92–98
Schirmacher MR, Castleberry SB, Ford WM, Miller KV (2007) Habitat associations of bats in south-central West Virginia. Proc Annu Conf Southeast Assoc Fish Wildl Agen 61:46–52
Sparks DW, Ritzi CM, Duchamp JE, Whitaker JO Jr (2005) Foraging habitat of Indiana myotis (*Myotis sodalis*) at an urban/rural interface. J Mammal 86:713–718
Swengel AB (2001) A literature review of insect responses to fire, compared to other conservation managements of open habitat. Biodivers Conserv 10:1141–1169
Swift LW Jr, Burns RG (1999) The three Rs of roads. J For 97:40–44
Swystun MB, Psyllakis JM, Brigham RM (2001) The influence of residual tree patch isolation on habitat use by bats in central British Columbia. Acta Chiropt 3:197–201
Szymanski JA, Runge MC, Parkin MJ, Armstrong M (2009) White-nose syndrome management: report on structured decision making initiative. USDI Fish and Wildlife Service, Fort Snelling
Thomas DW, Dorais M, Bergeron J-M (1990) Winter energy budgets and cost of arousals for hibernating little brown bats, *Myotis lucifugus*. J Mammal 71:475–479
Tibbels AE, Kurta A (2003) Bat activity is low in thinned and unthinned stands of red pine. Can J For Res 33:2436–2442
USDI Fish and Wildlife Service (1984) A recovery plan for the Ozark big-eared bat and Virginia big-eared bat. USDI Fish and Wildlife Service, Fort Snelling, 119 p
USDI Fish and Wildlife Service (2007) Indiana bat (*Myotis sodalis*) draft recovery plan: first revision. USDI Fish and Wildlife Service, Fort Snelling, 258 p
Veilleux JP, Whitaker JO Jr, Veilleux SL (2003) Tree-roosting ecology of reproductive female eastern pipistrelles, *Pipistrellus subflavus*, in Indiana. J Mammal 84:1068–1075
Veilleux JP, Whitaker JO Jr, Veilleux SL (2004) Reproductive stage influences roost use by tree roosting female eastern pipistrelles, *Pipistrellus subflavus*. Ecoscience 11:249–256
Vepakomma U, St-Onge B, Kneeshaw D (2008) Spatially explicit characterization of boreal forest gap dynamics using multi-temporal LIDAR data. Remote Sens Environ 112:2326–2340

Verboom B, Spolestra K (1999) Effects of food abundance and wind on the use of tree lines by an insectivorous bat, *Pipistrellus pipistrellus*. Can J Zool 77:1393–1401

Verboom B, Boonman AM, Limpens HJGA (1999) Acoustic perception of landscape elements by the pond bat (*Myotis dasycneme*). J Zool 248:59–66

Vonhof MJ, Barclay RM (1997) Use of tree stumps as roosts by the western long-eared bat. J Wildl Manage 61:674–684

Walsh AL, Harris S (1996) Foraging habitat preferences of verspertilionid bats in Britain. J Appl Ecol 33:508–518

Wear DN, Greis JG (eds) (2002) Southern forest resource assessment. Gen Tech Rep SRS-53, USDA Forest Service Southern Research Station, Asheville

Weller TJ (2007) Assessing population status of bats in forests: challenges and opportunities. In: Lacki MJ, Hayes JP, Kurta A (eds) Bats in forests: conservation and management. Johns Hopkins University Press, Baltimore, pp 264–291

Whitaker JO Jr, Hamilton WJ Jr (1998) Mammals of the eastern United States. Cornell University Press, Ithaca

Whitaker DM, Carroll AL, Montevecchi WA (2000) Elevated numbers of flying insects and insectivorous birds in riparian buffer strips. Can J Zool 75:740–747

Willis CKR (2006) Daily heterothermy by temperate bats using natural roosts. In: Zubaid A, McCracken GF, Kunz TH (eds) Functional and evolutionary ecology of bats. Oxford University Press, New York, pp 38–55

Willis CKR, Brigham RM (2005) Physiological and ecological aspects of roost selection by reproductive female hoary bats (*Lasiurus cinereus*). J Mammal 86:85–94

Yates MD, Muzika RM (2006) Effect of forest structure and fragmentation on site occupancy of bat species in Missouri Ozark forests. J Wildl Manage 70:1238–1248

Zimmerman GS, Glanz WE (2000) Habitat use by bats in eastern Maine. J Wildl Manage 64:1032–1040

Chapter 11
Reptile and Amphibian Response to Hardwood Forest Management and Early Successional Habitats

Christopher E. Moorman, Kevin R. Russell, and Cathryn H. Greenberg

Abstract Herpetofauna responses to forest management and early successional habitats are influenced by species-specific adaptations to historical disturbance regimes. It can take decades for woodland salamander diversity to recover after heavy overstory removal for even-aged forest regeneration or hot fires that yield higher light, drier microclimates, and reduced leaf litter cover, but some frog and toad species may tolerate or even increase after disturbances. In particular, disturbances that retain some canopy cover, such as selection harvests or low intensity burns, can mitigate effects on terrestrial salamanders. The same early successional conditions that are detrimental to salamanders can benefit many reptile species, such as fence lizards (*Sceloporus undulatus*). Maintaining stand age diversity across central hardwood forest landscapes, including retention of mature forest communities, should provide habitats for both early successional wildlife and mature forest species.

C.E. Moorman (✉)
Department of Forestry and Environmental Resources,
North Carolina State University, Raleigh, NC 27695, USA
e-mail: chris_moorman@ncsu.edu

K.R. Russell
College of Natural Resources, University of Wisconsin, Stevens Point, WI 54481, USA
e-mail: krussell@uwsp.edu

C.H. Greenberg
USDA Forest Service, Southern Research Station, Bent Creek Experimental Forest,
1577 Brevard Rd, Asheville, NC 28806, USA
e-mail: kgreenberg@fs.fed.us

C.H. Greenberg et al. (eds.), *Sustaining Young Forest Communities*,
Managing Forest Ecosystems 21, DOI 10.1007/978-94-007-1620-9_11,

11.1 Relevant Environmental Changes Following Disturbance

The extent and frequency of historical disturbances in central hardwood forests varied widely depending on slope position, aspect, stand age, and stand composition (White et al., Chap. 3). Gap phase disturbances following wind events, ice storms, and insect outbreaks were more common than the large-scale changes that followed hurricanes and wildfires in other regions of North America (White et al., Chap. 3). Amphibian and reptile species associated with mature hardwood forest presumably were common across much of the landscape, whereas those associated with early succession habitats were much more variable because they depended upon infrequent natural disturbance to create ephemeral patches of suitable habitat (Greenberg 2001).

Natural and anthropogenic disturbances that create young forest by removing or reducing canopy cover can greatly alter the microclimate at or just below the soil surface, where most amphibian and reptile species reside (but see Brooks and Kyker-Snowman 2008). Following overstory removal, light penetration increases, raising soil temperatures and evaporation rates and decreasing litter depth and moisture until it is replenished by leaf-fall and shade from the recovering vegetation (Greenberg et al., Chap. 8). Fire also can consume leaf litter and reduce leaf-fall input levels (Petranka et al. 1994). Increased light levels near the ground promote development of a grass and forb layer and establishment of shrubs or regenerating trees (Russell et al. 2004). These environmental changes can alter herpetofaunal movement patterns, survival rates, and prey abundance (Moseley et al. 2004).

Down wood or coarse woody debris (CWD) is used by many reptile and amphibian species for mating sites, egg-laying, feeding, and thermoregulation (Whiles and Grubaugh 1996). Down wood volume typically follows a U-shaped chrono-sequence in central hardwood forests, with highest levels in the 5–10 years following disturbance (i.e., downed trees following windthrow or logging slash following timber harvest) and again during late-succession or old growth stages when aging trees senesce (Gore and Patterson 1986). However, larger, more decayed logs may be more abundant in mature or old growth hardwood forest (Petranka et al. 1994). Webster and Jenkens (2005) reported that primary forests in the Southern Appalachians contained more large-diameter, highly decayed CWD compared to forests subjected to anthropogenic disturbances. Furthermore, among sites with similar disturbance histories, higher levels of CWD were associated with mesic conditions and higher elevations (Webster and Jenkens 2005; Keyser, Chap. 15). Therefore, reptile and amphibian species that use down wood heavily may be most abundant early (e.g., some reptiles) or late (e.g., salamanders) in stand development. However, the degree to which salamanders and other amphibians specifically rely on CWD is likely influenced by the availability of other surface cover. For example, salamanders may use cover objects less in undisturbed stands with intact leaf litter and vegetation cover compared to stands where leaf litter and vegetative cover is reduced from prescribed burning and herbivory (Ford et al. 2010).

Machinery associated with timber harvest operations can cause soil compaction or erosion. Disturbances of the subterranean environment, as occurs with most types of mechanical site preparation, can cause direct mortality or degrade habitat conditions for fossorial snakes and other species that spend portions of their life cycle below ground (Russell et al. 2004; Todd and Andrews 2008). However, mechanical site preparation and other forms of intensive forest management are uncommon in the Central Hardwood Region as compared to other regions such as the southeastern Coastal Plain (e.g., Russell et al. 2002).

Amphibians and reptiles often are generically lumped together as "herpetofauna," but in fact are as phylogenetically distinct from one another as are mammals and birds. Amphibians (class *Amphibia*) have permeable, moist skin that is used for respiration and increases vulnerability to desiccation. Amphibians have a two-stage or "biphasic" life cycle that includes morphologically distinct larval and adult stages. Most require water for egg deposition and development of larvae, which eventually metamorphose into adults that can be largely terrestrial (Duellman and Trueb 1986). Amphibian taxa vary considerably in their vulnerability to desiccation. For example, some frogs and toads can tolerate higher temperatures (Stebbins and Cohen 1995) and can store and reabsorb larger amounts of water in their bladders than salamanders (Zug 1993). Some salamanders are lungless, and some are completely terrestrial (deMaynadier and Hunter 1995). Many amphibian species have small home ranges (Duellman and Trueb 1986) and poor dispersal capabilities (Sinsch 1990). Conversely, most reptiles (class *Reptilia*) require warm temperatures (associated with higher light levels) for egg incubation and successful development of hatchlings (Deeming and Ferguson 1991). Reptiles have dry scaly skin that protects them from desiccation. Clearly, response to disturbance and early successional habitats might be expected to differ between the two taxonomic classes, and among species within them. Within *Amphibia*, salamanders tend to decline following disturbances that reduce canopy cover because of their increased risk of desiccation, whereas some toad and frog species may tolerate higher temperatures and lower moisture in early successional habitats (Russell et al. 2004). Many reptile species increase in recently disturbed areas, likely because of improved opportunity for thermoregulation and foraging (Russell et al. 2004).

11.2 Amphibian and Reptile Response to Timber Harvest

11.2.1 Amphibian Response

Heavy overstory removal for forest regeneration treatments (e.g., clearcut or shelterwood regeneration harvests) can adversely affect amphibians, especially terrestrial salamanders (Pough et al. 1987; Petranka et al. 1993, 1994; deMaynadier and Hunter 1995; Ash 1997; Harpole and Haas 1999; Reichenbach and Sattler 2007). Canopy removal results in higher light levels, a warmer, drier microclimate, and reduced leaf litter cover, which could cause salamanders to desiccate

Table 11.1 Estimated recovery periods for terrestrial plethodontid salamander populations following timber harvest

Authors	Recovery period	Disturbance	Comments
Ash 1997	20–24 years	Clearcut	Monitored salamanders in 3 clearcuts using night searches on 225-m^2 plots for 15 years post-harvest and recovery times estimated from regression curves
Harper and Guynn 1999	13–39 years	Clearcut	Used a terrestrial vacuum to sample leaf litter and associated fauna in 120, 0.04-ha plots in 3 stand age classes (0–12, 23–39, and ≥ 40 years old)
Pough et al. 1987	<60 years	Clearcut	Conducted nighttime surveys for salamanders in 50- × 2-m transects in 4 disturbed stand types of different ages and in 4 paired old-growth sites
Homyack and Haas 2009	>60 years	Various Harvests	Conducted nighttime searches of 15- × 2-m transects for 13 years following 7 canopy removal treatments and estimated population recovery from demographic models
Petranka et al. 1993	50–70 years	Clearcut	Surveyed salamanders in 50- × 50-m plots at 47 sites ranging in age from 2 to 120 years old
Herbeck and Larsen 1999	>80 years	Regeneration cut	Conducted area- and time-constrained searches for salamanders in 21 144-m^2 plots located in 3 age classes (<5, 70–80, >120 years old)
Ford et al. 2002a	>85 years	Clearcut	Captured salamanders in drift fence arrays in 13 cove hardwood stands ranging in age from 15 years old to >85 years old
Petranka et al. 1994	120 years	Clearcut	Conducted daytime searches for salamanders in 50 × 50-m plots at 52 forest sites ranging from <5 years old to approximately 200 years old

(deMaynadier and Hunter 1995; Renken 2006). In the Southern Appalachians, terrestrial salamander abundance declines following clearcutting (Ash 1988, 1997; Petranka et al. 1993, 1994; but see Adams et al. 1996).

There has been considerable debate about the time that it takes salamander populations to recover to pre-disturbance levels following canopy removal (Ash and Pollock 1999; Petranka 1999). Estimates range from approximately 20 years to more than 100 years (Table 11.1). Discrepancies in documented recovery periods likely are related to differences in study designs, salamander communities, and site and landscape characteristics. But, research suggests that post-disturbance recovery of salamander abundance is closely correlated with litter layer recovery (Pough et al. 1987; Ash 1997; Crawford and Semlitsch 2008a). Longer recovery periods may be required on drier aspects and ridge tops than on mesic sites where soil moisture remains relatively high even after disturbance (Harper and Guynn 1999;

Petranka 1999). However, the former sites generally are poorer sites for woodland salamanders. Disturbances that retain heavy canopy cover such as midstory removal, selection harvest, firewood cutting, thinning, and heavy browsing by white-tailed deer (*Odocoileus virginianus*) are less likely to affect salamander abundance (Pough et al. 1987; Adams et al. 1996, Messere and Ducey 1998; Brooks 1999; Ford et al. 2000; Harpole and Haas 1999; Moseley et al. 2003; Knapp et al. 2003; Homyack and Haas 2009; Semlitsch et al. 2009). Yet, salamander density may decline following partial canopy reduction (e.g., Duguay and Wood 2002), and reductions in canopy cover by as little as 41% can cause local declines in salamander abundance (Knapp et al. 2003).

The exact mechanisms for the disappearance of terrestrial salamanders from disturbed sites remain in question. Semlitsch et al. (2008) proposed three hypotheses to explain amphibian declines following timber harvest: (1) retreat to underground refugia; (2) mortality from desiccation or starvation; and (3) evacuation to adjacent forest. Although a percentage of pond-breeding Ambystomatid salamanders may disperse out of disturbed environments, it is not known how they fare once they reach adjacent forest (Semlitsch et al. 2008). Mortality is the most likely cause of declines in terrestrial salamander density following clearcutting because plethodontid salamanders primarily are surface feeders and individuals eventually would starve unless they came to the surface where they could desiccate. Adult plethodontid salamanders lack lungs and depend on cutaneous respiration for gas exchange. Because moist skin is necessary to facilitate respiration, salamanders are most active where the forest floor is moist or at night when relative humidity is highest (Petranka et al. 1993). Salamander desiccation results from reduced leaf litter cover and depth, and higher ground temperatures following clearcutting, rather than changes in soil moisture (Pough et al. 1987; Ash 1997; Rothermel and Luhring 2005). Rothermel and Luhring (2005) showed that salamander survival was 100% in uncut forest, but individuals could survive in clearcuts only by gaining access to protective underground burrows. Some researchers have speculated that salamanders are unlikely to evacuate to adjacent forested areas that already are saturated with territorial adults (e.g., Petranka 1999). For example, Bartman et al. (2001) did not detect any post-harvest emigration of plethodontid salamanders from sites that had been subjected to shelterwood harvests in western North Carolina. Interestingly, Ash (1997) speculated that adult salamanders disperse into early successional habitats such as clearcuts to avoid competition from smaller or immature salamanders that are restricted to mature forests with abundant, moist litter.

Juvenile frogs and salamanders typically exhibit higher rates of mortality than adults following canopy removal because their high surface:volume ratios make them prone to desiccation (Jaeger 1980; Ash et al. 2003; Marsh and Goicochea 2003). Additionally, the high adult:juvenile ratio of salamanders in clearcuts indicates low reproduction by adults or higher rates of mortality in juveniles (Ash 1997; Ash et al. 2003). Adults of some salamander species are better adapted to withstand the hot, dry conditions of recently disturbed sites or more exposed ridge top environments (Ash 1997; Ash et al. 2003; Ford et al. 2010). For example, Ford et al. (2010) reported that larger-bodied slimy salamanders (*Plethodon glutinosis*) were less affected by leaf litter reduction following fire than smaller-bodied red-backed salamanders (*P. cinereus*) or mountain dusky salamanders (*Desmognathus ocropheaus*). Riedel et al. (2008) documented high densities of both adult and juvenile eastern red-backed salamanders within former deciduous forests of West Virginia that had been converted to silvopastures, traditional pastures, and ungrazed meadows, indicating that this species may be more resilient to the creation of early successional habitats than previously thought. Interestingly, the physiological condition and sex ratios of salamanders within these open, early successional habitats were similar to those of salamanders in adjacent mature forest, although adults were significantly more abundant than juveniles (Riedel 2006). Riedel et al. (2008) suggested that the presence of artificial cover in these open, early successional habitats, in combination with moisture trapped by dense herbaceous vegetation, facilitated woodland salamander persistence. In addition, Marsh et al. (2004) showed that dispersal of *P. cinereus* was not limited by the presence of forest cover, and suggested that this species may be relatively insensitive to the creation of small, intervening, open habitats within deciduous forests such as fields, power line corridors, and even small residential areas. Accordingly, at least some species of woodland salamanders may tolerate the creation of small patches of early successional habitats within mature deciduous forests (Marsh et al. 2004; Riedel et al. 2008; Moseley et al. 2009), yet others can be highly sensitive to forest road edges (Semlitsch et al. 2007). However, individuals forced to forage in areas with reduced cover may be more exposed to predation (Moseley et al. 2004).

Timber harvest can affect stream-breeding salamanders by eliminating terrestrial habitat for adults and by degrading aquatic habitats required for larval development (Perkins and Hunter 2006; Crawford and Semlitsch 2008a; Peterman and Semlitsch 2009). Adult stream-breeding salamanders (e.g., *Desmognathus* and *Eurycea*) use terrestrial habitats some distance away from streams for foraging and overwintering habitat (Ashton and Ashton 1978; Crawford and Semlitsch 2007). Similar to terrestrial salamanders, adult stream salamander (e.g., Blue Ridge two-lined salamander [*E. wilderae*]) abundance may be reduced following timber harvest because of decrease in leaf litter depth, soil moisture, and overstory cover (Crawford and Semlitsch 2008a, b). Increased water temperatures and reduced litter input following canopy removal and sedimentation from logging roads (Vose and Ford, Chap. 14) are detrimental to larval salamanders that occur in the streams (Semlitsch 2000; Peterman and Semlitsch 2009). Stream sedimentation can fill interstitial spaces between rocks at the stream bottom, thus potentially reducing abundance of salamanders that use the spaces for cover, such as *Eurycea* and *Desmognathus* species (Lowe and Bolger 2002; Miller et al. 2007; Moseley et al. 2008; Peterman and Semlitsch 2009). However, retention of an uncut riparian buffer may mitigate the effects of clearcut harvests on larval salamanders (Peterman and Semlitsch 2009).

Frogs and toads tend to be more tolerant of canopy removal and elevated ground temperatures than salamanders (Gibbs 1998; Ross et al. 2000; Russell et al. 2004; Patrick et al. 2006). Additionally, tadpoles of some frog species may develop faster or survive better in ponds within clearcuts (Semlitsch et al. 2009; Felix et al. 2010). Some anuran species likely are attracted to the higher coverage of herbaceous vegetation around ponds in open environments (Felix et al. 2010). Response to canopy removal around breeding ponds differs among anuran species. Species associated with open habitats, such as gray treefrogs (*Hyla versicolor*), deposit more eggs in ponds in areas with heavy canopy removal. In contrast, species that require cooler water temperatures, such as mountain chorus frogs (*Pseudacris brachyphona*) and spotted salamanders (*Ambystoma maculatum*), only deposit eggs where at least 75% of the canopy is retained (Semlitsch et al. 2009; Felix et al. 2010). However, gray treefrogs oviposited more eggs in ponds in clearcuts close to forest edge than in ponds 50 m into clearcuts (Hocking and Semlitsch 2007), because adult treefrogs require mature trees for foraging (Johnson et al. 2007, 2008). Adult wood frogs (*Rana sylvatica*) were able to travel through clearcuts when dispersing between breeding ponds and non-breeding habitats in mature forest, but their rate of travel increased in response to the degraded micro-climatic conditions (Rittenhouse and Semlitsch 2009). Some anurans, especially juveniles, may experience increased predation or desiccation risks following timber harvests (Patrick et al. 2006; Rittenhouse and Semlitsch 2009; Rittenhouse et al. 2009). Species response to the creation of young forest may vary regionally. For example, adult wood frogs did not use hot, dry clearcuts in Missouri but did use moist areas within clearcuts as non-breeding habitat in Maine (Patrick et al. 2006; Rittenhouse and Semlitsch 2009).

11.2.2 Reptile Response

The same conditions following timber harvest that may be detrimental to amphibians appear to benefit many reptiles (Greenberg 2002; Adams et al. 1996). Most reptile species require the warm temperatures associated with higher light levels for egg incubation and successful development of hatchlings (Goin and Goin 1971; Deeming and Ferguson 1991). The hotter, drier microclimate in open, disturbed sites also may facilitate movement and thermoregulation for many reptile species (Greenberg 2001). Lizards, particularly fence lizards (*Sceloporus undulatus*), generally increase following canopy reduction (McLeod and Gates 1998; Greenberg 2001; Renken et al. 2004). Following timber harvests, Renken et al. (2004) determined that juvenile abundance of *S. undulatus* was twice as high as that of adults, suggesting that the lizards experienced an immediate boost in reproductive rates in disturbed sites or that the recently disturbed sites were colonized primarily by juveniles. In predominantly forested landscapes in Pennsylvania, snake abundance and richness increased with decreasing tree basal area (Ross et al. 2000).

However, there is evidence that some forest-dwelling reptile species may decline following timber harvest (Russell et al. 2004). In Coastal Plain pine forests, abundance of several small-bodied leaf litter snake species was lower in clearcuts than unharvested and thinned pine stands, but snake abundance was highest in thinned stands where habitat heterogeneity and presumably prey abundance was highest (Todd and Andrews 2008). In contrast with the management of deciduous forests, the intensive mechanical site preparation associated with Coastal Plain pine management not only removes surface cover used by small-bodied snakes but also likely results in direct destruction of nest sites (Russell et al. 2002).

11.3 Response to Prescribed Fire

Over the past 500 years, fire was a common forest disturbance across much of the Central Hardwood Region (Spetich et al., Chap. 4). Fire effects on vegetation structure likely varied with fire intensity and frequency, which in turn was influenced by topography, weather conditions, and population distribution of Native Americans or European settlers who intentionally burned to promote forage for game or livestock (Spetich et al., Chap. 4). Hot fires certainly reduced leaf litter and often killed overstory trees, creating patchy, heterogeneous early successional conditions with some snags and trees remaining. In contrast, cool, patchy burns likely had minimal impact on overstory trees or leaf litter depth and cover, but reduced shrub cover or killed midstory trees where it burned. In ecosystems such as longleaf pine-wiregrass or sand pine-scrub where lightning-ignited fires created and maintained "fire climax" habitat conditions, many species of reptiles and amphibians are behaviorally adapted to survive wildfire or prescribed burns, and require fire maintained habitat conditions (Russell et al. 1999; Greenberg 2002). Less is known about fire effects on herpetofauna of upland hardwood forest, where the majority of fires were historically human-caused. Fire is thought to have little direct effect on amphibians and reptiles, but the likelihood of individual mortality during a fire depends on the species' behavior, fire intensity, and season of burn (Russell et al. 1999). Negative indirect impacts of prescribed fire likely are most severe for species that require leaf litter or other forest debris that is consumed (Russell et al. 1999).

Relatively few studies have addressed fire effects on herpetofauna in hardwood forests (Russell et al. 2004; Renken 2006). Several studies have reported no difference between amphibian populations on prescribed burned sites and unburned controls (Ford et al. 1999; Floyd et al. 2001; Moseley et al. 2003; Keyser et al. 2004; Greenberg and Waldrop 2008; Ford et al. 2010; Matthews et al. 2010). Others have indicated that toad abundance may increase following fire (Kirkland et al. 1996; Greenberg and Waldrop 2008). Conversely, intense prescribed fires that cause immediate or delayed reduction in canopy cover following overstory tree mortality can produce micro-habitat changes near the forest floor (e.g., reduced leaf litter cover and depth, more sunlight, higher ground temperatures) that negatively impact salamander populations (Matthews et al. 2010).

Reptiles, lizards in particular, may increase after prescribed burns, especially after hot fires that reduce canopy cover (Moseley et al. 2003; Keyser et al. 2004; Greenberg and Waldrop 2008; Matthews et al. 2010). Litter removal, midstory and canopy reduction, and higher ground temperatures following intense fires likely create thermoregulatory conditions favorable for lizards (Moseley et al. 2003). Overstory mortality following intense fires also generates down wood that may be used as basking sites by lizards and large-bodied snakes (Matthews et al. 2010). However, it is not known whether these same changes negatively affect small-bodied fossorial snakes that depend on leaf litter.

Prescribed fire effects on wetland and stream-associated amphibians in central hardwood forests have not been well studied (Renken 2006). Intense fires that kill trees and reduce canopy cover in the uplands adjacent to streams or amphibian breeding ponds could result in higher water temperatures, increased sedimentation rates, or runoff of ash that changes water pH, potentially killing amphibian adults, eggs, or larvae (Renken 2006). However, other temperature and sediment-sensitive aquatic vertebrates in the Appalachians, such as brook trout (*Salvelinus fontinalis*), have been reported to respond positively to adjacent forest disturbances, presumably in response to greater abundance of macroinvertebrate prey after partial canopy removal (Nislow and Lowe 2006). In short, more research is needed on the effects of fire and other forest disturbances on aquatic and riparian reptiles and amphibians in central hardwood forests.

11.4 Mitigation Strategies

11.4.1 Stream and Wetland Buffers

Riparian buffers between upland timber harvests and adjacent streams or wetlands have been recommended to mitigate impacts on sensitive amphibian species (Semlitsch 2000). Buffers shade water, contribute leaf litter to streams, filter sediment, provide terrestrial habitats for biphasic amphibians and reptiles, and possibly provide refuge for individuals dispersing out of harvested areas (Mitchell et al. 1997; Semlitsch 2000; Perkins and Hunter 2006). Crawford and Semlitsch (2007) recommended a 92-m buffer adjacent to Southern Appalachian streams to provide core habitat free of edge effects for the widest ranging stream salamander species. Effects of timber harvest on sensitive amphibian species may extend at least 25 m into adjacent mature forest, possibly because of the reduced canopy and litter cover along edges created by timber harvests (deMaynadier and Hunter 1998). To provide the core biphasic habitat needs, Semlitsch and Bodie (2003) recommended 159–290 m buffers for amphibians and 127–289 m buffers for reptiles around wetlands and streams. However, it has been speculated that narrower 30-m buffers may provide adequate protection to larval salamanders (Peterman and Semlitsch 2009). Alternatively, a two-tiered approach has been recommended to protect aquatic herpetofauna, with unharvested 10–25 m buffers around streams surrounded by a wider partial harvest zone (deMaynadier and Hunter 1995). To date, however, the actual community and demographic responses of stream-dwelling herpetofauna to adjacent forest disturbance remain poorly characterized. Therefore, few data are available to evaluate the efficacy of specific buffer widths recommended to protect herpetofauna within deciduous forests of the Central Hardwood Region.

11.4.2 Coarse Woody Debris Retention

Salamander populations are positively linked to CWD abundance, especially on drier sites and where leaf litter cover is sparse, so retention of CWD may help mitigate the effects of disturbance on amphibians and provide critical habitat or refuge to a number of reptile species (Pough et al. 1987; Petranka et al. 1994; Brooks 1999; Herbeck and Larsen 1999; Russell et al. 2004). Retention of CWD and brush piles in clearcuts may decrease the proportion of salamanders leaving clearcuts and could contribute to increased juvenile amphibian survival by providing cool, moist refugia (Patrick et al. 2006; Rittenhouse et al. 2008; Semlitsch et al. 2008). Todd and Andrews (2008) captured more small snakes in clearcuts with CWD retention than in clearcuts without retention. However, CWD retention appears to provide only short-term benefits to sensitive amphibians by providing refuge from desiccating conditions immediately post-harvest, and may not prevent declines (Moseley et al. 2004; Semlitsch et al. 2009). Coarse woody debris diameter and degree of decay is generally much lower, and thus not used by salamanders, in recently harvested sites than in old growth stands (Herbeck and Larsen 1999). Additionally, several studies failed to show benefits of CWD retention for amphibians (Greenberg 2001; Ford et al. 2002a; Rothermel and Luhring 2005; Rittenhouse and Semlitsch 2009). Similarly, higher abundance of lizards and snakes in small canopy gaps was not related to CWD abundance (Greenberg 2001).

11.4.3 Overstory Retention

Small forest openings such as group selection harvests and wind-created downburst gaps with multiple treefalls, or partial harvests that retain a large percentage of the overstory, can mitigate the negative effects of timber harvest on amphibians by maintaining shade and leaf litter input and providing refuge and recolonization sources (Pough et al. 1987; Ford et al. 2000; Greenberg 2001; Lowe and Bolger 2002; Homyack and Haas 2009). Overstory retention adjacent to wetlands can be critical to maintaining connectivity between aquatic reproduction sites and other habitat features required by amphibians, as many, especially salamanders, avoid timber harvests when emigrating from breeding pools (Todd et al. 2009). In Maine, partial harvests adjacent to headwater streams had less effect on amphibian communities than clearcuts (Perkins and Hunter 2006). Increased growth of herbaceous plants or shrubs near the forest floor following small overstory reductions might improve habitat conditions for some herpetofaunal species and mitigate changes to the microclimate that are problematic for disturbance-sensitive species such as salamanders (Ross et al. 2000; Semlitsch et al. 2009). Retention of at least 50% of the overstory is recommended to minimize negative effects on amphibian populations (Ross et al. 2000; Semlitsch et al. 2009). However, as little as 41% reduction in the overstory may result in declines in the abundance of plethodontid woodland salamanders similar to clearcuts (Knapp et al. 2003). Group-selection harvests require more frequent stand entries across a larger land base to extract the same amount of wood fiber as a clearcut (Homyack and Haas 2009). We suggest that the relationships between partial overstory reduction and response by amphibian populations require more study.

11.4.4 Small Stand Sizes and Longer Rotations

Smaller harvest units may help to minimize the deleterious effects of timber harvest on wood frogs and other sensitive amphibians, especially juveniles (Patrick et al. 2006; Rittenhouse and Semlitsch 2009). The distance that dispersing individuals must traverse across smaller clearcuts could lessen the risks of desiccation and predation. Additionally, small timber harvests may facilitate evacuation by individuals from harvested areas into adjacent uncut areas (Semlitsch et al. 2008). Ford et al. (2002a) demonstrated that the amount of cove hardwood habitat surrounding harvested patches is an important determinant of woodland salamander population response to the disturbance, so designated no-harvest areas on the landscape could serve as sources for repopulating nearby harvest units. Additionally, breeding pools in small timber harvest openings could provide ideal locations for rapid larval development for larvae of some disturbance-adapted or early successional amphibians and be in close proximity to the mature forest required by adults (Barry et al. 2008; Semlitsch et al. 2009). Further, small harvest openings (<2 ha) provide habitat for lizards and other reptiles (Greenberg 2001). Similar to group-selection harvests, however, harvest of the same timber volume in smaller units requires more roads, potentially leading to sediment loading in streams and disturbance to a larger percentage of the land base.

Increasing the rotation length of managed forest stands would ensure that a portion of the landscape contained large trees, high accumulations of large diameter CWD, and other structural characteristics associated with late-seral forest (Herbeck and Larsen 1999). Alternatively, employing forest management practices that retain and enhance structural components of habitats important for herpetofauna (e.g., retention of CWD, green and legacy tree retention, selection harvest systems) may provide suitable conditions for these species while contributing to economic and other resource objectives. Additionally, management practices that mimic historical disturbance regimes may be used to promote a diversity of cover types across the landscape, which in turn would provide habitat for a variety of reptiles and amphibians. Examples of historical disturbance conditions include more frequent prescribed fires on xeric ridge tops in the Southern Appalachians and less disturbance on moist, north-facing slopes and ravines.

11.5 Research Challenges

More focus on reptile response to disturbance. Reptile response to disturbance from forest management has been studied much less than amphibian response. For example, a database search of journal articles using the keywords *salamander* and *clearcut* generated 64 citations; conversely, a search using the same database with the keyword *lizard* in place of *salamander* generated three citations and replacement of *salamander* with *snake* generated one citation. We can only speculate that the cause for the discrepancy is due in large part to the direction of response by amphibians and reptiles in previous studies. Because amphibians, especially woodland plethodontid salamanders, typically decline locally following disturbance, they have received the majority of research emphasis in the past two decades. However, some reptile species such as small fossorial snakes similarly show negative response; other reptile species, such as fence lizards increase in abundance following disturbance. We suggest that there may be a bias in the scientific literature attributable to a greater attraction by scientists to studying taxa that respond negatively to forest management, and journals to accept manuscripts that report significant results.

Longer study durations. Deleterious effects of canopy reduction on salamanders and other amphibians may be delayed for up to 5 years after timber harvest (Ash 1988; Reichenbach and Sattler 2007; Homyack and Haas 2009). Some species may experience a greater time lag in the demographic changes that occur following disturbance (Homyack and Haas 2009). Greenberg and Waldrop (2008) reported that a single prescribed burn that killed trees and reduced canopy cover did not reduce the relative abundance of terrestrial salamanders (*Plethodon* spp.), but salamander abundance was lower in the same treatment units compared to control plots after a second burn 5 years later in the same study area (Matthews et al. 2010). The delayed changes in salamander abundance following the fuel reduction treatments could either have been a result of additive effects of the treatments on environmental conditions, or the result of delayed changes in demographic parameters (Matthews et al. 2010). Lastly, long-term studies also should address the effects of forest management on population demography at large spatial scales (Homyack and Haas 2009).

More accurate assessment of detection bias. Most reptile and amphibian studies assume that sampled individuals represent the entire population (deMaynadier and Hunter 1995). This assumption is unlikely for salamanders because surface populations represent only a small percentage of the total population (Bailey et al. 2004a). Additionally, detection probabilities often differ among treatment areas because of variable habitat conditions, which in turn could influence abundance estimates for reptile and amphibian populations (Bailey et al. 2004b). For example, reduction of leaf litter from prescribed fire or timber harvest could cause individual salamanders to move more frequently and for longer periods (Moseley et al. 2004), or cause them to aggregate under coverboards being used to assess population response to burning or other disturbances (Ford et al. 2010). Few studies of reptile and amphibian response to forest management have accounted for detection bias (except see Bailey et al. 2004b; Ford et al. 2010). Mark-recapture methodologies can be used to account for detection probability, but recapture rates, especially with terrestrial salamanders, can be low and capture-recapture methods can be costly when used in large-scale field experiments (Bailey et al. 2004a). In the case of large-scale studies, researchers can use a double-sampling design that uses capture-recapture analysis on a subset of sites to estimate detection probability and calibrate counts for the complete set of sampling locations (Bailey et al. 2004a, c).

More focus on site conditions, landscape position, and abiotic features. There is evidence that elevation, slope, concavity, and other landform characteristics may be important determinants of woodland salamander occurrence and abundance in central hardwood forests (Ford et al. 2002a, b). Many studies of amphibian response to forest management do not account for landscape position and associated conditions such as moist, concave, lower-slope positions with a thicker leaf litter layer and drier, warmer ridge tops or south-facing slopes that could influence amphibian or reptile species composition and their response to disturbance. When compared to other vertebrates, patterns of amphibian distribution across landscape scales remain poorly known (Johnson et al. 2002; Dillard et al. 2008a). Because amphibians have limited dispersal abilities and small home ranges, site-specific habitat factors often are assumed to have an overriding influence on patterns of amphibian distribution. However, there is increasing evidence that abiotic habitat characteristics measured at broad spatial scales are important predictors of amphibian occurrence and abundance within forest ecosystems. Although disturbance and succession of vegetation exert a strong influence on amphibian distribution and abundance (deMaynadier and Hunter 1995, Russell et al. 2004), recent research indicates that the importance of abiotic habitat features such as geology, topography, and climate have not been sufficiently recognized (Russell et al. 2005, Harper 2007, Dillard et al. 2008a, b). For example, Dillard et al. (2008a, b) showed that elevation, slope, aspect, and parent geology were better predictors of the occurrence of the threatened Cheat Mountain salamander (*P. nettingi*) in deciduous forests of West Virginia than were the composition or successional stage of overstory vegetation. Moseley et al. (2009) determined that the effects of canopy openings (e.g., edge effects) on woodland salamanders within deciduous forests of West Virginia depended on site aspect.

Landscape-level population effects. Most studies of amphibian and reptile response to forest management have been conducted at the scale of an individual stream, forest stand, or wetland. Therefore, more research is needed to assess the persistence of reptile and amphibian communities at the landscape or watershed scale (Perkins and Hunter 2006). Renken et al. (2004) recorded similar responses by reptiles and amphibians to clearcuts as in other studies, but the researchers failed to detect larger-scale impacts given the relatively small percentage of the landscape that was harvested. Ford et al. (2002a) suggested that salamander populations in small, isolated cove hardwood stands might be more vulnerable to extirpation by timber harvests than populations in larger, less isolated coves. Because juvenile amphibians are more susceptible to habitat change, management activities that fragment habitats likely will have the greatest impact on species for which juveniles conduct the majority of dispersal among breeding and non-breeding locations (Patrick et al. 2008). Some amphibian species avoid roads likely because of reduced soil moisture and cover, so landscape-level conservation strategies should account for these increasingly prominent movement barriers (Gibbs 1998; Marsh and Beckman 2004; Semlitsch et al. 2007). In contrast, anecdotal evidence indicates that secondary forest roads and trails with little use may not have negative impacts on herpetofauna and in some cases be used as habitat (e.g., Dillard et al. 2008c). More information is needed to better understand how landscape factors influence amphibian and reptile response to the creation of early successional habitats in upland hardwood forest (Ford et al. 2002a).

Literature Cited

Adams JP, Lacki MJ, Baker MD (1996) Response of herpetofauna to silvicultural prescriptions in the Daniel Boone National Forest, Kentucky. Proc Annu Conf Southeast Assoc Fish Wildl Agen 50:312–320

Ash AN (1988) Disappearance of salamanders from clearcut plots. J Elisha Mitchell Sci Soc 104:116–122

Ash AN (1997) Disappearance and return of plethodontid salamanders to clearcut plots in the southern Blue Ridge Mountains. Conserv Biol 11:983–989

Ash AN, Pollock KH (1999) Clearcuts, salamanders, and field studies. Conserv Biol 13:206–208

Ash AN, Bruce RC, Castanet J, Francillon-Vieillot H (2003) Population parameters of *Plethodon metcalfi* on a 10-year-old clearcut and in nearby forest in the southern Blue Ridge Mountains. J Herpetol 37:445–452

Ashton RE, Ashton PS (1978) Movements and winter behavior of *Eurycea bislineata* (Amphibia, Urodela, Plethodontidae). J Herpetol 12:295–298

Bailey LL, Simons TR, Pollock KH (2004a) Estimating detection probability parameters for Plethodon salamanders using the robust capture-recapture design. J Wildl Manage 68:1–13

Bailey LL, Simons TR, Pollock KH (2004b) Spatial and temporal variation in detection probability of Plethodon salamanders using the robust capture-recapture design. J Wildl Manage 68: 14–24

Bailey LL, Simons TR, Pollock KH (2004c) Estimating site occupancy and species detection probability parameters for terrestrial salamanders. Ecol Appl 14:692–702

Barry DS, Pauley TK, Maerz JC (2008) Amphibian use of man-made pools on clearcuts in the Allegheny Mountains of West Virginia, USA. Appl Herpetol 5:121–128

Bartman CE, Parker KC, Laerm J, McCay TS (2001) Short-term response of Jordan's salamander to a shelterwood timber harvest in western North Carolina. Phys Geogr 22:154–166

Brooks RT (1999) Residual effects of thinning and high white-tailed deer densities on northern redback salamanders in southern New England oak forests. J Wildl Manage 63:1172–1180

Brooks RT, Kyker-Snowman TD (2008) Forest floor temperature and relative humidity following timber harvesting in southern New England, USA. For Ecol Manage 254:65–73

Crawford JA, Semlitsch RD (2007) Estimation of core terrestrial habitat for stream-breeding salamanders and delineaton of riparian buffers for protection of biodiversity. Conserv Biol 21: 152–158

Crawford JA, Semlitsch RD (2008a) Post-disturbance effects of even-aged timber harvest on stream salamanders in southern Appalachian forests. Anim Conserv doi:10.1111/j.1469-1795. 2008.00191.x

Crawford JA, Semlitsch RD (2008b) Abiotic factors influencing abundance and microhabitat use of stream salamanders in southern Appalachian forests. For Ecol Manage 255:1841–1847

Deeming DC, Ferguson MWJ (1991) Physiological effects of incubation temperature on embryonic development in reptiles and birds. In: Deeming DC, Ferguson MWJ (eds) Egg incubation: its effects on embryonic development in birds and reptiles. Cambridge University Press, New York

deMaynadier PG, Hunter ML Jr (1995) The relationship between forest management and amphibian ecology: a review of the North American literature. Environ Rev 3:230–261

deMaynadier PG, Hunter ML Jr (1998) Effects of silvicultural edges on distribution and abundance of amphibians in Maine. Conserv Biol 12:340–352

Dillard LO, Russell KR, Ford WM (2008a) Macrohabitat models of occurrence for the threatened Cheat Mountain salamander, *Plethodon nettingi*. Appl Herpetol 5:201–224

Dillard LO, Russell KR, Ford WM (2008b) Site level habitat models for the endemic, threatened Cheat Mountain salamander: the importance of geophysical and biotic attributes for predicting occurrence. Biodivers Conserv 17:1475–1492

Dillard LO, Russell KR, Ford WM (2008c) *Plethodon nettingi* habitat use and ovoposition. Herp Rev 39:335–336

Duellman WE, Trueb L (1986) Biology of amphibians. McGraw-Hill, New York

Duguay JP, Wood PB (2002) Salamander abundance in regenerating forest stands on the Monongahela National Forest, West Virginia. Forest Sci 48:331–335

Felix ZI, Wang Y, Schweitzer CJ (2010) Effects of experimental canopy manipulation on amphibian egg deposition. J Wildl Manage 74:496–503

Floyd TM, Russell KR, Moorman CE, Van Lear DH, Guynn, DC Jr, Lanham JD (2001) Effects of prescribed fire on herpetofauna within hardwood forests of the upper Piedmont of South Carolina: a preliminary analysis. In: Outcalt KW (ed) Proceedings of the 11th biennial southern silvicultural research conference. Gen Tech Rep SRS-48, USDA Forest Service Southern Research Station, Asheville

Ford WM, Menzel MA, McGill DW, Laerm J, McCay TS (1999) Effects of a community restoration fire on small mammals and herpetofauna in the southern Appalachians. For Ecol Manage 114:233–243

Ford WM, Menzel MA, McCay TS, Gassett JW, Laerm J (2000) Woodland salamander and small mammal responses to alternative silvicultural practices in the southern Appalachians of North Carolina. Proc Annu Conf Southeast Assoc Fish Wildl Agen 54:241–250

Ford WM, Chapman BR, Menzel MA, Odom RH (2002a) Stand age and habitat influences on salamanders in Appalachian cove hardwood forests. For Ecol Manage 155:131–141

Ford WM, Menzel MA, Odom RH (2002b) Elevation, aspect and cove size effects on woodland salamanders in the southern Appalachians. Southeast Nat 1:315–324

Ford WM, Rodrigue JL, Rowan EL, Castleberry SB, Schuler TM (2010) Woodland salamander response to two prescribed fires in the central Appalachians. For Ecol Manage 260:1003–1009

Gibbs JP (1998) Amphibian movements in response to forest edges, roads, and streambeds in southern New England. J Wildl Manage 62:584–589

Goin CJ, Goin OB (1971) Introduction to herpetology. W.H. Freeman and Company, San Francisco

Gore JA, Patterson WA III (1986) Mass of downed wood in northern hardwood forests in New Hampshire: potential effects of forest management. Can J Forest Res 16:335–339

Greenberg CH (2001) Response of reptile and amphibian communities to canopy gaps created by wind disturbance in the southern Appalachians. For Ecol Manage 148:135–144

Greenberg CH (2002) Fire, habitat structure and herpetofauna in the Southeast. In: Ford WM, Russell KR, Moorman CE (eds) The role of fire in nongame wildlife management and community restoration: traditional uses and new directions. Gen Tech Rep NE-288, USDA Forest Service Northeastern Forest Experiment Station, Newtown Square

Greenberg CH, Waldrop TA (2008) Short-term response of reptiles and amphibians to prescribed fire and mechanical fuel reduction in a southern Appalachian upland hardwood forest. For Ecol Manage 255:2883–2893

Harper EB (2007) The role of terrestrial habitat in the population dynamics and conservation of pond-breeding amphibians. Dissertation, University of Missouri, Columbia

Harper CA, Guynn DC Jr (1999) Factors affecting salamander density and distribution within four forest types in the southern Appalachian mountains. For Ecol Manage 114:245–252

Harpole DN, Haas CA (1999) Effects of seven silvicultural treatments on terrestrial salamanders. For Ecol Manage 114:349–356

Herbeck LA, Larsen DR (1999) Plethodontid salamander response to silvicultural practices in Missouri Ozark forests. Conserv Biol 13:623–632

Hocking DJ, Semlitsch RD (2007) Effects of timber harvest on breeding site selection by gray treefrogs (*Hyla versicolor*). Biol Conserv 138:506–513

Homyack JA, Haas CA (2009) Long-term effects of experimental forest harvesting on abundance and reproductive demography of terrestrial salamanders. Biol Conserv 142:110–121

Jaeger RG (1980) Microhabitats of a terrestrial forest salamander. Copeia 1980:265–268

Johnson CM, Johnson LB, Richards C, Beasley V (2002) Predicting the occurrence of amphibians: an assessment of multiple-scale models. In: Scott JM, Heglund PJ, Haufler JB, Morrison ML, Raphael MG, Wall WB, Sampson F (eds) Predicting species occurrences: issues of accuracy and scale. Island Press, Washington DC

Johnson JR, Knouft JH, Semlitsch RD (2007) Sex and seasonal differences in the spatial distribution of gray treefrog (*Hyla versicolor*) populations. Biol Conserv 140:250–258
Johnson JR, Mahan RD, Semlitsch RD (2008) Seasonal terrestrial microhabitat use by gray treefrogs (*Hyla versicolor*) in Missouri oak-hickory forests. Herpetologica 64:259–269
Keyser PD, Sausville DJ, Ford WM, Schwab DJ, Brose PH (2004) Prescribed fire impacts to amphibians and reptiles in shelterwood-harvested oak-dominated forests. Va J Sci 55:159–168
Kirkland GL Jr, Snoddy HW, Miller TL (1996) Impact of fire on small mammals and amphibians in a central Appalachian deciduous forest. Am Midl Nat 135:253–260
Knapp SM, Haas CA, Harpole DN, Kirkpatrick RL (2003) Initial effects of clearcutting and alternative silvicultural practices on terrestrial salamander abundance. Conserv Biol 17:752–762
Lowe WH, Bolger DT (2002) Local and landscape-scale predictors of salamander abundance in New Hampshire headwater streams. Conserv Biol 16:183–193
Marsh DM, Beckman NG (2004) Effects of forest roads on the abundance and activity of terrestrial salamanders. Ecol Appl 14:1882–1891
Marsh DM, Goicochea MA (2003) Monitoring terrestrial salamanders: biases cause by intense sampling and choice of cover objects. J Herpetol 37:460–466
Marsh DM, Thakur KA, Bulka KC, Clarke LB (2004) Dispersal and colonization through open fields by a terrestrial, woodland salamander. Ecology 85:3396–3405
Matthews CE, Moorman CE, Greenberg CH, Waldrop TA (2010) Response of herpetofauna to repeated fires and fuel reduction treatments. J Wildl Manage 74:1301–1310
McLeod RF, Gates JE (1998) Response of herpetofaunal communities to forest cutting and burning at Chesapeake Farms, Maryland. Am Midl Nat 139:164–177
Messere M, Ducey P (1998) Forest floor distribution of northern redback salamanders, *Plethodon cinereus*, in relation to canopy gaps: first year following selective logging. For Ecol Manage 107:319–324
Miller JE, Hess GE, Moorman CE (2007) Southern two-lined salamanders in urbanizing watersheds. Urban Ecosys 10:73–85
Mitchell JC, Rineharat SC, Pagels JF, Buhlmann KA, Pague CA (1997) Factors influencing amphibian and small mammal assemblages in central Appalachian forests. For Ecol Manage 96:65–76
Moseley KR, Castleberry SB, Schweitzer SH (2003) Effects of prescribed fire on herpetofauna in bottomland hardwood forests. Southeast Nat 2:475–486
Moseley KR, Castleberry SB, Ford WM (2004) Coarse woody debris and pine litter manipulation effects on movement and microhabitat use of *Ambystoma talpoideum* in a *Pinus taeda* stand. For Ecol Manage 191:387–396
Moseley KR, Ford WM, Edwards JW, Schuler TM (2008) Long-term partial cutting impacts on Desmognathus salamander abundance in West Virginia headwater streams. For Ecol Manage 254:300–307
Moseley KR, Ford WM, Edwards JW (2009) Local and landscape factors influencing edge effects on woodland salamanders. Environ Monit Assess 151:425–435
Nislow KH, Lowe WH (2006) Influences of logging history and riparian forest characteristics on macroinvertebrates and brook trout (*Salvelinus fontinalis*) in headwater streams (New Hampshire, U.S.A.). Freshwater Biol 51:388–397
Patrick DA, Hunter ML Jr, Calhoun AJK (2006) Effects of experimental forestry treatments on a Maine amphibian community. For Ecol Manage 234:323–332
Patrick DA, Calhoun AJK, Hunter ML Jr (2008) The importance of understanding spatial population structure when evaluating the effects of silviculture on spotted salamanders (*Ambystoma maculatum*). Biol Conserv 141:807–814
Perkins DW, Hunter ML Jr (2006) Effects of riparian timber management on amphibians of Maine. J Wildl Manage 70:657–670
Peterman WE, Semlitsch RD (2009) Efficacy of riparian buffers in mitigating local population declines and the effects of even-aged timber harvest on larval salamanders. For Ecol Manage 257:8–14

Petranka JW (1999) Recovery of salamanders after clearcutting in the southern Appalachians: a critique of Ash's estimates. Conserv Biol 13:203–205

Petranka JW, Eldridge ME, Haley KE (1993) Effects of timber harvesting on southern Appalachian salamanders. Conserv Biol 7:363–370

Petranka JW, Brannon MP, Hopey ME, Smith CK (1994) Effects of timber harvesting on low elevation populations of southern Appalachian salamanders. For Ecol Manage 67:135–147

Pough FH, Smith EM, Rhodes DH, Collazo A (1987) The abundance of salamanders in forest stands with different histories of disturbance. For Ecol Manage 20:1–9

Reichenbach N, Sattler P (2007) Effects of timbering on *Plethodon hubrichti* over twelve years. J Herpetol 41:622–629

Renken RB (2006) Does fire affect amphibians and reptiles in eastern U.S. oak forests? In: Dickinson MB (ed) Fire in eastern oak forests: delivering science to land managers. Gen Tech Rep NRS-P-1, USDA Forest Service Northern Research Station, Newtown Square

Renken RB, Gram WK, Fantz DK, Richter SC, Miller TJ, Ricke KB, Russell B, Wang X (2004) Effects of forest management on amphibians and reptiles in Missouri Ozark forests. Conserv Biol 18:174–188

Riedel BL (2006) Habitat relationships of red-backed salamanders (*Plethodon cinereus*) in Appalachian grazing systems. Thesis, University of Wisconsin, Stevens Point

Riedel BL, Russell KR, Ford WM, O'Neill KP, Godwin HW (2008) Habitat relationships of eastern red-backed salamanders (*Plethodon cinereus*) in Appalachian agroforestry and grazing systems. Agr Ecosyst Environ 124:229–236

Rittenhouse TAG, Semlitsch RD (2009) Behavioral response of migrating wood frogs to experimental timber harvest surrounding wetlands. Can J Zool 87:618–625

Rittenhouse TAG, Harper EB, Rehard LR, Semlitsch RD (2008) The role of microhabitats in the desiccation and survival of anurans in recently harvested oak-hickory forest. Copeia 2008:807–814

Rittenhouse TAG, Semlitsch RD, Thompson FR III (2009) Survival costs associated with wood frog breeding migrations: effects of timber harvest and drought. Ecology 90:1620–1630

Ross B, Fredericksen T, Ross E, Hoffman W, Morrison ML, Beyea J, Lester MB, Johnson BN, Fredericksen NJ (2000) Relative abundance and species richness of herpetofauna in forest stands in Pennsylvania. For Sci 46:139–146

Rothermel BB, Luhring TM (2005) Burrow availability and desiccation risk of mole salamanders (*Ambystoma talpoideum*) in harvested versus unharvested forest stands. J Herpetol 39:619–626

Russell KR, Van Lear DH, Guynn DC Jr (1999) Prescribed fire effects on herpetofauna: review and management implications. Wildl Soc Bull 27:374–384

Russell KR, Hanlin HH, Wigley TB, Guynn DC Jr (2002) Responses of isolated wetland herpetofauna to upland forest management. J Wildl Manage 66:603–617

Russell KR, Wigley TB, Baughman WM, Hanlin HG, Ford WM (2004) Responses of southeastern amphibians and reptiles to forest management: a review. In: Rauscher HM, Johnsen K (eds) Southern forest science: past, present, and future. Gen Tech Report SRS-75, USDA Forest Service Southern Research Station, Asheville

Russell KR, Mabee TJ, Cole MB, Rochelle MJ (2005) Evaluating biotic and abiotic influences on torrent salamanders in managed forests of western Oregon. Wildl Soc Bull 33:1413–1424

Semlitsch RD (2000) Principles for management of aquatic-breeding amphibians. J Wildl Manage 64:615–631

Semlitsch RD, Bodie JR (2003) Biological criteria for buffer zones around wetlands and riparian habitats for amphibians and reptiles. Conserv Biol 17:1219–1228

Semlitsch RD, Ryan TJ, Hamed K, Chatfield M, Drehman B, Pekarek N, Spath M, Watland A (2007) Salamander abundance along road edges and within abandoned logging roads in Appalachian forests. Conserv Biol 21:159–167

Semlitsch RD, Conner CA, Hocking DJ, Rittenhouse TAJ, Harper EB (2008) Effects of timber harvesting on pond-breeding amphibian persistence: testing the evacuation hypothesis. Ecol Appl 18:283–289

Semlitsch RD, Todd BD, Blomquist SM, Clhoun AJK, Gibbons JW, Gibbs JP, Graeter GJ, Harper EB, Hocking DJ, Hunter ML Jr, Patrick DA, Rittenhouse RAG, Rothermel BB (2009) Effects of timber harvest on amphibian populations: understanding mechanisms from forest experiments. BioScience 59:853–862

Sinsch U (1990) Migration and orientation in anural amphibians. Ethol Ecol Evol 2:65–79

Stebbins RC, Cohen NW (1995) A natural history of amphibians. Princeton University Press, Princeton

Todd BD, Andrews KM (2008) Response of a reptile guild to forest harvesting. Conserv Biol 22:753–761

Todd BD, Luhring TM, Rothermel BB, Gibbons JW (2009) Effects of forest removal on amphibian migrations: implications for habitat and landscape connectivity. J Appl Ecol 46:554–561

Webster CR, Jenkens MA (2005) Coarse woody debris dynamics in the southern Appalachians as affected by topographic position and anthropogenic disturbance history. For Ecol Manage 217:319–330

Whiles MR, Grubaugh JW (1996) Importance of coarse woody debris to southern forest herpetofauna. In: McMinn JW, Crossley DA Jr (eds) Biodiversity and coarse woody debris in southern forests. Proceedings of the workshop on coarse woody debris in southern forests: effects on biodiversity. Gen Tech Rep SE-94, USDA Forest Service Southeastern Forest Experiment Station, Asheville

Zug GR (1993) Herpetology: an introductory biology of amphibians and reptiles. Academic, San Diego

Chapter 12
Managing Early Successional Habitats for Wildlife in Novel Places

J. Drew Lanham and Maria A. Whitehead

Abstract Utility rights-of-way stretch for thousands of kilometers across the North American landscape. In deciduous forests of the Central Hardwood Region, rights-of-way provide opportunities for conserving early successional species, including a broad array of songbirds and butterflies. Although the millions of hectares managed by the utility industry to provide electricity, natural gas, and other services are not usually viewed by the public as beneficial for wildlife conservation, we suggest that rights-of-way can be valuable early succession habitats in addition to more "traditionally" created areas like clearcut harvests.

12.1 Introduction

Even as the amount of early successional habitats in the central hardwood forests of the USA (McNab, Chap. 2; Fig. 2.1) diminishes, some even-aged forest management practices (e.g. clearcutting) used to promote disturbance-dependent species are declining (see Loftis et al., Chap. 5; Shifley and Thompson, Chap. 6). Implementation of less intensive silvicultural systems in their stead, combined with "clean" agricultural practices that discourage early successional growth in weedy, ruderal areas, is

J.D. Lanham (✉)
Department of Forestry and Natural Resources, Clemson University, 261 Lehotsky Hall, Clemson, SC 29634–0331, USA
e-mail: lanhamj@clemson.edu

M.A. Whitehead
South Carolina Nature Conservancy, 960 Morrison Drive, Suite 100, Charleston, SC 29403, USA
e-mail: mwhitehead@tnc.org

C.H. Greenberg et al. (eds.), *Sustaining Young Forest Communities*, Managing Forest Ecosystems 21, DOI 10.1007/978-94-007-1620-9_12,

resulting in loss of early successional habitats (Confer 1992; LeGrand and Schneider 1992; Whitman and Hunter 1992; Trani et al. 2001). Despite conservation efforts to include early successional habitats and promote habitat diversity, these management trends and habitat losses are unlikely to be reversed in the foreseeable future. In this chapter, we explore novel means for creating early successional habitats and managing the disturbance-dependent species that require them. Specifically, we focus on utility rights-of-way (hereafter referred to as ROW).

Utility rights-of-way are ubiquitous anthropogenic landscape features that occupy millions of hectares in the USA. They lie across a diversity of landscapes within the Central Hardwood Region. Because utility companies must maintain transmission line ROWs to deliver electricity, natural gas, or other products and services, they manage these linear landscape features to keep them clear of obstacles and hazards that could interfere with construction, operation, and maintenance of facilities. Efforts to keep trees from growing into ROWs result in large expanses of land that are managed in a state of "perpetual" early succession, from grass-dominated areas to shrub-scrub habitats.

Although the primary function of ROW is to distribute service, these areas increasingly are being enhanced as early successional wildlife habitats. Extensive early work compiled by Lancia and McConnell (1976), Arner (1977), Bramble, and a host of co-workers (1972, 1979a, b, 1985, 1986, 1990, 1991) forms a basis for our understanding potential benefits of ROW for wildlife. These and several other investigators have shown that early successional habitats within ROW are used by vertebrate species such as songbirds (Hanowski et al. 1993; Confer and Pascoe 2003), game birds (Arner et al. 1993), raptors (Denoncour and Olson 1984; Bridges and Lopez 1993), white-tailed deer (*Odocoileus virginiana*) (Harlow 1991; Harlow et al. 1993), small mammals (Cavanaugh et al. 1976; Lauzon et al. 2002) and even sensitive, threatened or endangered species (Lowell and Lounsberry 2002; McLoughlin 2002; Thomas 2002). Birds are perhaps the most thoroughly studied wildlife that use ROW habitats. Highly charismatic organisms with important ecological functions, they are more easily observed than other vertebrate wildlife and the potential benefits of managing for them carries significant "green" capitol for industrial landowners.

Unlike birds, insects are rarely the focus of landscape-scale management activities that will promote their proliferation. The potential exceptions are butterflies (Lepidoptera). Arguably the most charismatic and publicly accepted insects, butterflies are critical pollinators and prey in temperate terrestrial ecosystems. Many species require features of early successional habitats (e.g. openness, abundant sunlight, flowering plants, bare ground) that occur in many ROW. Like birds, butterflies are often viewed as indicators of ecological health (Ries et al. 2001). Managing for these insects therefore also carries benefits for industries and natural resource managers striving to conserve biodiversity at all levels.

Many bird species, especially passerines, are inextricably linked to insects, especially butterflies, because Lepidoteran larvae (caterpillars) can make up much of their diet during breeding and migratory periods. As many songbirds and butterflies share requirements for early successional habitats, we will link the two groups in this chapter as we present the case for considering ROW as an option for managing some early successional species. In the case of birds, we present a brief review of the state of knowledge regarding avian use of ROW, including species diversity, ecological costs and benefits, and some potential management implications. For butterflies, there is scant literature for review, so we present a case study that provides compelling evidence for considering ROW as opportunities for managing this important, but frequently overlooked, group.

12.2 Songbirds and Rights-of-Way Management

Over the past several decades, many early successional (scrub/shrub, shrubland) songbird populations have declined (Robbins et al. 1989; Petit et al. 1995; Hagan et al. 1992; Hussell et al. 1992; Hunter et al. 2001; Askins 1994, 1998, 2000). The National Audubon Society lists several early successional, shrub-scrub birds among its highest priority species for conservation attention (http://www.audubon.org/bird/stateofthebirds/shrublands.htm, April 1, 2010). As habitat loss due to chronic conversion and aforestation can be a primary factor in these declines (Askins 1998, 2000), areas managed intensively to maintain vegetation in a state of arrested succession could contribute significantly to conservation of shrubland birds.

12.2.1 Songbird Species Diversity in Rights-of-Way

Most studies investigating shrub-scrub bird use of powerline ROW have been conducted in the northeastern USA (e.g. Bramble et al. 1992, 1994; Confer et al. 1998; Confer 2002; Marshall et al. 2002). The list of species using these habitats is impressively diverse (Table 12.1). It includes species ranked high on conservation prioritization lists such as the 2007 Audubon Society Watch List (http://web1.audubon.org/science/species/watchlist/browseWatchlist.php) and species with declining state or region-wide population trends (see Sauer et al. 1999). Several species of conservation concern, such as the Golden-winged Warbler (*Vermivora chrysoptera*), Blue-winged Warbler (*V. pinus*), Prairie Warbler (*Dendroica discolor*), Chestnut-sided Warbler (*D. pensylvanica*), Black-and-white Warbler (*Mniotilta varia*), Field Sparrow (*Spizella pusilla*), and Eastern Towhee (*Pipilo erythropthalmus*), are listed frequently in the ROW and bird literature as breeders in utility line corridors (Bramble et al. 1992, 1994; Confer 2002; Marshall et al. 2002).

As researchers and natural resources managers have come to understand the importance of managing early successional habitats for wildlife in general and songbirds in particular, recent studies highlight the potential for shrub-scrub (and other) bird conservation in ROW. Confer and Pascoe (2003) documented high bird species diversity in northeastern (New Hampshire, New York, Maine and Massachusetts) ROW, with many high priority conservation species occurring in ROW variably managed with herbicides, mowing, and prescribed fire. On the Cumberland Plateau in Tennessee, Bulluck and Buehler (2006) found that among three early successional

Table 12.1 Rights-of-way bird diversity in Pennsylvania and Tennessee (1982–2004)

Diurnal Raptors and Gallinaceous

Ruffed Grouse *Bonasa umbellus*[a]
American Kestrel *Falco sparverius*[a]
Northern Bobwhite *Colinus virginianus*[b]

Doves and Cuckoos

Mourning Dove *Zenaida macroura*[a]
Yellow-billed Cuckoo *Coccyzus americanus*[a]
Black-billed Cuckoo *C. erythopthaimus*[a]

Woodpeckers, Hummingbirds

Hairy Woodpecker *Picoides villosus*[a]
Downy Woodpecker *P. pubescens*[a]
Northern Flicker *Colaptes auratus*
Ruby-throated Hummingbird *Archilochus colubris* [a]

Flycatchers

Least Flycatcher *Empidonax minimus*[a]
Eastern Kingbird *Tyrannus tyrannus*[a]
Eastern Phoebe *Sayornis phoebe*[a]

Vireos

Red-eyed Vireo *Vireo olivaceus*[a]
Yellow-throated Vireo *V. flavifrons*[b]
White-eyed Vireo *V. griseus*[a, b]

Chickadees, Titmice and Nuthatches

Black-capped Chickadee *Poecile atricapillus*[a]
Tufted Titmouse *Baeolophus bicolor*[a]
White-breasted Nuthatch *Sitta carolinensis*[a]

Wrens

Carolina Wren *Thryothorus ludovicianus*[b]
House Wren *Troglodytes aedon*[a]

Thrushes

Veery *Hylocichla fucescens*[a]
Wood Thrush *H. mustelina*[a]
Eastern Bluebird *Sialia sialis*[a]
American Robin *Turdus migratorius*[a]

Mimic Thrushes, Wrens, Waxwings

Gray Catbird *Dumetella carolinensis*
Northern Mockingbird *Mimus polyglottos*[a]
Cedar Waxwing *Bombycilla cedrorum*[a]

Warblers

Chestnut-sided Warbler *Dendroica pensylvanica*
Prairie Warbler *D. discolor*[a]
Blackburnian Warbler *D. fusca*[a]
Common Yellowthroat *Geothlypis trichas*[a, b]
American Redstart *Setophaga ruticella*[a]
Black-and-white Warbler *Mniotilta varia*[a]
Canada Warbler *Wilsonia canadensis*[a]
Ovenbird *Seiurus aurocapillus*[a]
Blue-winged Warbler *Vermivora pinus*[a]
Golden-winged Warbler *V. chrysoptera*[a]
Yellow-breasted Chat *Icteria virens*[a, b]

Sparrows, Buntings, Grosbeaks, Finches

Dark-eyed Junco *Junco hyemalis*[a]
Field Sparrow *Spizella pusilla*[a, b]
Eastern Towhee *Pipilo erythrophthalmus*[a, b]
Song Sparrow *Melospiza melodia*[a]
Chipping Sparrow *Spizella passerina*[a]
Indigo Bunting *Passerina cyanea*[a, b]
Rose-breasted Grosbeak *Pheucticus ludovicianus* [a]
American Goldfinch *Carduelis tristis*[a]
Northern Cardinal *Cardinalis cardinalis*[a, b]

Tanagers and Blackbirds

Brown-headed Cowbird *Molothrus ater*
Common Grackle *Quiscalus quiscula*[a]
Scarlet Tanager *Piranga olivacea*[a]
Summer Tanager *Piranga rubra*[b]
Northern Oriole *Icterus galbula*[a]

[a]Pennsylvania (Bramble et al. 1992; Yahner et al. 2004)
[b]Kroodsma (1982)

habitat types (regenerating clearcuts, reclaimed surface mines, ROW), ROW were intermediate in bird diversity and harbored higher bird species richness than regenerating clearcuts. Of particular note, ROW sites were more likely to harbor the Kentucky Warbler (*Oporonis formosus*), a high conservation priority forest interior species across much of its range. Early successional species, including Chestnut-sided Warblers, Eastern Towhees, and Song Sparrows (*Melospiza melodia*) were more closely allied with ROW than either clearcuts or reclaimed mines.

12.2.2 Ecological Benefits and Costs of Rights-of-Way Occupancy

Creation of early successional habitats, whether by forest management or maintenance of ROW, is often controversial. Aesthetically, the dense, jungle-like vegetation in a regenerating clearcut or a power line ROW may not be appealing. Beyond appearances, edges between forested and early successional habitats can have both positive and negative impacts on songbirds. Avian species diversity and density are greater along edges (MacArthur and MacArthur 1961; Odum 1971; Roth 1976; Hansson 1983); ecotones between forests and ROW could contain more species and more individuals than pure communities if species requiring both ecosystems occur together with more specialized species from each ecosystem (Odum 1971). Additionally, edges can act as boundaries of individual breeding territories and concentrate birds (Anderson et al. 1977). Increased primary productivity, insect species richness and density (Hansson 1983), vegetation density and structure, and light intensity (Strelke and Dickson 1980) are other factors thought to attract birds to edges. Hanowski et al. (1993) found differences in composition of bird communities over 200 m into the forest interior from a ROW edge. Additionally, Small and Hunter (1989) contrasted two permanent ecotones (transmission-line corridors and river edges) and found that edges bordering transmission-line corridors supported more avian species than river edges – probably due to abundant brushy cover, which is required by many edge species. Although use of ROW in the breeding season by early successional obligates and edge species might be expected, the number of forest interior birds occurring in these areas is noteworthy. Forest interior neotropical migrants like the Rose-breasted Grosbeak (*Pheuticus ludovicianus*), Scarlet Tanager (*Piranga rubra*), Red-eyed Vireo (*Vireo olivaceous*), Eastern Wood-Pewee (*Contupus virens*), Ovenbird (*Seirus aurocapillus*) and Wood Thrush (*Hylocichla mustelina*) were recorded in upland hardwood forest ROW in Pennsylvania (Bramble et al. 1992; Yahner et al. 2004). Several studies (Pagen et al. 2000; Marshall et al. 2003; Yahner 2004; Vitz and Rodewald 2006) have revealed that early successional habitats provide important cover and foraging for mature forest birds in the post-fledging period. Thus, the appearance of mature forest birds might be explained by the high levels of vegetative productivity in ROW attracting a diverse arthropod assemblage that provides potential prey items such as adult and larval Lepidoptera (see Greenberg et al., Chap. 8).

Transmission-line ROW may also serve as movement corridors and provide important stopover habitats for migrating land birds, (as passerines are demonstrated to use some early successional habitats) during migration (Moore et al. 1990; Mabey et al. 1992). Moore et al. (1990) found that scrub-shrub areas were selected as stopover habitats by migratory birds on an island in the Gulf of Mexico. Relative to four additional habitat types available to the migrating songbirds, scrub/shrub areas harbored both the greatest species diversity and number of individuals. The "value-added" facets of early successional habitats as areas used for fledgling foraging, migration stopover, and species associated with other habitat types (e.g., forest interior species) may place utility-line corridor management for songbirds in a new and more fully informed light.

Although the potential benefits of shrubby vegetation in ROW and elsewhere to songbirds are known, creation of early successional vegetation structure and edges as practiced in traditional wildlife management (Leopold 1933) is controversial. Studies of forest fragmentation and its impacts on forest interior species reveal some potential negatives of creating too much edge that may outweigh the benefits (Harris 1984). For example, nest productivity may decrease near habitat edges. Chasko and Gates (1982) showed that higher fledging success was associated with increased distance from ROW edge. Gates and Gysel (1978) suggested that forest species' breeding habitat suitability decreases as the number of nests increases toward narrow field-forest edges. The isolation of mature forests and impacts of nest predation and brood parasitism on edges are well documented and have remained an issue of concern among some researchers (Ratti and Reese 1988; Robinson 1988). Suarez et al. (1997) found that nest predation rates were higher along exterior agricultural edges than forest-interior edges and were also higher along abrupt, permanently-maintained edges than along gradual edges. Rights-of-way edges may also attract potential nest predators due to their greater prey density and natural travel lanes created by the change in structure of vegetation (Gottfried and Thompson 1978; Ratti and Reese 1988). By attracting songbirds as well as brood parasites and nest predators, edges may act as "ecological traps" for some disturbance-dependent species while adversely affecting forest-interior songbirds in adjacent forests (Wilcove 1985; Pulliam 1988; Ratti and Reese 1988; Robinson 1988).

The impacts of ROW on brood parasitism in Brown-headed Cowbirds (*Molothrus ater*) exemplify the variable effects of ROW on parasitism and nest predation. Confer and Pascoe (2003) reported Cowbird parasitism was relatively low (5.3%) in some northeastern USA study sites and ROW did not reduce forest bird productivity. Although Suarez et al. (1997) found that cowbird parasitism rates in southern Illinois did not differ among several types of edge habitats (agricultural, forest interior, stream, wildlife openings, and treefall gaps), predation rates were higher and nest productivity was lower next to agricultural edges. Chasko and Gates (1982) and Gates and Griffen (1991) found Brown-headed Cowbird parasitism rates were higher near power-line corridors than within adjacent forest. These contradictory results illustrate the complex influences within variably managed areas and landscape context. For example, managers might consider impacts of an agricultural landscape on parasitism when creating habitats such as wildlife food plots for game species. Rights-of-way embedded within different matrices (e.g., forest versus agriculture versus riparian zones) may have different effects on parasitism and predation.

There are often strong associations between mature forest bird species and forest fragmentation (see Gates and Gysel 1978; Wilcove 1985; Yahner and Wright 1985; Angelstam 1986; Robinson 1988; Ratti and Reese 1988; Keyser et al. 1998); yet, some area-sensitive species require large, contiguous patches of early successional habitats (Robbins et al. 1989; Robinson et al. 1992). In contrast, most early successional species in the northeastern USA breed in a wide range of habitat patch sizes, and patch size appears not to be a limiting factor for their conservation

(Dettmers 2003). However, an increasing number of researchers (Askins 1998; Confer 1992; Lanham 1997; Lanham and Guynn 1998; Kremmentz and Christie 2000; Rodewald and Vitz 2005; Schlossberg and King 2008) provide strong evidence that early successional species may also be sensitive to variations in habitat area. For example, Yellow-breasted Chats (*Icteria virens*) were absent from shrubby patches smaller than 2 ha (Dennis 1958). Similarly, Golden-winged Warblers are probably an area-sensitive species with 10 ha early successional patches offering the minimum area for nesting in New York (Confer and Knapp 1981; Confer 1992). Lanham (1997) and Lanham and Guynn (1998) described Prairie Warblers and Yellow-breasted Chats as area-sensitive shrub-scrub species in because neither species was ever observed in small (<2 ha), regenerating, clearcut patches in the South Carolina mountains and upper piedmont. Although linear in configuration, some ROW may provide adequate area for area–sensitive, disturbance-dependent species. Powerline corridors vary in width by the voltage the lines they carry, with 500-kv transmission-lines typically among the widest at 45.7 m (150 ft) or greater (American Transmission Company; http://www.atcllc.com/IT6.shtml, April 5, 2010). Over sufficient lengths, ROW may provide adequate area as evidenced by the occurrence of Prairie Warblers and Yellow-breasted Chats in 500-kv transmission corridors in Pennsylvania (Yahner et al. 2004) and Prairie Warblers, Yellow-breasted Chats, and Golden-winged Warblers in 230-kv corridors in that same region (Bramble et al. 1992).

12.2.3 Songbird Responses to Transmission-Line Maintenance

Mowing, cut/stump treatments, and broadcast or selectively applied herbicides are some common methods used to manage ROW. In Tennessee, densities of several bird species were positively associated with presence of blackberry (*Rubus* spp.), which often increases in managed ROW. The numbers of Yellow-breasted Chats, White-eyed Vireos (*Vireo griseus*), Common Yellowthroat (*Gymnothylpis trichas*) and Prairie Warbler all increased with density of blackberry patches. Field Sparrows showed the opposite trends as the grass-dominated habitats the species prefers were apparently not abundant (Kroodsma 1982). In New York, bird diversity and density were highest in ROWs where trees received cut and stump or basal treatments with herbicides (Malefyt 1984). Bird density and diversity also were lowest on brush-mowed ROWs in the year after the mowing. Similarly, in central Pennsylvania, songbird species diversity decreased immediately after hand-cutting and herbicide treatments were applied to ROWs (Bramble et al. 1986). However, Bramble et al. (1992) observed higher bird densities on those areas sprayed with herbicides (basal, stem foliage and foliage sprayed) relative to those that received mechanical treatments and herbicide treatments.

In a comparison of two transmission line corridors, one maintained in an early successional stage (grass and forbs) by annual mowing and the other maintained in a later seral stage (small trees and shrubs) by selective herbicides, Chasko and Gates

(1982) found that bird use of a grassland stage corridor was minimal (limited to two mixed-habitat species) and did not include grassland birds characteristic of the region. Songbird diversity and breeding success were highest in the shrubland-stage corridor. Maintenance by selective herbicide application may create more vertical structure and habitat heterogeneity (Bramble and Byrnes 1979a; Lawson and Gates 1981) than annual mowing. Yahner et al. (2004) assessed the trends in bird usage of Pennsylvania ROW managed by the "wire/border zone method" (Bramble et al. 1992) that creates lower strata of grass, forbs and shrubs in the central wire zone and higher shrub zones at the forest edge. During 15 years of mowing and selective herbicide management, bird community composition remained relatively stable throughout the study units, with highest densities (birds/100 ha/day) observed in mowed and herbicide-treated units. Somewhat in contrast to other research, Kroodsma (1982, 1984) found bird density was lower in mowed, grass-dominated corridors than in forb-*Rubus*-dominated transmission-line corridors in Tennessee.

12.3 Butterfly Diversity in Rights-of-Way: A Case Study of an Ecological Benefit

Butterfly watching and gardening have become popular outdoor activities. Because many butterfly species are brightly colored, have interesting life histories, and even allow close observation, the interest in conserving them and their habitats has increased. Beyond their aesthetic appeal to nature enthusiasts, butterflies are key components of terrestrial ecosystems. They pollinate many species of flowering plants (Webb and Bawa 1983) and are prey for a diverse array of vertebrates, including many species of birds. The lives of many bird species are closely linked to the presence and abundance of arthropods (Rodenhouse and Holmes 1992; Marra and Holberton 1998; Marshall et al. 2002) including butterfly larvae (caterpillars) that are important food resources for both nestlings and adults. Because of the important role that they play as insect prey, the abundance and diversity of butterflies may impact local bird populations dramatically (Holmes and Sherry 1988; Sherry and Holmes 1995).

Many nectar-producing flowering plants and other species important in the life cycles of butterflies occur in early successional habitats. For example, Monarch butterflies (*Danaus plexippus*) prefer milkweeds (*Asclepias* spp.), which grow primarily in disturbed, open sites as an egg-laying substrate and larval food source. From this and multiple other examples of early-successional plants that proliferate in ROW, it should follow that these areas may have great potential as butterfly conservation areas. However, relatively little work has been done to determine the suitability of ROW for butterflies (see Yahner 2004). Nichols and Lanham (2002) surveyed butterflies and skippers in six South Carolina ROW and found 101 species occupying those sites (Table 12.2). They also recorded a diverse community of nectar and larval host plants as well as an abundance of other butterfly habitat requisites (e.g. open areas, bare ground, and moist puddling areas; see Nichols and Lanham 2002).

Table 12.2 Butterflies and skippers on six South Carolina ROW[a]

Papilionidae - Parnassians and Swallowtails	**Nymphalidae - Brushfoots**	**Hesperiidae - Skippers**
Pipevine Swallowtail *Battus philenor*	***Snouts (Libytheinae)***	***Spread-wing Skippers (Pyrginae)***
Black Swallowtail *P. polyxenes*	Snout *Libytheana bachmanii*	Silver-spotted Skipper *Epargyreus clarus*
Giant Swallowtail *P. cresphontes*	***Milkweed Butterflies (Danainae)***	Golden Banded-Skipper *Autochton cellus*
Eastern Tiger Swallowtail *P. glaucus*	Monarch *Danaus plexippus*	Hoary Edge *Achalarus lyciades*
Spicebush Swallowtail *P. troilus*	***Longwings (Heliconiinae)***	Southern Cloudywing *Thorybes bathyllus*
Pieridae - Whites and Sulphurs	Gulf Fritillary *Agraulis vanillae*	Northern Cloudywing *T. pylades*
Whites (Pierinae)	Variegated Fritillary *Euptoieta claudia*	Dreamy Duskywing *Erynnis icelus*
West Virginia White *Pieris virginiensi*	Great Spangled Fritillary *Speyeria cybele*	Horace's Duskywing *E. horatius*
Cabbage White *P. rapae*	***True Brushfoots (Nymphalinae)***	Mottled Duskywing *E. martialis*
Great Southern White *Ascia monuste*	Silvery Checkerspot *Chlosyne nycteis*	Wild Indigo Duskywing *E. baptisiae*
Sulphurs (Coliadinae)	Gorgone Crescent *Charidryas gorgone*	Common Checkered-Skipper *Pyrgus communis*
Clouded Sulphur *Colias philodice*	Pearl Crescent *Phyciodes tharos*	***Grass Skippers (Hesperiinae)***
Orange Sulphur *C. eurytheme*	Common Buckeye *Junonia coenia*	Clouded Skipper *Lerema accius*
Southern Dogface *C. cesonia*	Question Mark *Polygonia interrogationis*	Least Skipper *Ancyloxypha numitor*
Cloudless Sulphur *Phoebis sennae*	Eastern Comma *P. comma*	Fiery Skipper *Hylephila phyleus*
Barred Yellow *Eurema daira*	Mourning Cloak *Nymphalis antiopa*	Leonard's Skipper *Hesperia leonardus*
Little Yellow *Pyrisitia lisa*	Red Admiral *Vanessa atalanta*	Cobweb Skipper *H. metea*
Sleepy Orange *Abaeis nicippe*	Painted Lady *V. cardui*	Sachem *Atalopedes campestris*
Lycaenidae - Gossamer-wings	American Lady *V. virginiensis*	Tawny-edged Skipper *Polites themistocles*
Harvesters (Miletinae)	***Admirals and Relatives (Limenitidinae)***	Crossline Skipper *P. origenes*
Harvester *Feniseca tarquinius*	Red-spotted Purple *Limenitis arthemis astyanax*	Whirlabout *P. vibex*
Coppers (Lycaeninae)	Viceroy *L. archippus*	Southern Broken-Dash *Wallengrenia otho*
American Copper *Lycaena phlaeas*	***Emperors (Apaturinae)***	Northern Broken-Dash *W. egeremet*
Hairstreaks (Theclinae)	Tawny Emperor *Asterocampa clyton*	Little Glassywing *Pompeius verna*
Great Purple Hairstreak *Atlides halesus*	***Satyrs and Wood-Nymphs (Satyrinae)***	Delaware Skipper *Anatrytone logan*
Banded Hairstreak *Satyrium calanus*	Southern Pearly Eye *Enodia portlandia*	Aaron's Skipper *Poanes aaroni*

(continued)

Table 12.2 (continued)

Striped Hairstreak *S. liparops*	Northern Pearly Eye *E. anthedon*	Hobomok Skipper *P. hobomok*
Coral Hairstreak *S. titus*	Creole Pearly Eye *E. creola*	Zabulon Skipper *P. zabulon*
Olive Hairstreak *Mitoura grynea*	Appalachian Brown *Satyrodes appalachia*	Yehl Skipper *P. yehl*
Gray Hairstreak *Strymon melinus*	Gemmed Satyr *Cyllopsis gemma*	Broad-winged Skipper *P. viator*
White M Hairstreak *Parrhasius m-album*	Carolina Satyr *Hermeuptychia sosybius*	Byssus Skipper *Problema byssus*
Red-banded Hairstreak *Calycopis cecrops*	Hermes Satyr *H. hermes*	Dion Skipper *Euphyes dion*
Northern Hairstreak *Euristrymon ontario*	Little Wood Satyr *Megisto cymela*	Dun Skipper *E. vestris*
Brown Elfin *Callophrys augustinus*	Common Wood-Nymph *Cercyonis pegala*	Dusted Skipper *Atrytonopsis hianna*
Eastern Pine Elfin *C. niphon*		Pepper and Salt Skipper *Amblyscirtes hegon*
Blues (Polyommatinae)		Carolina Roadside-Skipper *A. carolina*
Eastern Tailed-Blue *Cupido comyntas*		Reversed Roadside-Skipper *A. reversa*
Spring Azure *Celastrina ladon*		Lace-winged Roadside-Skipper *A. aesculapius*
Silvery Blue *Glaucopsyche lygdamus*		Common Roadside-Skipper *A. vialis*
		Bell's Roadside Skipper *A. belli*
		Eufala Skipper *Lerodea eufala*
		Ocola Skipper *Panoquina ocola*
		Sickle-winged Skipper *Achylodes thraso*[b]

[a]Common and scientific names from http://www.carolinanature.com/butterflies/ (1/27/2011)
[b]Rare or extralimital species

Their work showed that the ROW were botanically diverse, structurally heterogeneous, early successional habitats that harbored an impressive diversity of butterflies and skippers. The high butterfly/skipper habitat suitability was in part due to the combination of selective herbicide application and mowing that resulted in a rich mosaic of shrubs, grasses, and bare ground.

12.4 The Future for Songbirds, Butterflies and ROW Management

Permanently maintained openings on public lands (e.g., ROW, wildlife openings, reclaimed landfills, and strip mines) may become increasingly important to disturbance-dependent birds. Bulluck and Buehler (2006) provide some of the most compelling evidence for the value of ROW and reclaimed areas as early successional forest habitats relative to "traditional" regenerating forest habitats. Many surface (e.g., mine) reclamation efforts focus on grassland habitats (Bajema and Lima 2001; Bajema et al. 2001; Cox and Maehr 2004; Scott et al. 2002), and few appear to be focused on "restoring" shrub-scrub habitats. Thus, ROW may provide the best opportunities for managing early successional shrub-scrub habitats.

Rights-of-way have been largely ignored by conservationists and frequently derided as landscape scars. However, in light of the declines in early successional habitats and the concurrent declines in disturbance-dependent bird and butterfly species, there are numerous opportunities to maximize the benefits of a system that essentially arrests succession and provides habitats that would otherwise disappear. Again, the growing body of evidence linking early successional habitats to the important post-fledgling stage of some songbirds (Pagen et al. 2000; Marshall et al. 2003; Yahner 2004; Vitz and Rodewald 2006) provides strong evidence that managing early successional areas like ROW can benefit many disturbance-dependent species that include both predators (songbirds) and prey (caterpillars).

Because landscape-level habitat restoration or alteration (e.g., "ecosystem management") is often regarded as the best management tool for maintenance of songbird communities, perhaps the ubiquitous nature of ROW offers opportunities to manage for some early successional songbirds on a large scale. Development and implementation of landscape-level management requires intimate knowledge of population-level requirements and limitations, as well as community-level interactions. Transmission-line position on the landscape, maintenance regimes, and successional stage may ultimately determine ROW use by shrub-scrub songbird species.

Future research to evaluate transmission-line corridors as songbird habitats should continue to focus on: (1) impacts and benefits for both disturbance-dependent and forest interior species, (2) differences among power line corridors and other early successional areas (e.g., ephemeral timber harvest openings), and (3) effects of landscape level characteristics (e.g., shape, size, configuration, and context) and maintenance regimes. Perhaps most importantly, these studies should include some measure of fitness (e.g., nesting success, predation rates, or fledgling and adult survival) along with density and diversity estimates when addressing ROW quality as songbird habitats. Petit et al. (1995) claim that "the landscape context in which the habitat is imbedded can be an important influence on [songbird] demography and, hence, population health." By understanding how transmission line ROWs and other forest openings, within the context of the landscape, impact songbird density, diversity, and reproductive rates, we can better devise management alternatives at a landscape level for the maintaining declining, disturbance-dependent songbird populations in ROW.

The ecological linkages and public appeal of birds and butterflies can provide opportunities for natural resource managers and conservationists to move advocacy for early successional habitats forward. Some utility companies have recognized the potential for enhancing their reputations on environmental issues, and have implemented programs to enhance the habitat suitability of their ROW. Likewise, as the use of selective herbicides has become the common means of managing ROW vegetation, those companies providing the chemicals have also taken advantage of the opportunities to "green" their image by promoting the highly selective nature of the herbicides and advancing their use in promoting or restoring wildlife habitats (Hurst 1997). Beyond the ecological benefits to early successional species and the marketing advantages to corporate utility and chemical entities, there are social benefits to be gained. Rights-of-way teeming with bird life and butterflies within forests that cross urban and suburban landscapes can provide opportunities for educating audiences heretofore uninformed or with negative impressions of early successional habitats - novel opportunities indeed for wildlife conservation.

Literature Cited

Anderson SH, Mann K, Shugart HH Jr (1977) The effect of transmission-line corridors on bird populations. Am Midl Nat 97:216–221

Angelstam P (1986) Predation on ground-nesting birds' nests in relation to predator densities and habitat edge. Oikos 47:365–373

Arner DH (1977) Transmission line rights-of-way management. FWS/OBS-76/20.2, USDI Fish and Wildlife Service, Laurel, 12 pp

Arner DH, Jones J, Hering B (1993) A cooperative program for the management of utility line rights-of-way. In: Doucet GJ, Séguin C, Giguère M (eds) Proceedings of the fifth international symposium on environmental concerns in rights-of-way management, Montreal, pp 259–261

Askins RA (1994) Open corridors in a heavily forested landscape: impact on shrubland birds and forest-interior birds. Wildl Soc Bull 22:339–347

Askins RA (1998) Restoring forest disturbance to sustain populations of shrubland birds. Restor Manage Notes 16:166–173

Askins RA (2000) Restoring North America's birds: lessons from landscape ecology. Yale University Press, New Haven

Bajema RA, Lima SL (2001) Landscape level analysis of Henslow's Sparrow (*Ammodramus henslowii*) abundance in reclaimed coal mine grasslands. Am Midl Nat 145:288–298

Bajema RA, DeVault TL, Scott PE, Lima SL (2001) Reclaimed coal mine grasslands and their significance for Henslow's sparrows in the American Midwest. Auk 118:422–431
Bramble WC, Byrnes WR (1972) A long-term ecological study of game food and cover on a sprayed utility right-of-way. Res Bull 885, Purdue Univ Agri Exp Stn, Lafayette, 20 pp
Bramble WC, Byrnes WR (1979a) Effect of an electric transmission right-of-way on forest wildlife habitat. In: Tillman R (ed) Proceedings of the second international symposium on environmental concerns in rights-of-way management, University of Michigan, Ann Arbor, pp 1–11
Bramble WC, Byrnes WR (1979b) Evaluation of the wildlife habitat values of rights-of-way. J Wildl Manage 43:642–649
Bramble WC, Byrnes WR, Hutnik RJ (1985) Effects of a special technique for right-of-way maintenance on deer habitat. J Arboricult 11:278–284
Bramble WC, Byrnes WR, Schuler MD (1986) Effects of special right-of-way maintenance on an avian population. J Arboricult 12:219–226
Bramble WC, Byrnes WR, Hutnik RJ (1990) Resistance of plant cover types to tree seedling invasion on an electric transmission right-of-way. J Arboricult 16:21–25
Bramble WC, Byrnes WR, Hutnik RJ, Liscinsky SA (1991) Prediction of cover types on rights-of-way after maintenance treatments. J Arboricult 17:38–43
Bramble WC, Yahner RH, Byrnes WR (1992) Breeding bird population changes following right of-way maintenance treatments. J Arboricult 18:316–321
Bramble WC, Yahner RH, Byrnes WR (1994) Nesting of breeding birds on an electric utility right-of-way. J Arboricult 20:124–129
Bridges JM, Lopez R (1993) Reducing large bird electrocutions 12.5 kV distribution line originally designed to minimize electrocutions. In: Tillman R (ed) Proceedings of the first national symposium on environmental concerns in rights-of-way management, Mississippi State University, Starkville, pp 263–265
Bulluck LP, Buehler DA (2006) Avian use of early successional habitats: are regenerating forests, utility rights-of-ways and reclaimed surface mines the same? For Ecol Manage 236:76–84
Cavanaugh JB, Olson DP, Macrigeanis S (1976) Wildlife use and management of power line rights-of-way in New Hampshire. In: Tillman R (ed) Proceedings of the first national symposium on environmental concerns in rights-of-way management, Mississippi State University, Starkville, pp 275–285
Chasko GG, Gates JA (1982) Avian habitat suitability along a transmission-line corridor in an oak-hickory forest region. Wildl Monogr 82:1–41
Confer JL (1992) Golden-winged warbler, *Vermivora chrysoptera*. In: Schneider KJ, Pence DM (eds) Migratory nongame birds of management concern in the Northeast. USDI Fish and Wildlife Service, Newton Corner, pp 369–383
Confer JL (2002) Management, vegetative structure and shrubland birds of rights-of way. In: Goodrich-Mahoney JW, Mutrie D, Guild C (eds) Proceedings of the seventh international symposium on environmental concerns in rights-of-way management, Calgary, pp 373–381
Confer JL, Knapp K (1981) Golden-winged warblers and blue-winged warblers: the relative success of a habitat specialist and a generalist. Auk 98:108–114
Confer JL, Pascoe SM (2003) Avian communities on utility rights-of-way and other managed shrublands in the northeastern United States. For Ecol Manage 185:193–205
Confer JL, Gebhards J, Yrizarry J (1998) Golden-winged and blue-winged warblers at Sterling Forest: a unique circumstance. Kingbird 39:50–55
Cox JJ, Maehr DS (2004) Surface mining and wildlife resources: addition and subtraction on the Cumberland Plateau. Transnati Wildl Am Nat Res 69:236–250
Dennis JV (1958) Some aspects of the breeding ecology of the Yellow-breasted Chat (*Icteria virens*). Bird Band 29:169–183
Denoncour JE, Olson DP (1984) Raptor utilization of power line rights-of-way in New Hampshire. In: Crabtree AF (ed) Proceedings of the third international symposium on environmental concerns in rights-of-way management, San Diego, pp 527–553
Dettmers R (2003) Status and conservation of shrubland birds in the northeastern US. For Ecol Manage 185:81–93

Gates JE, Griffen NR (1991) Neotropical migrant birds and edge effects in a forest-stream ecotone. Wilson Bull 103:204–217

Gates JE, Gysel LW (1978) Avian nest dispersion and fledging success in field-forest ecotones. Ecology 59:871–883

Gottfried BM, Thompson CF (1978) Experimental analysis of nest predation in an old field habitat. Auk 95:304–312

Hagan JM III, Lloyd-Evans TL, Atwood JL, Wood DS (1992) Long-term changes in migratory landbirds in the northeastern United States: evidence from migration capture data. In: Hagan JM, Johnston DW (eds) Ecology and conservation of neotropical migrant landbirds. Smithsonian Inst Press, Washington, DC, pp 115–130

Hanowski JG, Niemi JN, Blake JG (1993) Seasonal abundance and composition of forest bird communities adjacent to a right-of-way in northern forests USA. In: Doucet GJ, Séguin C, Giguère M (eds) Proceedings of the fifth international symposium on environmental concerns in rights-of-way management, Montreal, pp 276–282

Hansson L (1983) Bird numbers across edges between mature conifer forest and clearcuts in Central Sweden. Ornis Scand 14:97–103

Harlow RF (1991) The effect of management treatments on the biomass, nutritive quality and utilization of deer forages on utility rights-of-way. Dissertation, Clemson University, Clemson

Harlow RF, Guynn DC Jr, Davis JR (1993) The effect of management treatments on the biomass, nutritive quality and utilization of deer forages on utility rights-of-way. In: Doucet GJ, Séguin C, Giguère M (eds) Proceedings of the fifth international symposium on environmental concerns in rights-of-way management, Montreal, pp 284–289

Harris LD (1984) The fragmented forest: island biogeography theory and the preservation of biotic diversity. Chicago University Press, Chicago

Holmes RT, Sherry TW (1988) Assessing population trends of New Hampshire forest birds: local vs. regional patterns. Auk 105:756–768

Hunter WC, Buehler DA, Canterbury RA, Confer JL, Hamel PB (2001) Conservation of disturbance-dependent birds in eastern North America. Wildl Soc Bull 29:440–455

Hurst G (1997) Project habitat: ROW management to enhance wildlife habitat and utility image. In: Proceedings of the sixth international symposium on environmental concerns in rights-of-way management, New Orleans, pp 311–315

Hussell DJT, Mather MH, Sinclair PH (1992) Trends in numbers of tropical- and temperate-wintering migrant landbirds in migration at Long Point, Ontario, 1961–1988. In: Hagan JM, Johnston DW (eds) Ecology and conservation of neotropical migrant landbirds. Smithsonian Inst Press, Washington, DC, pp 101–114

Keyser AJ, Hill GE, Soehren EC (1998) Effects of forest fragmentation size, nest density, and proximity to edge on the risk of predation to ground-nesting passerine birds. Conserv Biol 12:986–994

Kremmentz DG, Christie JS (2000) Clearcut stand size and scrub-successional bird assemblages. Auk 117:913–924

Kroodsma RL (1982) Bird community ecology on power-line corridors in east Tennessee. Biol Conserv 23:79–94

Kroodsma RL (1984) Effects of power-line corridors on the density and diversity of bird communities in forested areas. In: Crabtree AF (ed) Proceedings of the third international symposium on environmental concerns in rights-of-way management, San Diego, pp 551–561

Lancia RA, McConnell CA (1976) Wildlife management on utility company rights-of-way: results of a national survey. In: Tillman R (ed) Proceedings of the first national symposium on environmental concerns in rights-of-way management, Mississippi State University, Starkville, pp 177–184

Lanham JD (1997) Attributes of avian communities in early-successional clearcut habitats in the mountains and upper piedmont of South Carolina. Dissertation, Clemson University, Clemson

Lanham JD, Guynn DC Jr (1998) Habitat-area relationships of birds in regenerating clearcuts in South Carolina. Proc Annu Conf Southeast Fish Wildl Agen 52:222–231

Lauzon RD, Grindal SD, Hornbeck GE (2002) Ground squirrel re-colonization of a pipeline right-of-way in southern Alberta. In: Goodrich-Mahoney JW, Mutrie D, Guild C (eds) Proceedings of

the seventh international symposium on environmental concerns in rights-of-way management, Calgary, pp 439–445
Lawson BA, Gates JE (1981) Habitat niche discrimination of passerines along a transmission-line corridor. In: Tillman R (ed) Proceedings of the second international symposium on environmental concerns in rights-of-way management. University of Michigan, Ann Arbor, pp 1–8
LeGrand HE Jr, Schneider KJ (1992) Bachman's Sparrow, (*Aimophila aestivales*). In: Schneider KJ, Pence DM (eds) Migratory nongame birds of management concern in the Northeast. USDI Fish and Wildlife Service, Newton Corner, pp 299–313
Leopold A (1933) Game management. Charles Scribner's Sons, New York, p 481
Lowell F, Lounsberry S (2002) Karner blue butterfly habitat restoration on pipeline right-of-way in Wisconsin. In: Goodrich-Mahoney JW, Mutrie D, Guild C (eds) Proceedings of the seventh international symposium on environmental concerns in rights-of-way management, Calgary, pp 345–354
Mabey SE, McCann J, Niles LJ, Bartlett C, Kerlinger P (1992) Neotropical migratory songbirds regional coastal corridor study. Final report, National Oceanic and Atmospheric Administration, contract # NA90AA-H-CZ839
MacArthur RH, MacArthur JW (1961) On bird species diversity. Ecology 42:594–598
Malefyt JW (1984) Effect of vegetation management on bird population along electric transmission rights-of-way. In: Crabtree AF (ed) Proceedings of the third international conference on environmental concerns, San Diego, pp 570–580
Marra PP, Holberton RL (1998) Corticosterone levels as indicators of habitat quality: effects of habitat segregation in a migratory bird during the non-breeding season. Oecologia 116:284–292
Marshall MR, Cooper RJ, DeCecco JA, Strazanca J, Butler L (2002) Effects of experimentally reduced prey abundance on the breeding ecology of the Red-eyed Vireo. Ecol Appl 12:261–280
Marshall MR, DeCecco JA, Williams AB, Gale GA, Cooper RJ (2003) Use of regenerating clearcuts by late-successional bird species and their young during the post-fledging period. For Ecol Manage 183:127–135
McLoughlin K (2002) Endangered and threatened species and ROW vegetation management. In: Goodrich-Mahoney JW, Mutrie D, Guild C (eds) Proceedings of the seventh international symposium on environmental concerns in rights-of-way management, Calgary, pp 319–326
Moore FR, Kerlinger P, Simons TR (1990) Stopover on a Gulf coast barrier island by spring trans-Gulf migrants. Wilson Bull 102:487–500
Nichols M, Lanham JD (2002) Butterfly and skipper fauna on utility line rights-of-way in the upper Piedmont of South Carolina. In: Goodrich-Mahoney JW, Mutrie D, Guild C (eds) Proceedings of the seventh international symposium on environmental concerns in rights-of-way management, Calgary, pp 337–343
Odum EP (1971) Fundamentals of ecology, 3rd edn. WB Sanders Company, Philadelphia
Pagen RW, Thompson FR III, Burhans DE (2000) Breeding and post-breeding habitat use by forest migrant songbirds in the Missouri Ozarks. Condor 102:738–747
Petit LJ, Petit DR, Martin TE (1995) Landscape level management of migratory birds: looking past the trees to see the forest. Wildl Soc Bull 23:420–429
Pulliam HR (1988) Sources, sinks, and population regulation. Am Nat 132:652–661
Ratti JT, Reese KP (1988) Preliminary test of the ecological trap hypothesis. J Wildl Manage 52:484–491
Ries L, Debinski DM, Wieland ML (2001) Conservation value of roadside prairie restoration to butterfly communities. Conserv Biol 15:401–411
Robbins CS, Sauer JR, Greenberg RS, Droege S (1989) Population declines in North American birds that migrate to the neotropics. Proc Nat Acad Sci 86:7658–7662
Robinson SK (1988) Reappraisal of the costs and benefits of habitat heterogeneity for nongame wildlife. N Am Wildl Nat Res Conf 53:145–155
Robinson SK, Rothstein SI, Brittingham MC, Petit LJ, Grzybowski JA (1992) Ecology and behavior of cowbirds and their impact on host populations. In: Martin TE, Finch DM (eds) Ecology and management of neotropical migratory birds. Oxford University Press, New York, pp 428–460
Rodenhouse NL, Holmes RT (1992) Result of experimental and natural food reductions for breeding Black-throated Blue Warblers. Ecology 73:357–372

Rodewald A, Vitz A (2005) Edge and area sensitivity of shrubland birds. J Wildl Manage 69: 681–688
Roth RR (1976) Spatial heterogeneity and bird species diversity. Ecology 57:773–782
Sauer JR, Hines JE, Thomas I, Fallon J, Gough G (1999) The North American breeding bird survey, results and analysis 1966–1998. Version 98.1. USGS Patuxant Wildlife Research Center, Laurel
Schlossberg S, King DI (2008) Are shrubland birds edge specialists. Ecol Appl 18:1325–1330
Scott PE, DeVault TL, Bajema RA, Lima SL (2002) Grassland vegetation and bird abundances on reclaimed midwestern coal mines. Wildl Soc Bull 30:1006–1014
Sherry TW, Holmes RT (1995) Summer versus winter limitation of populations: conceptual issues and evidence. In: Martin TE, Finch DM (eds) Ecology and management of neotropical migratory birds: a synthesis and review of the critical issues. Oxford Univ. Press, New York, pp 85–120
Small MF, Hunter ML (1989) Response of passerines to abrupt forest-river and forest-powerline edges in Maine. Wilson Bull 101:77–83
Strelke WK, Dickson JG (1980) Effect of forest clear-cut edge on breeding birds in east Texas. J Wildl Manage 44:559–567
Suarez A, Pfennig KS, Robinson SK (1997) Nesting success of a disturbance-dependent songbird on different kinds of edges. Conserv Biol 11:928–935
Thomas DP (2002) Recruitment of gopher tortoises (*Gopherus polyphemus*) to a newly constructed pipeline corridor in Mississippi. In: Goodrich-Mahoney JW, Mutrie D, Guild C (eds) Proceedings of the seventh international symposium on environmental concerns in rights-of-way management, Calgary, pp 465–469
Trani MK, Brooks RT, Schmidt TL, Rudis VA, Gabbard CM (2001) Patterns and trends of early successional forest in the eastern United States. Wildl Soc Bull 29:413–424
Vitz A, Rodewald A (2006) Can regenerating clearcuts benefit mature-forest songbirds? an examination of post-breeding ecology. Biol Conserv 127:477–486
Webb CJ, Bawa KS (1983) Pollen dispersal by hummingbirds and butterflies: a comparative study of two lowland tropical plants. Evolution 37:1258–1270
Whitman JW, Hunter ML Jr (1992) Population trends of neotropical migrant landbirds in northern coastal New England. In: Hagan JM, Johnston DW (eds) Ecology and conservation of neotropical migrant landbirds. Smithsonian Inst Press, Washington, DC, pp 85–95
Wilcove DS (1985) Nest predation in forest tracts and the decline of migratory songbirds. Ecology 66:1211–1214
Yahner RH (2004) Wildlife response to more than 50 years of vegetation maintenance on a Pennsylvania, US right-of-way. J Arboricult 30:123–126
Yahner RH, Wright AL (1985) Depredation on artificial ground nests: effects of edge and plot age. J Wildl Manage 49:508–513
Yahner RH, Ross BD, Yahner RT, Hutnik RJ, Liscinsky S (2004) Long term effects of rights-of-way maintenance via the wire-border zone method on bird nesting ecology. J Arboricult 30:288–294

Chapter 13
Conservation of Early Successional Habitats in the Appalachian Mountains: A Manager's Perspective

Gordon S. Warburton, Craig A. Harper, and Kendrick Weeks

Abstract The plight of species dependent upon disturbed or early successional habitats of Appalachian Mountain forests has been documented by wildlife managers across the region. We conducted surveys of managers and examined State Wildlife Action Plans and initiatives aimed at addressing conservation of these species and their habitats. Although the decline of disturbance-dependent species and the types of habitats they need are well documented, determining the amount and quality of existing early successional habitats is difficult. Few managers have clear goals of how much early successional habitat is needed and where it should be located, although most agree that much more is needed. Recently developed game bird and songbird plans represent some of the best efforts to date to address levels of habitat needed. Managers have prescriptions but face serious social and cultural barriers to establishing early successional habitats on the ground. Ecological forestry and collaborative management approaches may be the best solutions to overcome these barriers.

G.S. Warburton (✉)
North Carolina Wildlife Resources Commission, 783 Deepwoods Drive,
Marion, NC 28752, USA
e-mail: gordon.warburton@ncwildlife.org

C.A. Harper
University of Tennessee, 280 Ellington Plant Sciences, 2431 Joe Johnson Drive,
Knoxville, TN 37996–4563, USA
e-mail: charper@utk.edu

K. Weeks
North Carolina Wildlife Resources Commission, 346 South Mills River Road,
Mills River, NC 28759, USA
e-mail: Kendrick.weeks@ncwildlife.org

C.H. Greenberg et al. (eds.), *Sustaining Young Forest Communities*,
Managing Forest Ecosystems 21, DOI 10.1007/978-94-007-1620-9_13,

13.1 Introduction

> The greatest burden on bird populations in the Appalachian Mountains may be reduced structural diversity and spatial heterogeneity due to insufficient acreage of both older age classes and early successional conditions.
>
> Partners in Flight
>
> *Conservation of the Land Birds of the United States* (Rich et al. 2004)

A diversity of wildlife habitats produces a diversity of wildlife species. This is one of the basic tenets of wildlife management. Landscapes of the Central Hardwood Region, and more specifically the Appalachian Mountains, contain large amounts of mature hardwood forests with trees of similar age and poor understory development. Although this presents a visually appealing experience for those traveling roads such as the Blue Ridge Parkway, the situation for many wildlife species is less than satisfactory. Interspersed are urban and other residential areas complete with manicured lawns and more big trees. The amounts and diversity of habitats needed to support certain guilds of wildlife species are either not available or are in such short supply that viability of populations is a concern. These populations are often subject to frequent local extinctions with little recolonization opportunities (i.e. weak metapopulation dynamics).

Southern Appalachian forests have recovered from the massive disturbances of the last 300 years, particularly the major cutting events of the early 1900s. Since that time, birds associated with mature forests have increased in numbers. Of 60 species in the eastern USA that are not obviously dependent on early successional habitats, over 85% show either stable or increasing populations (Fig. 13.1a, Dessecker in press). None of these species is listed as endangered, threatened or special concern. Such is not the case for 128 species (98 excluding species that use canopy gaps) that depend on disturbance (Fig. 13.1b, Dessecker in press). Fourteen of these species are federally listed. Many scientists and wildlife managers are concerned about the plight of species that depend on early successional habitats in the eastern USA (Askins 2001; Litvaitis 2001; North Carolina Wildlife Resources Commission 2005; Appalachian Mountains Joint Venture Management Board 2008).

As a result of logging at the turn of the last century, 80% of Southern Appalachian forests are now in a mid-successional stage (80–100 years old; Fig. 13.5a, b). This is a challenging situation for wildlife managers because the large area makes effective management treatments difficult. On top of this recent history, pre-European historical and "natural" disturbance regimes are either absent or very altered (White et al., Chap. 3; Spetich et al., Chap. 4). Loss of soil layers and invasive exotic plant species are two key influences that severely hinder our ability to restore ecosystems to previous historical conditions. Ecological relationships are severely altered and keystone species are no longer present. Compounding the problem is ever increasing human population and urbanization, which superimpose a myriad of problems for wildlife in areas that appear to be providing suitable habitats on private lands.

Wildlife biologists once considered early successional wildlife species to be habitat generalists. Research over the last decade has refuted this hypothesis; indeed,

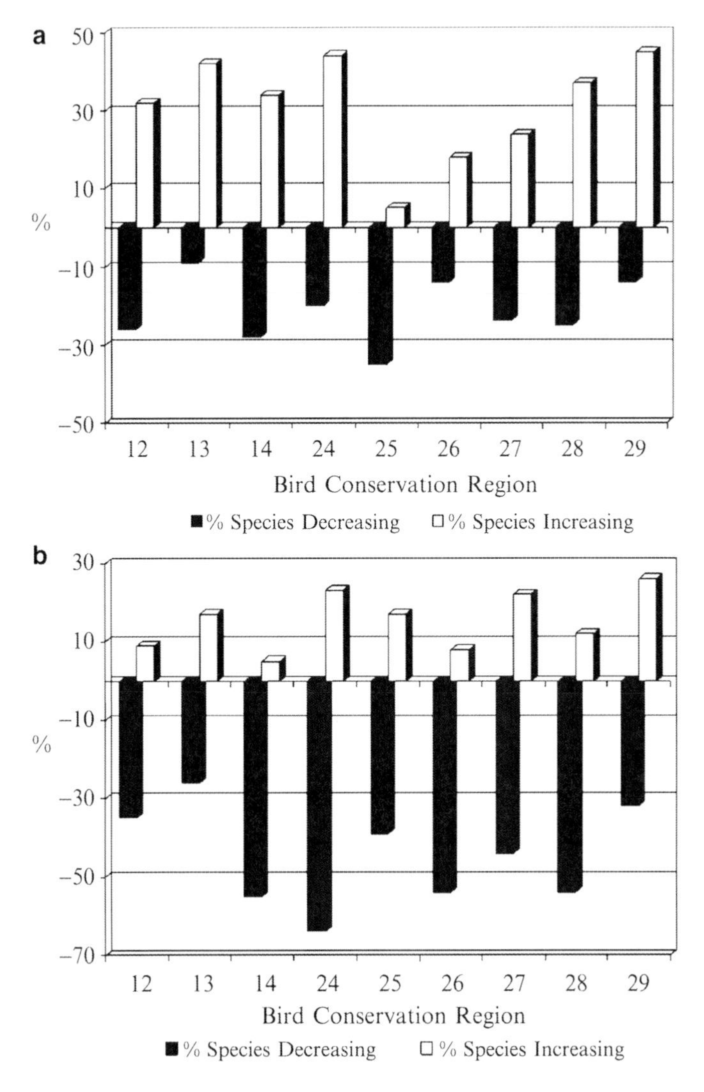

Fig. 13.1 (**a**) Percent of bird species characteristic of mature forest habitats that are experiencing population increases exceeds the percent that are decreasing based on Breeding Bird Survey data (1966–2007) for Bird Conservation Regions in the eastern USA (including BCR 28 – Appalachian Mountains). Graph is from Dessecker (in press). (**b**) The percent of bird species characteristic of early successional habitats that are experiencing population declines exceeds the percent that are increasing based on Breeding bird survey data (1966–2007) for Bird Conservation Regions in the eastern United States (including BCR 28 – Appalachian Mountains). Graph is from Dessecker (in press)

many early successional habitat species are specialists (DeGraaf and Yamasaki 2003). Although areas may be in an early successional stage, they may lack necessary structural complexity, composition, size, or spatial characteristics important for many early successional species (Askins 2001). Many early successional habitat patches have become isolated in either forested areas or within urban areas where domestic predators (e.g., house cats), automobile kills, and other mortality or disturbance agents make them unsuitable for habitation (Litvaitis 2003).

The plight of disturbance-dependent species has drawn the attention of wildlife biologists, Non-Government Organizations (NGOs), bird enthusiasts, and hunters. The American Bird Conservancy listed early successional habitat in eastern deciduous forest as one of the top 20 threatened habitats in North America. The Audubon Society's Watch List emphasizes the plight of grassland and shrub/scrub species. Increasingly throughout the last decade, the shortage of early successional habitats and its impacts on disturbance-dependent species has been the focus of numerous special issues of professional publications (e.g., The Wildlife Society Bulletin 2001, vol. 29; Forest Management and Ecology 2003, vol. 185), workshops, research and involvement of NGOs.

This chapter examines early successional habitats from a manager's perspective. Wildlife managers are charged with maintaining healthy and viable populations of wildlife species. We first examine which species and habitats need management attention. Next, we examine ways in which managers are determining how much early successional habitat is on the landscape, how biologists are determining the amounts and types needed, and where it should be placed at the local level. This leads to a discussion of how we create or develop these habitats on the landscape. Finally and perhaps most importantly, we discuss challenges managers face in creating these important habitats over large areas given the social and ecological concerns with creating disturbances in our forested landscape.

13.2 Methods and Study Area

We focused on the Appalachian Mountains Bird Conservation Region (BCR) (Fig. 13.2) and looked more closely at the Southern Appalachians in many cases. Early successional habitats were defined as young forest (0–20 years old) and shrub-scrub habitats typically found in upland forested areas. Forest types corresponded roughly with the Eastern Mesic Subregion described by McNab (Chap. 2). We conducted informal surveys of wildlife managers in eight states and examined Wildlife Action Plans in ten states in the Appalachian Mountain BCR. Wildlife Action Plans are comprehensive conservation plans developed by state wildlife agencies to be eligible for federal funding under the State Wildlife Grants program. We also reviewed regional and state initiatives and plans developed by the Appalachian Mountain Joint Ventures (Appalachian Mountains Joint Venture Management Board 2008) and other cooperatives.

13.3 What Species? Which Habitats

Hunter et al. (2001) listed 128 species of birds that depend on some type of early successional habitat. Every State Wildlife Action Plan (SWAP) we reviewed included disturbance-dependent species and their habitats as priorities. The number

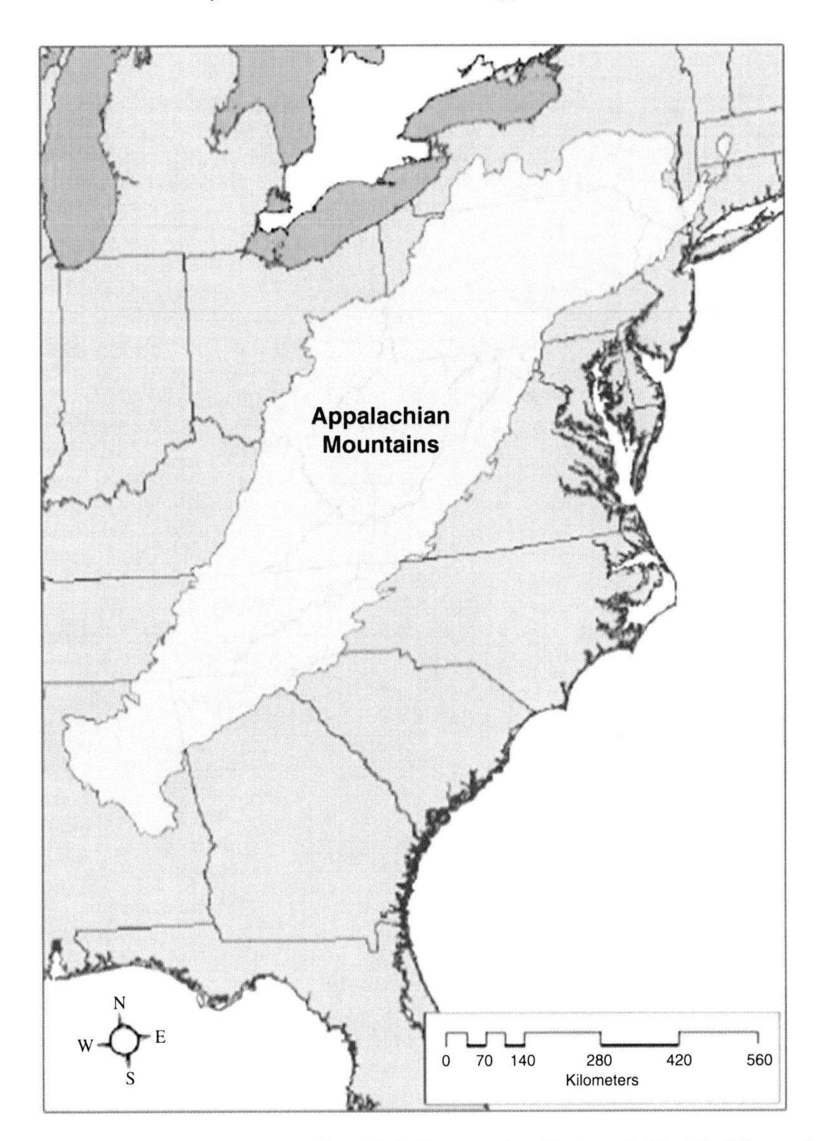

Fig. 13.2 The Appalachian Mountains Bird Conservation Region (BCR 28) (Figure from AMBCRP Appalachian Mountains Bird Conservation Region Partnership 2005)

of species listed ranged from 7 to 80 and included birds, mammals, herpetofauna and game species (Table 13.1). As seen in Table 13.1, the different SWAPs used a variety of terminology to define types of early successional habitat. There is a need to develop specific definitions of early successional habitat types. It is clear that managers across the Appalachians consider many disturbance-dependent species and their habitats to be priorities for management.

Table 13.1 Priority early successional wildlife species and habitats identified in State Wildlife Action Plans for selected states in the Appalachian Mountains

State	Priority species using early successional habitats	Types of early successional habitat
GA	Golden-winged Warbler, Grasshopper Sparrow, Bewick's Wren, Appalachian cottontail, least weasel, star-nosed mole, and pygmy shrew	High-elevation early successional habitats. Includes a variety of vegetation types found at high elevations that are maintained by periodic natural or anthropogenic disturbance, mountain bogs, wet meadows
KY[a]	American Woodcock, Bachman's Sparrow, Barn Owl, Bell's Vireo, Bewick's Wren, Blue-winged Warbler, Golden-winged Warbler, Least Flycatcher, Long-eared Owl, Northern Bobwhite, Prairie Warbler, Red-headed Woodpecker, Willow Flycatcher, Appalachian cottontail, evening bat, Indiana bat, Kentucky red-backed vole, Rafinesque's big-eared bat, southeastern myotis, swamp rabbit, Virginia big-eared bat, coal Skink, corn snake, eastern coachwhip, eastern slender glass lizard, northern pine snake, northern scarlet snake, scarlet kingsnake, six-lined racerunner, southeastern crowned snake, southeastern five-lined skink, timber rattlesnake, western pygmy rattlesnake	Savannah/shrub-scrub terrestrial habitat and "transitional zones" from grassland to forest (both savannahs and woodlands), early successional stages of forest (e.g., regenerating stands, reforestation projects, latter stages of "old fields" where shrubs and young trees dominate), and previously mined areas currently in the shrub-scrub successional stage
NC	Grasshopper Sparrow,Whip-poor-Will, Common Nighthawk, Northern Bobwhite, Prairie Warbler, Chestnut-sided Warbler, Bobolink, Alder Flycatcher, Willow Flycatcher, Horned Lark, American Kestrel, Orchard Oriole, Savannah Sparrow, Vesper Sparrow, American Woodcock, Field Sparrow, Eastern Meadowlark, Eastern Kingbird, Barn Owl, Golden-winged Warbler, Blue-winged Warbler, rock vole, meadow vole, least weasel, Appalachian cottontail, meadow jumping mouse, timber rattlesnake, coal skink, smooth greensnake, eastern box turtle	Ancient grassy balds on or adjacent to broad ridgetops (containing a variety of unique grass and herb species), shrub-dominated heath balds (alder, rhododendron and mountain laurel are common dominant species), lower elevation fields, meadows, pastures, and clear cuts resulting from agriculture or forestry activities
VA	Golden-winged Warbler, loggerhead shrike, smooth greensnake, Northern Harrier, Barn Owl, timber rattlesnake, chuck wills widow, whip poor will, Northern Bobwhite, Prairie Warbler, Yellow-Breasted Chat, Eastern Towhee, Field Sparrow, Eastern Meadowlark, Appalachian cottontail, American Woodcock	Open vegetated habitats – only delineated into herbaceous and shrub/scrub
WV	Northern Harrier, Golden-winged Warbler, Appalachian cottontail, Grasshopper Sparrow, Bobolink, Horned Lark, Vesper Sparrow, Blue-winged Warbler, Prairie Warbler, Field Sparrow, Appalachian Bewick's Wren	Successional conifer forests and woodlands, successional deciduous forest, old fields, anthropogenic grassland, heath/grass barrens and balds, marshes and wet meadows

GA Georgia, *KY* Kentucky, *NC* North Carolina, *VA* Virginia, *WV* West Virginia
[a]Omitted grassland-agricultural species from table

Table 13.2 Types of databases used to delineate early successional habitats by state agencies

- National Land Cover Database
- Forest Inventory and Analysis
- Gap Analysis Program (GAP)
- National Vegetation Classification Standard
- NatureServe Vista
- Aerial photos
- Continuous Inventory of Stand Condition (FSVeg used now)
- State Forest Inventory
- National Wetland Inventory
- National Vegetation Classification System

13.4 How Much Is Out There?

Managers throughout the Appalachian Mountains use a variety of methods to determine how much early successional habitat is on the landscape (Table 13.2). The methods employed, however have shortcomings and limitations. In many cases, agencies use gross estimates with no specific field verification or ground-truthing to validate these approximations. Because early successional habitats are ephemeral, and in many cases occur in small patches, they are difficult to quantify, and databases need frequent updating. In addition, early successional habitats can be further divided into classifications that have relevance to specific species (i.e. grass/forbs, shrub/scrub, sapling, or mosaics of these structural elements). The two primary databases used to estimate the extent of early successional habitats are the National Land Cover Dataset (NLCD) and the Forest Inventory and Analysis (FIA).

13.4.1 National Land Cover Dataset

In 1992, a consortium of federal agencies purchased Landsat 5 imagery of the USA, which they used to develop the NLCD (MRLC 2010). A second generation version was produced in 2001 using Landsat 7 imagery. The NLCD contains 21 land cover categories. Figure 13.3 shows eight categories of the NLCD for the Appalachian Mountains. Although the NLCD gives managers a good overview of the landscape, it misses many details needed to quantify early successional habitats. In the Southern Appalachians, 30% of the landscape is characterized as early successional (SAMAB 1996); however, this is misleading. Over 26% of the early successional designation consists of pastures, croplands, water or developed areas. These areas are usually poor quality early successional habitats compared to recently disturbed hardwood forest (see Greenberg et al., Chap. 8) due to species composition and structure, poor agricultural management, and human disturbance (but see Lanham et al., Chap. 12). Estimates of semi-permanent early successional habitats, like balds and old fields, approximate 1% in the Southern Blue Ridge (North Carolina Wildlife Resources Commission 2005).

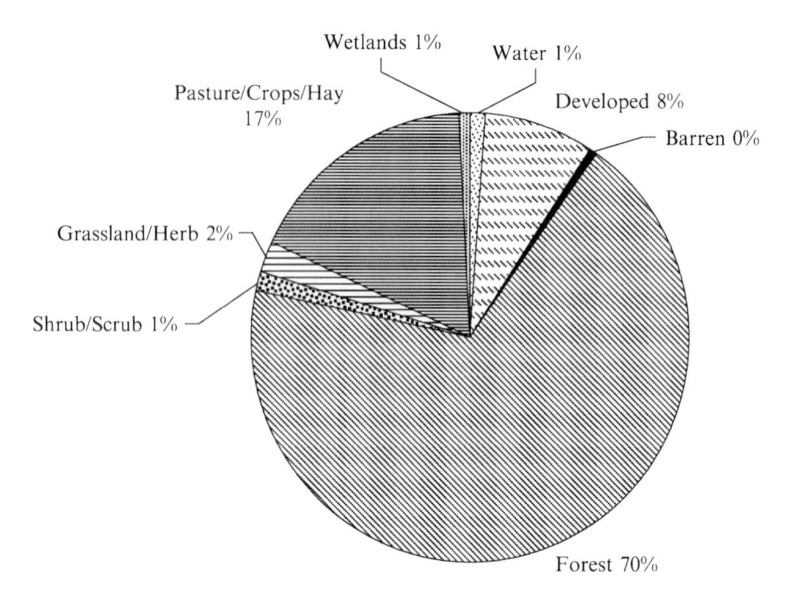

Fig. 13.3 Estimates of percent cover or habitat classes based on the National Land Cover Data (MRLC 2010) for the Appalachian Mountain Bird Conservation Region. The 21 classes of land cover data were combined into eight categories. Data do not provide information about the quality or condition of the habitat types (data summary from Appalachian Mountains Joint Venture Management Board 2008)

13.4.2 Forest Inventory and Analysis

The US Forest Service's FIA is based on periodic field surveys of total 1.0 acre plots stratified by county and state (Trani et al. 2001). They adhere to a strict set of national standards for accuracy (Hansen et al. 1992). Fixed radius and variable radius prism points are used to measure trees within plots, and area expansion factors are used to extrapolate measurement to the larger population. Typically, managers or researchers make inferences about early successional habitats using stand-level diameter size class. Diameter class is a surrogate for stand age and the size class of interest is seedling-sapling (trees less than 12.7 cm diameter and 30.5 cm in height) because it represents stands that are 0–20 years old. In some cases, non-stocked stands include some aspects of seedling-sapling and are included in analyses. An example of using these data to examine Ruffed Grouse (*Bonasa umbellus*) habitat is shown in Fig. 13.4 (also see Shifley and Thompson, Chap. 6; Franzreb et al., Chap. 9). Small diameter forests had declined nearly 40% in the Appalachian Mountains from 1980 to 2005.

Problems using FIA data can affect results. Data are summarized at the county level, and in cases where counties do not correspond to physiographic or ecological boundaries, results can be misinterpreted. Online tools are available however that allow a user to define an analysis area that does not correspond to county boundaries. As the survey is always ongoing, not all states are surveyed during the same year.

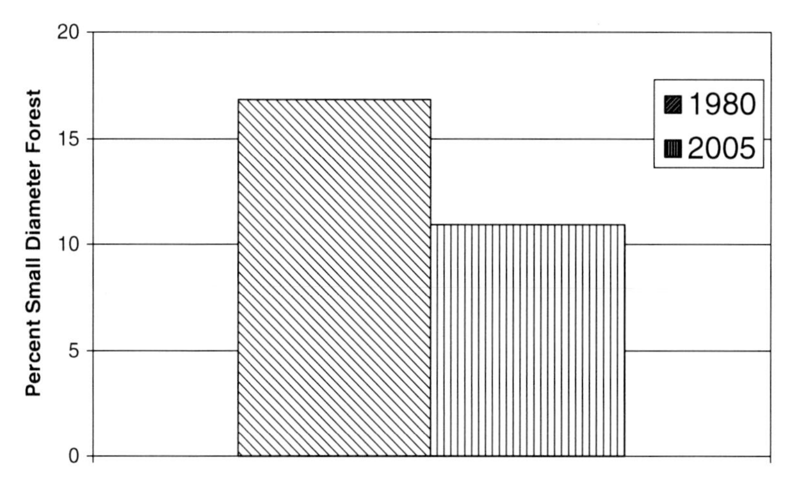

Fig. 13.4 Percent of small diameter forest in 1980 and 2005 in the Appalachian Mountain Bird Conservation Region using FIA data (compiled data from Dessecker et al. 2006)

This presents issues when combining states. Unfortunately, methodologies have changed over the years and some comparisons cannot be made between years in many cases. In addition, many early successional habitats may not be considered as "forestland" and therefore are not included in forestland totals.

13.4.3 US Forest Service Inventory Data

The US Forest Service used a Continuous Inventory of Stand Condition (CISC) system to track forest types, and that has been replaced recently by a system titled Field Sampled Vegetation (FSVeg). This intensive on-the-ground inventory can be used to track stand age and document forest age composition as illustrated for all the national forests in the Appalachian Mountains (Fig. 13.5a). Age class 0–10 is currently 1.1% of national forest land, and 11–20 year old stands comprise 4.5%. The 0–10 age class has declined from highs of nearly 5% in the early 1990s to just above 1% currently. Figure 13.5b shows the decline in acres that are regenerated each year on national forests in the Appalachian Mountains.

13.4.4 Problems with Measuring Early Successional Habitats

Many of the coarser-grained data sets miss important smaller-scale early successional habitats. In addition, there is no indication of their quality or structure. From a wildlife perspective, these quality issues can be critical. Inconsistent definitions of early successional habitat among databases becomes problematic; FIA values can be affected by

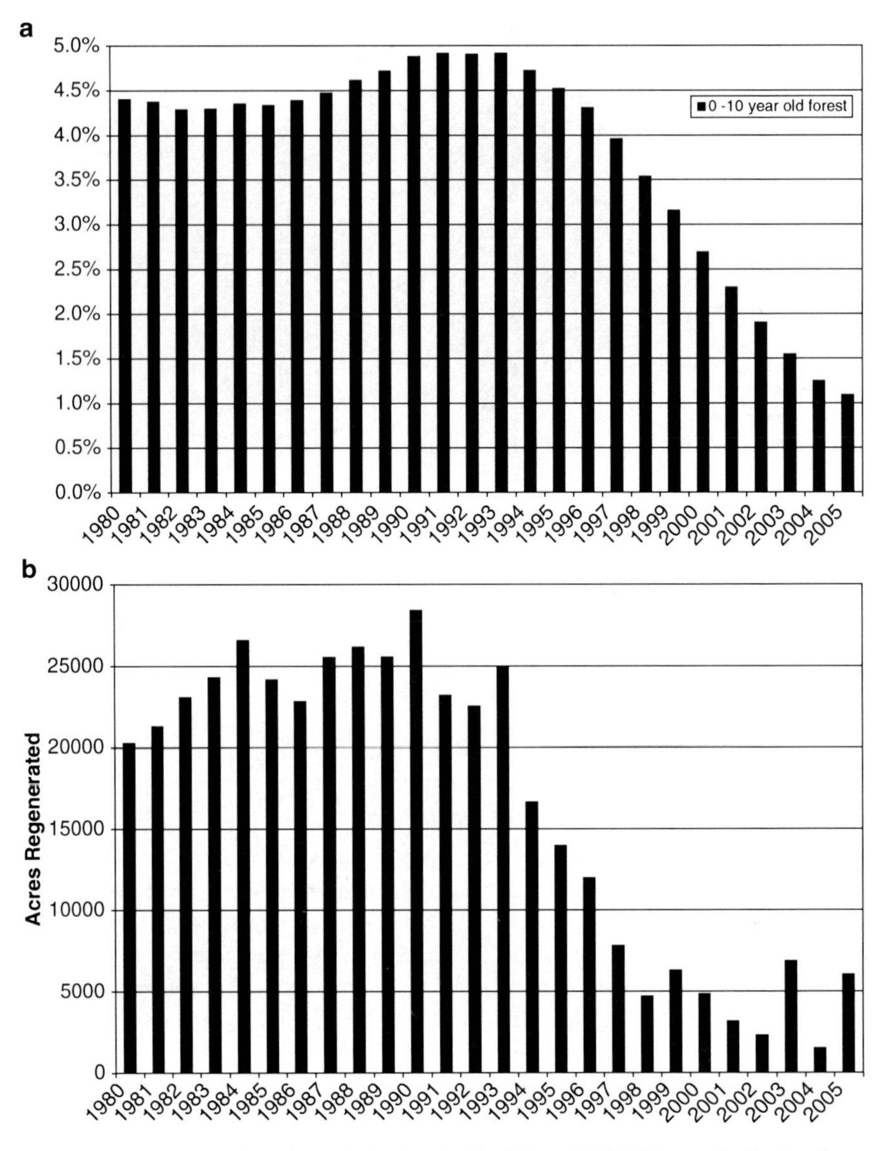

Fig. 13.5 (**a**) Percent of national forests in the Appalachian Mountain Bird Conservation Region that are in the 0–10 age class by year (data from FSVeg dataset, USDA Forest Service, Dave Casey 2011, *pers. comm.*). (**b**) Acres of regeneration in national forests in the Appalachian Mountain Bird Conservation Region by year (data from FSVeg dataset, USDA Forest Service, Dave Casey 2011, *pers. comm.*)

changes in methodologies and definitions from one sample period to the next (Trani et al. 2001). Finally, many databases do not measure these ephemeral early successional habitats at a frequency that allows managers to track them with any consistency or accuracy. There is clearly a need for new solution for measuring disturbed habitats.

13.4.5 LiDar

One of the most promising techniques for measuring early successional habitats is Light Detection and Ranging (LiDar). LiDar is a remote sensing system used to collect topographic data using aircraft-mounted lasers. LiDar takes elevation measurements by determining the time delay between transmission of laser pulses and detection of the reflected signal. A Global Positioning System (GPS) calculates location, with both the LiDar and GPS collectively assigning a series of x, y, and z coordinates. These data allow for the generation of a digital elevation model of the ground surface which can be used to determine the height of vegetation. For example, Fusting (2009) used LiDar to document the structure of a high elevation bald on Roan Mountain, North Carolina. He was able to classify and discern grasses from shrubs and extract blueberry bushes that measured 10′ × 10′ × 2′ (Fig. 13.6). Modern software (Fusion) developed by the US Forest Service now allows users to manipulate LiDar layers on personal computers (McGaughey 2010). The result is a map of user-defined canopy heights which are a much better indicator of forest structure than stand age. Further exploration of this technology is needed, but early indications are positive.

13.5 How Much Early Successional Habitat Is Needed and Where on the Landscape?

How much early successional habitat is needed and where it is needed are difficult questions for managers. Answers to these questions are being addressed at regional levels; however attempts to step down large-scale estimates to these local levels are virtually lacking. All states surveyed indicated that early successional habitats were important in their jurisdiction, and that regardless of whether the amounts were quantified, all agreed that there are not enough quality early successional habitats in the Appalachian Mountains. Most states have general goals to increase the amount of early successional habitats (Oehler 2003), but do not have specific targets for amounts or locations. State agencies typically have rather aggressive goals for increasing early successional habitats on state-owned areas. In general, in providing advice to the US Forest Service, many managers still recommend that between 8% and 12% of forests be in the 0–10 year age class. This recommendation is most likely related to maintaining balanced age classes in forests managed at 80–100 year rotations (see Shifley and Thompson, Chap. 6) but it is also related to considerations for species such as the Golden-winged Warbler (*Vermivora chrysoptera*) (D. Buehler 2010, *pers. comm.*). States are attempting to use regional plans to set goals for providing early successional habitats for both game and non-game species, and it is worth looking at some of these regional initiatives.

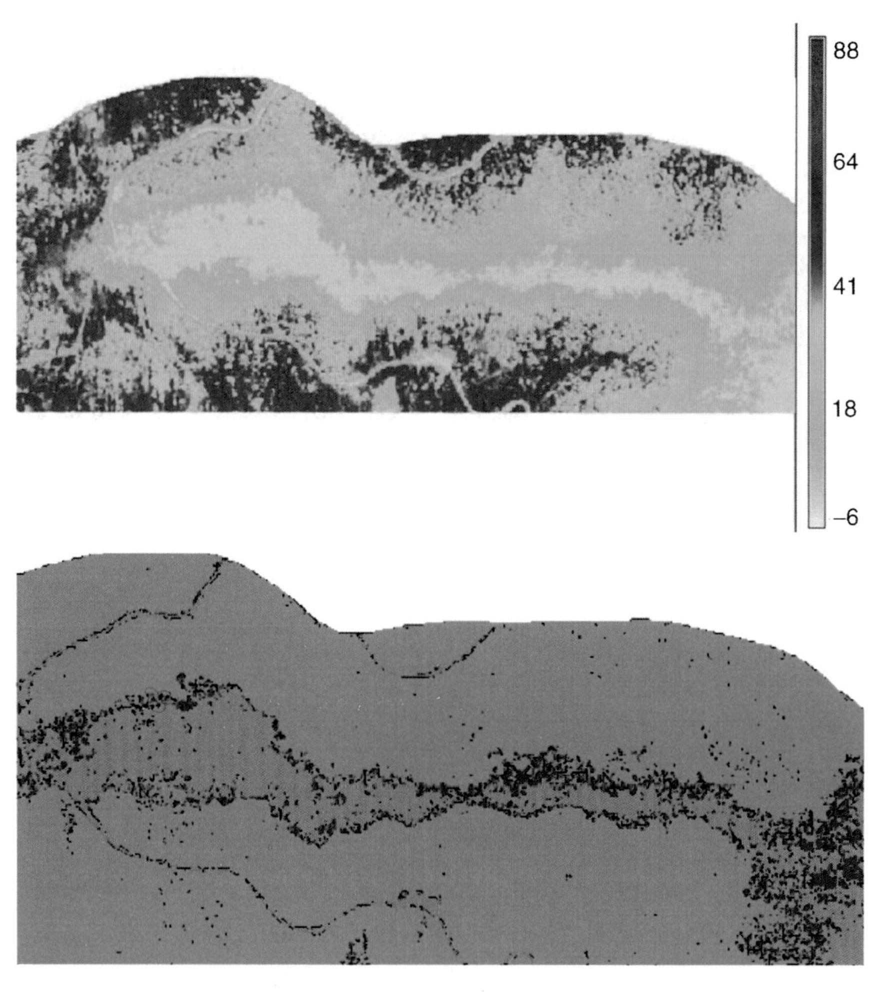

Fig. 13.6 Vegetation layer on Roan Mountain, North Carolina, created from LiDar data (Fusting 2009). The top figure shows the raw data showing cell heights. The lower figure shows these data reclassified into polygons. Green – Forest canopy; Brown – Bald; purple – shrubs (graphs are from Fusting 2009)

13.5.1 Single Species Plans

Regional plans for three major disturbance-dependent game species: Ruffed Grouse (Dessecker et al. 2006), American Woodcock (*Scolopax minor*) (Kelley et al. 2008) and Bobwhite Quail (*Colinus virginianus*) (Dimmick et al. 2002) have been developed. All three plans establish population and habitat goals for specific regions of the USA, and these are stepped down to individual states. Importantly, all plans reference numerous species associated with the featured game bird of interest. These regional initiatives all rely on multi-state and multi-agency cooperation, and they represent excellent starting points for setting goals that managers can use as guidelines for areas within their respective states or management areas.

The Ruffed Grouse management plan (Dessecker et al. 2006) is an example of the regional initiative process. The plan's major goal is to identify habitat availability and management objectives required to sustain or restore Grouse populations to 1980 levels. Recognizing that farm abandonment during the mid-twentieth century may have led to Grouse populations higher than historical norms, the plan authors selected 1980 as the year when abandoned lands moved past functional Grouse habitat. The target year for accomplishing plan objectives is 2025 with a goal of applying proactive habitat management at regular intervals to provide a steady supply of quality Grouse habitat on the landscape. This notion of "steady supply of quality habitat" implies active management to sustain species at desired levels as opposed to passive management schemes relying more on natural or random disturbance events.

The Ruffed Grouse management plan used data from the 1980 and 2005 FIA inventories, and BCRs as analysis units. The seedling-sapling or small diameter (<5″ dbh) size class was used as a surrogate for stand age. Drumming male density was estimated for forest types from published results, regional surveys and professional opinion. If stands were in optimal age classes for Grouse, density estimates were then doubled to account for habitat quality differences. By tying population levels to habitat conditions, the plan authors derived population estimates for 1980 and 2005 and then used these figures to determine habitat needed to restore Grouse to 1980 population levels. They established goals for each BCR and then stepped these results down to each state. Table 13.3 shows the amount of habitat needed to establish Grouse populations at 1980 levels for the Appalachian Mountains BCR. In this BCR, meeting plan objectives requires increasing the current proportion of small diameter forest by 10% (7.3 million acres), and maintaining this amount requires annual even aged treatments on 364,000 acres per year.

Table 13.3 An example of a "step down" process from a regional plan to individual states. The Ruffed Grouse Conservation Plan (Dessecker et al. 2006) outlines the current amount of annual even-age management, and the amount of small-diameter forest and annual management required to maintain or restore Ruffed Grouse populations to 1980 levels

State	Current annual even-age treatment (acres)	Small-diameter forest objective (acres)	Even-age management annual objective (acres)
Arkansas	7,000	159,100	8,000
Georgia	23,400	515,400	25,800
Kentucky	23,900	26,500	26,300
North Carolina	34,100	729,000	36,500
Ohio	46,300	1,013,400	50,700
Pennsylvania	91,100	1,992,100	99,600
Maryland	3,700	80,700	4,000
Tennessee	28,500	626,900	31,300
Virginia	45,200	962,400	48,100
West Virginia	48,200	1,060,400	53,000

Source: The Ruffed Grouse Conservation Plan (Dessecker et al. 2006)

Similar goals were established in the American Woodcock Conservation Plan (Kelley et al. 2008). The primary goal of this plan is to halt the decline of Woodcock and restore densities to levels that provide opportunity for use of the resource. The premise of the plan is that loss of young forest habitat is responsible for declines in Woodcock recruitment and overall population status. The baseline year established for restoration is 1970, as that is when Woodcock populations began to decline. The plan attempts to set goals for restoring Woodcock densities rather than absolute population size because many areas that were habitat in 1970 may never be developed into suitable Woodcock habitat. To restore Woodcock in the Appalachian Mountains BCR to 1970 levels, managers need to add 88,000 singing males to the population and create three million acres of new Woodcock habitat. Again, the Woodcock plan sets out both regional and state objectives.

13.5.2 Multi-Species Efforts

We have covered game species plans and recognized that many other species will likely benefit from accomplishing goals outlined in these plans. What about plans or initiatives for nongame species? Most SWAPs prioritize the need to increase the area of quality early successional habitats for a variety of species. By the very nature of addressing the conservation needs of many species, goals in SWAPs tend to be very general in nature and they tend to reference various regional initiatives for amounts of early successional habitats needed. We will look at initiatives related to bird conservation and examine how the various geographic levels are organized.

The North American Bird Conservation Initiative Committee is a collection of government agencies, NGOs, and other initiatives dedicated to coordinating bird conservation in North America. The Committee acts as an umbrella for various other ventures, initiatives, and cooperatives involved in bird conservation. Over the last 20 years, a number of bird conservation initiatives have produced national and international bird conservation plans. Typically, these plans include assessments of species status, population and habitat goals, threats, and monitoring needs. Partners in Flight is an example of such an initiative. Partners in Flight has produced both a North American Landbird Conservation Plan (Rich et al. 2004) and Landbird Conservation Plans for individual states and BCRs which provide more details than the continental plans. For example, the Partners in Flight Bird Conservation Plan for the Southern Blue Ridge sets a population goal to establish 5,000 pair of Golden-winged Warblers (Hunter et al. 1999), but habitat targets and locations are very general. These initiatives have spurred a number of regional plans that are more detailed and specific to BCRs. In order to implement plans and continue the step down process, Joint Ventures, which are regional partnerships between public and private agencies and groups, have taken a lead role in attempting to create on-the-ground translation of plan objectives (USDI Fish and Wildlife Service 2010a). There are a number of state-based bird conservation initiatives throughout the Appalachian Mountains, and many projects identified by the Joint

Ventures are implemented through State Wildlife Grants administered through state nongame programs and specified in SWAPs.

Although some progress has been made in determining amounts of early successional habitats needed and locations at the regional and state levels, there is still a need for further step-down at more local levels. This question is being addressed aggressively by the Joint Ventures and other conservation groups (TNC 2000). Specifically, the Appalachian Mountains Joint Venture is focused on making great strides in the area of conservation design. Conservation design "refers to identification of specific areas with landscape or habitat characteristics that will sustain viable populations at target levels, in this case, for priority bird species" (Appalachian Mountains Joint Venture Management Board 2008). The conservation design process uses biological information, and specifically species – habitat relationships and viability modeling, to evaluate landscapes currently and into the future. This process will provide tools for managers to determine how much habitat and where it is needed over large areas, and it will also evaluate trade-offs between species with contradictory habitat needs (e.g., early successional vs. mature forest species). Ultimately conservation design processes should produce a spatially explicit conservation blueprint of future desired condition to sustain populations (Appalachian Mountains Joint Venture Management Board 2008). State Wildlife Agencies would do well to support these efforts because they can be devised to not only test competing hypotheses of land management options, but also because they will allow for the most expedient use of limited resources to manage wildlife populations.

In an attempt to include all taxa, Land Conservation Cooperatives have been formed by the US Fish and Wildlife Service (USDI Fish and Wildlife Service 2010b). The Land Conservation Cooperatives are intended to support a landscape-scale, collaborative approach to conservation. Like the Joint Ventures, the Land Conservation Cooperatives are to assist with biological planning, conservation design, conservation delivery, monitoring, and research. As SWAPs are revised and rewritten, they will include both game and nongame in an attempt to be truly comprehensive.

13.5.3 Setting Goals for Early Successional Habitats – Use of Appropriate Benchmarks

In determining the quantity and location of early successional habitats, one must determine appropriate benchmarks for population or habitat goals. There are several options for selecting points of reference or targets. One could choose a time in recent history or a historical reference point before European settlement. The question becomes at what time in history does one "stop the car" (Dessecker 1997)?

The use of historical benchmarks for setting population or habitat goals can be problematic. Lorimer (2001, 2003) found the use of historical disturbance regimes to guide modern day systems to have two major issues. First, given that pre-European settlement disturbance regimes were variable in time and space, how does one define "natural" vegetation? Second, anthropogenic fire greatly shaped many ecosystems in ways we may not be able to document (see Spetich et al., Chap. 4), and "natural" is again difficult to discern (Frost 1998). Lorimer (2001) posed a case study involving open oak woodlands. This community type was present historically, but the fire disturbance regime to keep these areas open most likely was more intense than natural fires alone. The management question becomes do we burn oak woodlands to keep them open since they were here and many species are adapted to this habitat, or do we let nature take its course in which case they become a late successional forest dominated by red maple (*Acer rubrum*), operating with gap dynamics? There are fair arguments on both sides, but one could end up losing a unique ecosystem. Cultural considerations certainly come into play; are Native American activities considered "natural"?

A more useful approach for wildlife management may be a proactive or "wildlife first" approach. Under this system, managers look forward and identify desired future conditions, such as amount and types of early successional habitats needed to maintain viable populations of target disturbance-adapted wildlife species and then create them accordingly (see Shifley and Thompson, Chap. 6). In this way, we can plan for a steady supply of quality early successional habitats on the limited areas available for management on the landscape (see Ruffed Grouse management plan). Indeed, Lorimer (2001) seemed to recognize this approach, stating "a more clearly defined role for scientific input might be questions on amount of habitat needed to maintain viable populations of early successional species."

13.6 How Do We Get Early Successional Habitats on the Ground?

If we know the disturbance-dependent species of interest and habitat requirements and we have a sense of the total amount of habitat needed in an area, we now must set about actually getting the habitat created or developed on the ground. This task can be broken down into two parts: management prescriptions and actual techniques for manipulating vegetation and the land (see Loftis et al., Chap. 5). Ultimately, the success of getting these important early successional habitats on the ground will depend on our ability to overcome challenges to actually doing the work.

13.6.1 Establishing Early Successional Habitats

Table 13.4 lists the actions suggested in SWAPs and by state biologists for creating early successional habitats. Many of these are standard techniques (e.g., Loftis et al., Chap. 5) used by wildlife managers in the USA for over 80 years. Some recurring themes from the informal survey of wildlife managers were that managers have experience developing disturbed habitats, and there is large body of literature available about using these techniques. In addition, we have habitat prescriptions for

Table 13.4 Actions suggested in selected State Wildlife Action Plans to create early successional habitats for priority species

Acquisitions, prescribed fire, integrate other programs like Northern Bobwhite Conservation Initiative, invasive species control – Alabama
Acquisitions (connect blocks), regional land use planning, livestock BMPs, cost share incentives, encourage prescribed fire, encourage natives, control invasives, identify and protect areas at risk of urban development; also developed approach to prioritize areas to focus conservation efforts -Kentucky
"The needs of these [scrub-shrub and early successional] birds, including game species such as American Woodcock and Northern Bobwhite, should be considered within the context of forest habitat objectives." – Georgia
Habitat descriptions in the plan include "key factors" and actions that will restore habitat: Composition (addresses invasives), Fire Regime (seasonality and frequency), Remoteness (road density), and Spatial Ecology (patch proximity, block size, block number) – Arkansas
Acquisition of conservation lands (we need more *quality* early successional habitat, e.g., grass balds). Consider increasing the size of timber harvest areas where appropriate to support greater variety and density of early successional "area sensitive" species. Control of exotic species. Implement conservation measures on private lands through various programs and initiatives. Increased management of balds - North Carolina

many species, even though we may not have complete information for all species. There seems to be a sense of urgency among managers for the need to get going and creating a steady supply of early successional habitats for wildlife in many of our Appalachian Mountain landscapes. A common comment was "we have the prescriptions, now let's put the habitat on the ground". Harper et al. (in press) provide excellent habitat management prescriptions for Ruffed Grouse in the Appalachian Mountains and there are many other such prescriptions for other species.

States have a variety of programs on both private and public lands. One fairly common approach is to define focal areas, which are areas with reasonable potential for habitat development and with some minimum population level identified for the species of interest. Once defined, management efforts and resources are focused in these areas in an attempt to build strong core populations. The Pennsylvania Game Commission recommends using public lands as "hubs" for habitat development and then complimenting these efforts with adjoining private landowners, analogous to spokes on a wheel (B. Jones 2010, *pers. comm.*).

13.6.2 Barriers to Establishing Early Successional Habitats

Wildlife managers identified several significant barriers to developing early successional habitats on the ground (Table 13.5). The lack of clear goals and the lack of adequate funding to create early successional habitats are two important barriers. Although most biologists recognize that early successional habitats are in short supply, there are no clearly defined targets for the types needed or where they are needed, other than general regional initiatives. The increased "parcelization" of

Table 13.5 Barriers to creating quality early successional habitats on the ground as identified in State Wildlife Action Plans and a survey of wildlife managers in the Appalachian Mountains

- Lack of specific habitat goals
- Lack of adequate funding
- Increasing "parcelization" – size of private tracts decreasing
- Lack of market for roundwood
- Fire suppression
- Increasing development
  - Reduced management potential – fire
  - Negative urban effects – reducing quality of early successional habitats
- Non-native invasive species
- Public opposition to techniques used to create early successional habitats
- Assumption that only game species need habitat management
- Visual values (scenery) more important than ecological or wildlife values
- Poor agricultural practices
- Clean farming
- Unmanaged recreational uses
- Incompatible road and utility corridor management

private lands is another important barrier (Litvaitis 2003; see Wear and Huggett, Chap. 16). As tracts become more divided each time they are sold, potential management options become more limited. The ability and willingness of landowners to engage in forest management practices declines with tract size. A private landowner who owns 20 acres is most likely not going to cut 15 acres for Grouse habitat. Further, poor timber markets for smaller-sized trees were another barrier mentioned by state biologists (see Wear and Huggett, Chap. 16). This eliminates monetary incentives to pay for creation of certain early successional habitat types. Along with increased parcelization, increased rural land development is problematic not only for reducing the creation of early successional habitats (Wear and Huggett, Chap. 16), but also negatively impacting the quality of habitats that are actually out there. Disturbance, cats and dogs, automobile mortality, exotics and a host of other factors diminish the quality of early successional habitats in developed areas. Even the land management practices in undeveloped rural areas have become "modernized" providing only lower quality early successional habitats. Further, residential areas limit many management options, particularly the ability to use prescribed fire (see Spetich et al., Chap. 6). Smoke management has become a major issue in conducting controlled burns for wildlife and other reasons.

Almost unanimously, wildlife managers identified negative public perceptions about the techniques used to create early successional habitats as the major barrier or reason why agencies do not reach goals for creating these habitats. Many citizens believe that cutting trees is destruction and does not play a role in creating habitat for any species of wildlife. Fire is often viewed as destructive, and herbicides, even modern day, low toxicity and high selectivity ones, are still thought to have toxic effects on the environment. Certain groups often show logging done poorly in attempts to curtail or eliminate forestry practices on public lands. One respondent mentioned that some of the public recognize active management of habitat as a game issue, and they believe that nongame species do not need such management. The struggle to provide early successional habitats is sometimes referred to as the "challenge of managing ugly habitat" (Askins 2001). Although probably unintentional, many of the public today wish to manage for visual values rather than ecological values. Scenic and visual issues must be considered on most federal forest lands.

13.7 Solutions

Wildlife managers must find ways to overcome the many barriers to establishing levels and types of early successional habitat that are necessary to maintain viable or healthy populations of disturbance-dependent species. This will necessarily involve education, but most importantly, collaboration. It will also mean that traditional definitions of forestry may need to be expanded, especially on public lands, in order to satisfy competing human objectives.

Education always moves to the forefront when it comes to many management issues. The importance of getting accurate information to the public on the major points related to early successional habitats and species that depend on them is obvious. First, managers must educate the public as to the status of disturbance-dependent species and early successional habitats. Many look at the disturbed areas on private lands and do not understand habitat issues of quality, size, and location. The more difficult issues involve elucidating the positive aspects of cutting trees and burning areas to create wildlife habitat. A good example of efforts to educate the public about the benefits of fire is the US Fire Learning Network (FLN). The FLN is a joint project of The Nature Conservancy, the US Forest Service and several agencies of the US Department of the Interior. This effort has spawned many regional and state networks, including the Appalachian FLN. The US FLN seeks to "overcome barriers to implementing ecologically appropriate restoration projects using a facilitated conservation action planning approach" (TNC 2008). The FLN has fostered many successful burning projects and educated the public about the value of fire. Project Learning Tree, a program of the American Forest Foundation, is an environmental education program focusing on forest issues and other controversial environmental problems for school-aged children (Project Learning Tree 2004).

Ecological forestry is a concept that might allow movement past some of the controversy about using certain habitat management techniques (Corace and Goebel 2010). Ecological forestry attempts to link site conditions and natural disturbance regimes with silvicultural treatments that mimic outcomes of natural disturbances (Smith et al. 1996; Franklin et al. 2007). It requires a detailed understanding of the underlying disturbance regimes of a given ecosystem. One way to approach this is to create a "probability of disturbance" map where the probability of each of the major "natural" disturbance agents is mapped across the landscape (see White et al., Chap. 3). For example, fire probabilities would be highest on south and southwest-facing slopes, and wind disturbances would be more likely on areas with shallow soils that are more exposed to winds such as ridge tops (Lorimer and White 2003). The probabilities of beaver flooding would increase along low gradient waterways. In this way, prescriptions for watersheds or analysis areas could be crafted to mimic the most likely disturbance types for that site. Rather than waiting for chance events to create disturbances, managers could proactively establish a steady supply of early successional habitats in the areas they most likely occurred. Viewed in this way, ecological forestry is a win–win situation. It is good for wildlife, it uses resources wisely, and it fits in with the principles of ecological restoration.

Ecological forestry or any attempt to create critical early successional habitats for wildlife needs to involve partnerships, stakeholders, and other interested parties. One such paradigm involving these components is adaptive management. Adaptive management is a structured decision-making approach for improving resource management by learning from management outcomes (Conroy and Carroll 2009; Williams et al. 2009). It is a good way to explore alternatives but yet have action on the ground, especially in the face of uncertainty (Moore et al. 2005). There are opportunities for both public input and collaboration and for science to be used. Describing the problem, defining objectives and proposing actions are all policy decisions and are best done with stakeholders, decision makers and experts. Developing models, consequences, and optimization are left to biologists, modelers and other professionals. Monitoring and evaluation can involve citizen science and stakeholders as well as biologists and modelers (Nichols and Martin 2010). For many public agencies, input from stakeholders must be balanced against legislative mandates.

Within the context of structured decision-making, the early successional habitat issue can be viewed as a multi-species problem (Nichols and Martin 2010). Such an analysis requires a common currency for placing values on different species. This may mean determining the relative value of one species versus another one, and weighting the value of that species accordingly. Another approach is to use an integrated metric such as one of the diversity indices. In this case, the weighting process is done implicitly because the diversity index assigns greater values to species of lower abundance. Nichols and Martin (2010) created a theoretical objective function (Fig. 13.7) using the Shannon-Wiener diversity index. In this example, species abundance distributions are plotted over some type of habitat feature gradient such as stand age (or percent of area in a particular disturbed state), and the Shannon's diversity index is then calculated for each point along the gradient (*red line* in Fig. 13.7). This objective function then gives the manager an idea of where along the habitat gradient a piece of property should be to maximize the diversity index. It is important to note that in using an integrated metric such as a diversity index, there is a loss of some information that must be considered, so perhaps other metrics might be examined also. This is the type of analysis that can allow for an objective look at the tradeoffs for managing for multiple species. Moore et al. (2005) provide information on using optimization functions for examining endangered species recovery using red cockaded woodpeckers as an example. Habitat suitability models have also been used to examine large subsets of a species' habitat needs over a large landscape (Larson et al. 2003).

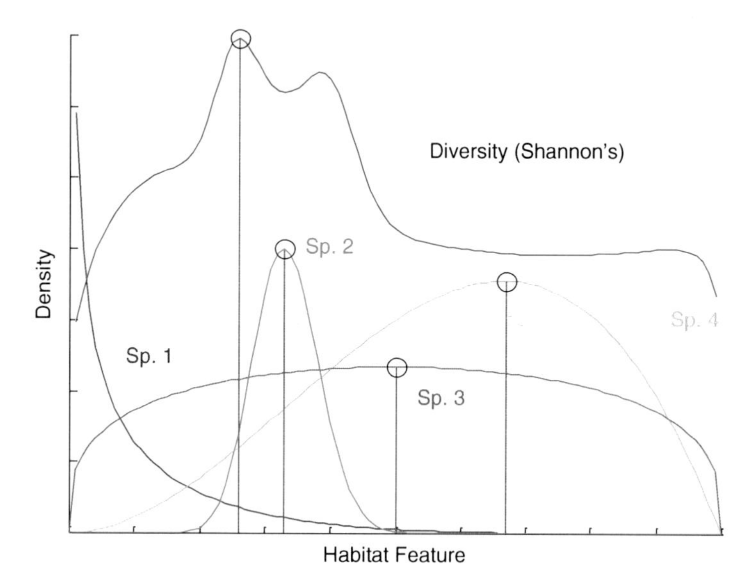

Fig. 13.7 A theoretical multi-species model of species density along a habitat feature gradient for four species. For each point along the habitat feature gradient, Shannon's Diversity is calculated. A manager then could use this type of analysis to select a desired future condition for an analysis area (graph from Nichols and Martin 2010)

Cooperative organizations that pull together people and resources are very important in getting habitat work done on the ground. The Appalachian Mountain Joint Venture has been very effective in overseeing and initiating various projects aimed at providing early successional habitats in many states. The Appalachian Mountain Joint Venture is a "self-directed partnership of agencies and organizations whose focus is to conserve (i.e., protect, restore, enhance) habitats for priority bird species in order to improve or sustain their populations" (Appalachian Mountains Joint Venture Management Board 2008). The group has focused on the American Woodcock and the Golden-winged Warbler because of their steep declines and because other species of concern also use similar habitats (e.g. Blue-winged Warbler (*Vermivora cyanoptera)*, Prairie Warbler (*Dendroica discolor*), Ruffed Grouse). The Appalachian Mountain Joint Venture is developing spatially explicit population and habitat goals in order to guide the conservation of early successional habitats in this BCR and have developed best management practices for Golden-winged Warblers. The Appalachian Mountain Woodcock initiative (Kelley et al. 2008), which recently developed best management practices, is working with partners to usher in demonstration areas (54 areas in 2009 in five states). Many cooperative projects for both woodcock and Golden-winged Warblers have been initiated on US Forest Service lands as the agency embarks in large scale prescribed fire projects. Finally, the Appalachian Mountain Joint Venture is very involved with state and NGO biologists in establishing early successional habitats on both state-owned lands and private lands, and more recently in assisting with land acquisition for

Table 13.6 Farm bill and other private lands incentive programs for wildlife

Conservation Reserve Program (CRP)
Conservation Reserve Enhancement Program (CREP)
Wetlands Reserve Program (WRP)
Wildlife Habitat Incentive Program (WHIP)
Conservation Stewardship Program (CSP)
Grassland Reserve Program (GRP)
Environmental Quality Incentives Program (EQIP)
Forest Stewardship
State Wildlife Tax Credits

priority bird species. It is clear that regional initiatives pulling together resources, stakeholders and money can do many good things for wildlife.

Sportsmen's groups will continue to play an important role in habitat conservation. Organizations like the Ruffed Grouse Society fund important field projects to directly develop young forest habitat. An important tool that sportsmen's groups are using on US Forest Service lands is Stewardship Contracting (USDA Forest Service 2009). Ranger districts may apply the value of timber removed against the cost of land management services received (goods for services). Receipts from the sale of wood products are not returned to the national treasury but rather may be used to fund service work such as habitat improvement or maintenance of existing habitat developments. The actual timber sale also contributes to the development of young forest stands. Groups like the National Wild Turkey Federation have become very active in this program, and it is one important way for sportsmen to become involved in creating wildlife habitat on public lands.

Sportsmen in many states are becoming more attuned to habitat issues. On many national forests, white-tailed deer (*Odocoileus virginianus*) harvests have declined (NCWRC, unpublished). Sportsmen are becoming more involved in forest policy issues. It would certainly be more effective if sportsmen joined with bird enthusiasts who recognize the need for active habitat management for songbirds.

Private lands have not been discussed here, but clearly programs to assist private landowners with the development of early successional habitats are going to be critical. There are a number of farm bill programs (Table 13.6) that are available to help these landowners. The Northern Bobwhite Quail Initiative actually sets state goals for grassland habitats, and funding from farm bill programs is coordinated with these goals (Dimmick et al. 2002). Connecting government assistance to regionally set goals is an outstanding way to tie incentives to productive and meaningful development of key wildlife habitats. The Forest Stewardship program is another program coordinated by the US Forest Service but operated by the states and it offers key technical advice to private forest land owners. In addition, many forest certification groups aim to make it worthwhile for landowners to keep lands in a forested condition. North Carolina enacted a wildlife incentive tax law that allows landowners who provide certain priority habitats identified in the SWAP to be taxed at a lower property tax rate. Early successional habitats of 20–100 contiguous acres qualify for the tax credit.

13.8 Triage Situations

The situation for many species dependent on disturbed habitats is critical. Many disturbance- dependent species and habitats are on various watch lists and lists of endangered habitat types. The Bewick's Wren (*Thryomanes bewickii*) is seemingly extirpated from much of the Central Hardwood Region and is now listed as endangered, threatened, or of special concern in many of the midwestern and all of the eastern states where it occurs (Powers 2001). Bewick's Wren's population levels were high during the 1800s at a time when farms and homesteads created favorable habitat–mixtures of thick scrubby vegetation near open woodlands (Powers 2001). With field abandonment and the maturing of the hardwood forests of the Central Hardwood Region, habitat is now very limited for this bird. And, in places where habitat is present near human development, competition from House Wrens (*Troglodytes aedon*) is too much for this bird. Many biologists feel it may be too late to bring the Bewick's Wren back.

Northern Georgia is home to the southernmost range of the Golden-winged Warbler. The bird was "quite common especially in open oak savannahs and second growth on hillsides" in the mountains of Georgia from 1850 to the 1950s (Brewster 1886). From 1966 to 1979, the Breeding Bird Survey population trend was −18.7%, and by 1979 no birds were detected on these routes. Klaus (2004) surveyed North Georgia and found Golden-winged Warblers on only five sites. The bird was listed by the Georgia Department of Natural Resources as a state endangered species in 2007, and by 2009, the bird population was reduced to one site with 12 breeding pairs. The remaining site is on US Forest Service land on Brawley's Mountain (Klaus 2011, *pers. comm.*). A proposed project to restore the open oak woodlands and shrublands of this area was immediately appealed despite support from the Audubon Society. The project was finally approved in 2010, and 400 acres of the original 1,400 that were planned will be developed to improve habitat for this bird (J. Wentworth 2011, *pers. comm.*).

These are just two examples of situations in the Appalachian Mountains. When a species gets to the point that Golden-winged Warblers are in north Georgia, biologists must begin "triage". Klaus (2004, 2011, pers. comm.) made several recommendations for increasing Golden-winged Warblers in Georgia. First, the existing population must be augmented by creating additional habitat within 1 mile of where the birds currently exist (he found that Golden-winged Warblers will not disperse great distances when at low population levels). New habitat should be created through a combination of fire and timber harvest as fire alone was not practical due to the required high intensity burns necessary to thin the forests with fires alone. Then, he suggested that connectivity with existing source or stronghold areas (southwestern North Carolina) be established via grouped patches of early successional habitats at high elevations (3,000–5,000 ft). When a population gets to this stage, options become limited. It is better to address species needs before population levels become too low.

13.9 Conclusion

Early successional habitats and the species dependent on them are a major conservation concern in the Appalachian Mountains. Current conditions present many challenges to getting these important habitats out on the landscape. Early successional habitats are ephemeral and require that managers constantly either create newly disturbed areas to replace those that succeed to older age classes, or treat those areas that are already disturbed to retard or set back succession. In order to conserve disturbance-dependent species, we need to develop reliable means of identifying and quantifying early successional habitats and assessing quality. Much work remains to be done in the area of conservation design so that managers know clearly how much and what types of early successional habitats are needed and where they should be located. Finally, mangers must step out of the field and engage the public through collaboration.

For the short term, we need to identify "triage" situations or situations and take action to keep the species in the region. Also in the short term, we need to identify two or three key species and begin the process of adaptive management to "learn by doing" as we develop habitat and monitor populations. These strategies both can be done in a collaborative setting.

Looking ahead long range, managers need to embrace the principles of conservation design and the requisite modeling that goes with that approach. On-the-ground mangers must move into this area and use the modern tools of habitat suitability assessment and population viability analyses (Millspaugh and Thompson 2009). These tools should be applied to multiple sets of species and move from coarse-grained to fine-grained analyses. A hierarchical method of moving from large regions to focal areas within the states is a valid approach, and a clear step down process from regional initiatives to local areas is much needed (Thompson and DeGraaf 2001). Right now, conservation design tools are the privy of researchers at universities and research institutions. But, they need to be adapted in useful forms for those charged with managing the landscape for wildlife. This "will necessitate a move in the practice of conservation design away from the domain of scientists and academicians and into the hands of 'quantitatively savvy practitioners'. The success of conservation design … will hinge on the success of this transition" (Thogmartin et al. 2009).

Hopefully we can find areas of concurrence with those philosophically opposed to human management of forestlands. Some points of agreement may include that the suite of early successional species are important and that we need to ensure their future existence. Also, we can all recognize the importance of both mature forest and early successional species in long term conservation planning, and the importance of maintaining a balance of various disturbed habitats as well as old growth within a forested landscape.

It takes much effort and action when working with disturbance-dependent species and early successional habitats. Managers may not have 100% of the information necessary when operating, but it is important to examine the costs of inaction. The public needs to understand the very real consequences to wildlife from decisions not to impart a habitat management action on a particular landscape. We must give equal consideration to both the long term implications of inaction and the short term implications of action (Dessecker in press). The very future of many species depends on our ability as a society to consider both sides of this issue.

> Are we not just another step in the natural process of human ecological interactions? We manage existing undeveloped landscapes for biodiversity... (we) maintain diversity and ensure the viability of all species (Delcourt and Delcourt 2004).

Literature Cited

Appalachian Mountains Bird Conservation Region Partnership (2005) Appalachian Mountains bird conservation initiative concept plan, p 26

Appalachian Mountains Joint Venture Management Board (2008) Implementation plan for the Appalachian mountains joint venture: a foundation for all-bird conservation in the region. In: Smith BW (ed) Appalachian Mountains Joint Venture, Frankfort

Askins RA (2001) Sustaining biological diversity in early successional communities: the challenge of managing unpopular habitats. Wildl Soc Bull 29:407–412

Brewster W (1886) An ornithological reconnaissance in western North Carolina. Auk 3:94–111

Conroy MJ, Carroll JP (2009) Quantitative conservation of vertebrates. Blackwell, London

Corace RG, Goebel PC (2010) Ecological forestry; integrating disturbance ecology patterns into forest treatments. Wildl Prof 4:38–40

DeGraaf RM, Yamasaki M (2003) Options for managing early-successional forest and shrubland bird habitats in the northeastern United States. For Ecol Manage 185:179–191

Delcourt PA, Delcourt HR (2004) Prehistoric native Americans and ecological change; human ecosystems in eastern North America since the Pleistocene. Cambridge University Press, Cambridge

Dessecker DR (1997) Back to the future – is the past a guide to a "healthy" forest landscape in the northern Great Lakes region? Trans N Am Wildl Nat Res 62:469–478

Dessecker DR (in press) Constant change: bird conservation on grassland and early successional forest landscapes. Trans N Am Wildl Nat Res 75:000–000

Dessecker DR, Norman GW, Williamson SJ (2006) Ruffed grouse conservation plan. Association of Fish & Wildlife Agencies, Resident Game Bird Working Group

Dimmick RW, Gudlin MJ, McKenzie DF (2002) The northern bobwhite conservation initiative. Misc Publ Southeast Association of Fish and Wildlife Agencies, South Carolina, p 96

Franklin JF, Mitchell RJ, Palik BJ (2007) Natural disturbance and stand development principles for ecological forestry. Gen Tech Rep NRS-19, USDA Forest Service Northern Research Station, Newtown Square

Frost CC (1998) Pre-settlement fire frequency regimes of the United States: a first approximation. In: Pruden TL, Brenman LA (eds) Proceedings of the 20th tall timbers fire ecology conference. Fire in wetlands: a management perspective. Tall Timbers Research Station, Tallahassee, pp 70–81

Fusting CW (2009) Roan Mountain LiDAR investigation; vegetation extraction. Unpublished final report, Warren Wilson College

Hansen MH, Frieswyk T, Glover JF, Kelly JF (1992) The eastwide forest inventory data base: users manual. Gen Tech Rep NC-151, USDA Forest Service North Central Forest Experiment Station, St. Paul

Harper CA, Jones BC, Whitaker DM, Norman GW, Banker MA, Tefft BC (2011) Habitat management. In: Stauffer DF, Edwards J, Giuliano WM, Norman GW (eds) Ecology and management of Appalachian ruffed grouse. Hancock House, pp 130–151

Hunter WC, Katz RT, Pashley DN, Ford RP (1999) Partners in flight bird conservation plan for the Southern Blue Ridge (Physiographic Area 23) Version 1.0

Hunter WC, Buehler DA, Canterbury RA, Confer JL, Hamel PB (2001) Conservation of disturbance-dependent birds in eastern North America. Wildl Soc Bull 29:440–455

Kelley J, Williamson S, Cooper T (2008) American woodcock conservation plan: a summary of and recommendations for woodcock conservation in North America. Woodcock Task Force, Association of Fish and Wildlife Agencies

Klaus NA (2004) Status of Golden-winged Warblers in North Georgia. Oriole 69:1–7

Larson MA, Dijak WD, Thompson FR, III, Millspaugh JJ (2003) Landscape-level habitat suitability models for twelve species in southern Missouri. Gen Tech Rep NC-233, USDA Forest Service North Central Research Station, St. Paul

Litvaitis JA (2001) Importance of early successional habitats to mammals in eastern forests. Wildl Soc Bull 29:466–473

Litvaitis JA (2003) Are pre-Columbian conditions relevant baselines for managed forests in the northeastern United States? For Ecol Manage 185:113–126

Lorimer CG (2001) Historical and ecological roles of disturbance in eastern North American forests: 9000 years of change. Wildl Soc Bull 29:425–439

Lorimer CG, White AS (2003) Scale and frequency of natural disturbances in the northeastern US: implications for early successional forest habitats and regional age distributions. For Ecol Manage 185:41–64

McGaughey RJ (2010) Fusion home page, USDA Forest Service, Pacific Northwest Research Station. http://www.fs.fed.us/eng/rsac/fusion/. Accessed 28 Jan 2010

Millspaugh JJ, Thompson FR III (eds) (2009) Models for planning wildlife conservation in large landscapes. Academic, Burlington

Moore CT, Plummer WT, Conroy MJ (2005) Forest management under uncertainty for multiple bird population objectives. In: Ralph CJ and Rich TD (eds) Bird conservation implementation and integration in the Americas. Proceedings of the third international partners in flight conference. Gen Tech Rep PSW-191, USDA Forest Service Pacific Southwest Research Station, Albany, pp 373–380

Multi-Resolution Land Characteristics Consortium (MRLC) (2010) National land cover datatbase home page. http://www.mrlc.gov/index.php. Accessed 10 Jan 2010

Nichols JD, Martin, J (2010) Webinar short course on adaptive management of natural resources, U.S. Geological Survey. http://www.fort.usgs.gov/brdscience/AdaptiveManagementCourse.htm. Accessed 8 Jun 2010

North Carolina Wildlife Resources Commission (2005) North Carolina wildlife action plan. Raleigh

Oehler JD (2003) State efforts to promote early-successional habitats on public and private lands in the northeastern United States. For Ecol Manage 185:169–177

Powers M. (2001) Decline of Bewick's Wren. Birdscope. 15 (3), http://www.birds.cornell.edu/Publications/Birdscope/Summer2001/Bewicks.html. Accessed 16 Dec 2010

Project Learning Tree (2004) Project learning tree home page. http://www.plt.org/cms/pages/21_19_1.html. Accessed 2 Mar 2010

Rich TD, Beardmore CJ, Berlanga H, Blancher PJ, Bradstreet MSW, Butcher GS, Demarest DW, Dunn EH, Hunter WC, Iñigo-Elias EE, Kennedy JA, Martell AM, Panjabi AO, Pashley DN, Rosenberg KV, Rustay CM, Wendt JS, Will TC (2004) Partners in flight North American landbird conservation plan. Cornell Lab of Ornithology, Ithaca

Smith DM, Larson BC, Kelty MJ, Ashton PMS (1996) The practice of silviculture: applied forest ecology, 9th edn. Wiley, New York, p 560

Southern Appalachian Man and the Biosphere (1996) The southern Appalachian assessment terrestrial technical report. Report 5 of 5 USDA Forest Service, Southern Region, Atlanta

The Nature Conservancy (2000) Conservation by design: a framework for mission success. The Nature Conservancy, Arlington

The Nature Conservancy (2008) Global fire initiative – the fire learning network. http://www.tncfire.org/training_usfln.htm. Accessed 1 Mar 2010

Thogmartin WE, Fitzgerald JA, Jones MT (2009) Conservation design: where do we go from here? In: Rich TD, Arizmendi C, Demarest D, Thompson C (eds) Tundra to tropics: connecting birds, habitats and people. Proceedings of the fourth international partners in flight conference, McAllen, pp 426–436

Thompson FR III, DeGraaf RM (2001) Conservation approaches for woody-early successional communities in the eastern USA. Wildl Soc Bull 29:483–494

Trani MK, Brooks RT, Schmidt TL, Rudis VA, Gabbard CM (2001) Patterns and trends of early successional forest in the eastern United States. Wildl Soc Bull 29:413–424

USDA Forest Service (2009) Stewardship contracting; basic stewardship contracting concepts. Brochure FS-893. http://www.fs.fed.us/fstoday/091106/03.0About_Us/stewardship_brochure.pdf. Accessed 16 Mar 2010

USDI Fish and Wildlife Service (2010a) Division of Bird Conservation; Joint Ventures. http://www.fws.gov/birdhabitat/jointventures/index.shtm. Accessed 3 Mar 2010

USDI Fish and Wildlife Service (2010b) Division of Bird Conservation; Landscape Conservation Cooperatives. http://www.fws.gov/science/shc/lcc.html. Accessed 3 Mar 2010

Williams BK, Szaro RC, Shapiro CD (2009) Adaptive management: the U.S. department of the interior technical guide. Adaptive Management Working Group, USDI, Washington. http://www.doi.gov/initiatives/AdaptiveManagement/. Accessed 16 Mar 2011

Chapter 14
Early Successional Forest Habitats and Water Resources

James M. Vose and Chelcy R. Ford

Abstract Tree harvests that create early successional habitats have direct and indirect impacts on water resources in forests of the Central Hardwood Region. Streamflow increases substantially immediately after timber harvest, but increases decline as leaf area recovers and biomass aggrades. Post-harvest increases in stormflow of 10–20%, generally do not contribute to downstream flooding. Sediment from roads and skid trails can compromise water quality after cutting. With implementation of Best Management Practices (BMPs), timber harvests are unlikely to have detrimental impacts on water resources, but forest conversion from hardwood to pines, or poorly designed road networks may have long lasting impacts. Changing climate suggests the need for close monitoring of BMP effectiveness and the development of new BMPs applicable to more extreme climatic conditions.

14.1 Introduction

Watershed management requires understanding the tight linkages among vegetation, soils, and water quantity and quality. Because of these linkages, forest management activities that alter vegetation, such as creation of early successional habitats, have the potential to impact water resources. From a hydrologic standpoint, we define early successional habitats by the structural and functional attributes that are created by disturbance and influence hydrologic processes. Early successional habitats can be created by either natural disturbances (e.g., hurricanes, tornados, severe wildfires), or human-mediated intentional (e.g., forest cutting)

J.M. Vose (✉) • C.R. Ford
USDA Forest Service, Southern Research Station, Coweeta Hydrologic Laboratory, Otto, NC 28763, USA
e-mail: jvose@fs.fed.us; crford@fs.fed.us

C.H. Greenberg et al. (eds.), *Sustaining Young Forest Communities*, Managing Forest Ecosystems 21, DOI 10.1007/978-94-007-1620-9_14,

and unintentional (e.g., invasive insects and disease introductions) disturbances (White et al., Chap. 3). Defining structural attributes of early successional forests include low leaf, stemwood and sapwood areas, high forest floor mass and coarse woody debris, and a high proportion of fast-growing, shade intolerant species (Keyser, Chap. 15). Defining functional attributes include high leaf-level C gain and low water use efficiencies, rapid organic matter decomposition, and accelerated nutrient cycling and accumulation (Keyser, Chap. 15). Although early successional forest attributes can be maintained with repeated disturbances, these attributes more often are transitional and recovery to pre-disturbance conditions occurs quickly (e.g., leaf area) or over several decades (e.g., species composition). Where disturbances are particularly severe, such as road building or loss of a dominant overstory species, structural and functional attributes may never recover to pre-disturbance conditions (Ellison et al. 2005). Combined, these changes in structural and functional attributes can impact water resources, and land managers need to consider those impacts when managing forests for multiple benefits. In particular, forest harvesting (with and without species conversion) and associated forest operations have the potential to substantially alter both water quantity and quality; in some cases, these changes persist long-term. In short, good land management is good watershed management.

Our understanding of the changes in water resources associated with creating early successional habitats is largely derived from a long history of paired watershed studies that have examined long-term streamflow and water quality responses to forest cutting (Calder 1993; Stednick 1996; Jones and Post 2004; Brown et al. 2005). Paired catchment studies have been critical to understanding how land management and other disturbances affect streamflow and quality. Accurate measurement of streamflow is at the core of paired watershed studies and this typically requires installation of a weir at the watershed outlet (Reinhart and Pierce 1964). Streamwater quality can be measured directly for some parameters (e.g., turbidity, pH, temperature, conductivity) using automated sensors, or water samples can be analyzed in a laboratory for these and other parameters such as nutrient concentration. The primary goal of the paired catchment method is to isolate streamwater response to cutting by accounting for the influences of climate or other factors. Using a paired untreated watershed that serves as a reference, streamflow response to cutting can be determined by examining the difference between expected streamflow (e.g., what would be expected if the watershed had not been treated) from observed streamflow. When measured streamflow differs from expected, the inference is that the treatment alone resulted in the streamflow response. Catchment scale manipulations at experimental watersheds such as the Coweeta Hydrologic Laboratory in the Southern Appalachians of North Carolina, the Fernow Experimental Forest in the Central Appalachians of West Virginia, and Hubbard Brook Experimental Forest in New Hampshire involve various intensities and types of management activities, as well as variation in watershed characteristics such as aspect, elevation, and size (Adams et al. 2008). These long-term watershed studies provide a powerful database from which we can examine the effects of managing for early successional habitats on streamflow amount, timing, and quality.

Annual streamflow generally increases for the first few years after forest canopy removal,, but the magnitude, timing, and duration of the response varies considerably among ecosystems. Using data from water yield studies across the globe, a general model suggests that for each percent of the forest removed streamflow increases 2.5–3.3 mm (Calder 1993; Stednick 1996); however, general models typically explain less than 50% of the variation of the streamflow increase (Stednick 1996) due to high variability in stand structure, pre- and post-harvest species composition, and the interaction between vegetation and climate. In some cases, streamflow returns to pre-harvest levels within 10–20 years. In others, streamflow remains higher, or can even be lower than pre-harvest flow, for several decades after cutting. This wide variation in temporal response patterns is attributable to the complex interactions between climate and vegetation, which can vary considerably from dry to wet to snow-dominated climatic regimes, and with differences in vegetation structure and phenology (coniferous vs. deciduous forest) (McNab, Chap. 2).

While gauged watershed studies provide the foundation for quantifying streamflow responses to forest disturbances, process-level studies are required to fully understand the structural and functional attributes that regulate the magnitude and duration of responses. For example, timber harvest simultaneously alters forest structure by reducing leaf area index, interception surface area, and vegetation height. Harvesting also alters forest function by changing the relative abundance of plant species (Loftis et al., Chap. 5; Elliott, Chap. 7), and the physical environment by changing the energy balance, wind environment, hydrologic flowpaths, and soil temperature and moisture. The topographic/edaphic complexity and high vegetation diversity of forest ecosystems in the Central Hardwood Region is likely to result in a wide range of streamflow response patterns. A more in depth understanding of the factors regulating these response patterns can help managers create and maintain early successional habitats and protect or enhance water resources.

Water quality can also be substantially affected by management activities that create early successional habitats and can have detrimental impacts on aquatic habitats and organisms (Moorman et al., Chap. 11). Research indicates that the harvest of forest biomass in itself has little or no measureable impact on sediment yield. Instead, the primary factors that determine sediment yield are the forest operations required to remove logs, such as roads and skid trails, and the implementation and effectiveness of Best Management Practices (BMPs) that either minimize erosion or prevent sediment from reaching the stream. Stream nutrients can also be impacted by creating and maintaining early successional habitats; however, response magnitude and duration vary considerably among chemical constituents, post-disturbance successional dynamics, and other silvicultural practices such as the use of herbicides or fertilizers.

In this chapter we focus on the first several years after harvesting to assess potential impacts of using forest harvests to create early successional habitats on water resources. To provide examples and illustrate concepts, we use data primarily from long-term studies in the Southern Appalachians, but also include and integrate results of studies from watershed experiments in other areas of the Central Hardwood Region. In addition, we include a discussion of the potential implications of climate change and how associated changes in precipitation regimes might interact with early successional habitats.

14.2 The Hydrologic Budget of Forested Watersheds

The three main components of the hydrologic budget of forested watersheds are **inputs** in the form of rain, snow, and ice (P); **outputs** in the form of transpiration, canopy interception, and soil and forest floor evaporation (evapotranspiration, ET), and groundwater recharge and streamflow (RO or runoff); and change in **soil water storage** (S). Thus, the hydrologic budget can be expressed in terms of a simple mass balance equation: $RO = P - ET \pm S$. Over the long-term, changes in soil water storage (S) are assumed to be negligible so that the storage component of the budget is usually ignored.

Understanding components of the water budget is useful for interpreting and predicting potential impacts of creating and maintaining early successional habitats. ET is the primary component influenced by forest cutting. However, significant alterations to hydrologic flowpaths due to compaction, roads, and other physical changes can influence runoff processes as well, especially stormflow. Timber harvesting alters ET by changing forest structure and function, and the micrometeorological factors that drive transpiration and evaporation. Structural changes include less leaf and stem surface area, and change in the distribution and arrangement of branch surface area. A major functional change that ensues when shifting from mature trees to seedlings, sprouts, and herbaceous vegetation is a decrease in abundance of plant species with conservative water use, resulting in increased transpiration per unit leaf area (Wallace 1988). The vegetation layer can also be more coupled to the atmosphere after forest harvest, thus changing energy balances and wind profiles (Swift 1976; Swank and Vose 1988). For example, Sun et al. (2010) found that net radiation of an 18-year old loblolly pine plantation was 20% higher than a younger stand (4–6 year old) in on the Coastal Plain of North Carolina, resulting in a 25% higher ET in the former.

14.3 Streamflow Responses to Forest Removal

14.3.1 Amount and Timing

Forest harvesting increases annual streamflow in almost all cases in the Central Hardwood Region (Jackson et al. 2004). For example, average increases (% increase relative to that expected based on flow in a reference watershed) in water yield for the first 2 years after cutting ranges from 9.1% at Hubbard Brook in New Hampshire, 14.3% at the Fernow in West Virginia, and 23.0% at the Coweeta Hydrologic

Table 14.1 Post-treatment streamflow response expressed as a percentage increase relative to expected streamflow (adapted from Vose et al. 2010)

Experimental forest	Average annual response (first 2 years post-cut)	Minimum	Maximum
Coweeta, NC ($n=6$)	23.0	10.3	44.1
Fernow, WV ($n=3$)	14.3	10.8	18.2
Hubbard Brook, NH ($n=3$)	9.1	1.7	18.9

Laboratory in North Carolina (Table 14.1). Comparing clearcut harvests with and without BMPs in hardwood forest in eastern Kentucky, Arthur et al. (1998) found a 138% (without BMPs) and a 123% (with BMPs) increase in streamflow during the initial 17 month post-cutting period. Water yield was still 15 to 12% greater 8 years after cutting for the BMP and without BMP watersheds, respectively (Arthur et al. 1998). Differences among regions are likely the result of a complex array of factors, but syntheses of worldwide data from watershed experiments suggest that absolute increases after cutting are greatest in high rainfall areas (Bosch and Hewlett 1982; Swank and Johnson 1994). Other factors include soil depth, the proportion of the annual water budget accounted for by ET, and annual snow fall. The amount of steamflow response is greatest during the first few years following treatment and can be estimated for upland hardwood forests using a model (Douglass and Swank 1975) where first year streamflow increase (water yield) is predicted as a function of the amount of basal area removed and an index of solar radiation inputs:

$$\text{Yield} = 0.00224 * (\text{BA} / \text{PI})^{1.4462},$$

where

Yield = first year increase in streamflow (cm),
BA = amount of basal area removed (%), and
PI = solar insolation index.

Highest yields are observed when 100% of the forest is harvested on north facing slopes. On south or west facing slopes where solar radiation inputs are greater, first year responses are lower because ET on harvested south facing slopes is not as responsive to the increased energy load as ET on harvested north facing slopes. The model also includes an equation to predict the exponential decline in streamflow response as the forest re-grows and LAI recovers (Swank and Douglass 1975). Applications of the model indicate good performance in the Southern Appalachians (Swank and Johnson 1994; Swank et al. 2001) and other eastern deciduous and coniferous forests (Douglass and Swank 1975; Douglass 1983).

Forest cutting can also impact streamflow timing throughout the year and alter storm hydrographs. For example, in areas with high snowfall and shallow soils, cutting increases the proportion of annual streamflow in the spring and summer months due to faster snowmelt and reduced transpiration. In areas with deeper soils and higher precipitation, typical of the Southern Appalachians, flow increases are greatest in the late summer and fall, and may extend into the winter months (Swank and Johnson 1994). For example, on a south facing clearcut watershed in the Southern Appalachians, streamflow increased by approximately 48% during August through October, a time when flows from mature forests are typically lowest (Swank et al. 2001). Storm hydrographs (i.e., a graphical analysis of stream flow vs. time during and after storm events) can also be impacted by cutting and the effects of timber harvesting on flooding have been a focus of intense debate and research for the past several decades (Lull and Reinhart 1972; Andreassian 2004; Eisenbies et al. 2007). Flooding is defined by hydrologic events that exceed bankfull. The linkage between timber harvesting, storm hydrographs, and flooding is complex, and can be better understood by examining the components of stormflow, and then dissecting how forest harvesting influences these components. Streamflow is comprised of baseflow and stormflow, with the latter being described by both the magnitude (peakflow) and duration (stormflow volume). Flooding occurrence and severity is determined largely by peakflow (essentially analogous to stage or the height of the stream) and stormflow volume (the amount of flow contributed by the storm). In forests of the Central Hardwood Region, peakflow and stormflow volume are primarily affected by forest operations that create soil disturbances that alter stormflow pathways; chief among these operations is the road network. For example, in the Southern Appalachians, stormflow volume was nearly double on a watershed logged with a high road density (Douglass and Swank 1976) compared to a watershed logged with a low road density (Swank et al. 2001). However, increases were still relatively minor (10% increase for the low road density watershed versus 17% increase for the high road density watershed). Peak discharges increased on the low road density watershed by up to 15% (Swank et al. 2001). In other sites where trees were felled, but no material removed and no roads were built, peakflow rates increased very little over all (<7%) although stormflow volume increased by 11% (Hewlett and Helvey 1970). In West Virginia, peak discharges after logging were up to four times greater during the growing season (Patric and Reinhart 1971) and they were up to 30% greater after cutting in New Hampshire (Hornbeck 1973).

If BMPs are implemented, most of the physical impacts related to harvest soil disturbances (e.g., skid trails, landing decks, etc.) are short-lived and have little impact on flood risk over the long-term. In contrast, construction of roads and associated engineering related to road surfacing, drainage, culvert design and location are much longer lasting. Depending on the design and surface area impacted, these can permanently alter hydrologic flow paths and storm hydrographs. In short, road design needs to focus on "disconnecting" the surface water draining from the road network to the stream network. Analyses of the impacts of cutting on downstream flooding suggests that many extreme flood events are unrelated to forest cutting and associated road networks and skid trails. Instead, they are primarily determined by storm size and intensity (Perry and Combs 1998; Kochendorfer et al. 2007) and occur regardless of forest management activities.

14.3.2 Duration of Streamflow Response

Among the biological and physical process changes that occur with timber harvest, the duration of streamflow response primarily depends on how quickly leaf and sapwood area recover, and the physiological and structural characteristics of the tree species that occupy the site after the cutting. Long-term streamflow responses for six watersheds in the Southern Appalachians illustrate the temporally variable nature of the response. The response depends on both the forest management objective (e.g., thinning, species conversion, clear cut, etc.,) and how the resulting vegetation responds to climate (Fig. 14.1). Few watershed treatments show no effect (e.g., zero line represents no difference between observed and expected flow based on flow from the reference watershed); and more importantly, few of the watersheds have returned to expected levels after 20 years. For example, where timber harvesting was followed by a species conversion (in this case, from deciduous hardwood to conifer, Fig. 14.1a–b), annual streamflow returned to reference levels after approximately 10 years, marking the point in time when canopy closure was complete. Thereafter, streamflow has been about 25% lower on the conifer dominated watershed (relative to the hardwood reference watershed) due to higher interception and year round transpiration by conifers (Swank and Douglass 1974; Ford et al. 2011).

Variation in sapwood area and species composition among hardwood species during succession can also play an important role in determining the magnitude and timing of streamflow responses after cutting (Ford et al. 2011). For example, transpiration rates for a given diameter yellow-poplar (*Liriodendron tulipifera*) are nearly twofold greater than hickory (*Carya* spp.) and fourfold greater than oaks (*Quercus* spp.). Yellow-poplar transpiration and stomatal conductance rates are also much more responsive to climatic variation compared to oaks and hickories (Ford et al. 2011) (Table 14.2). Xylem anatomy and resulting sapwood area are important determinants of stand transpiration (Wullschleger et al. 2001). For example, transpiration of trees with diffuse-porous, ring-porous, semi-ring-porous, and tracheid xylem anatomies vary more among these three xylem types than they do within a type by species (Fig. 14.2). Diffuse ring porous species have greater sapwood area than ring- or semi-ring porous species and as sapwood area increases, potential water transport increases (Enquist et al. 1998; Meinzer et al. 2005). Hence, if the early successional stand is dominated by diffuse porous species such as yellow-poplar, black birch (*Betula lenta*), or red maple (*Acer rubrum*), we would expect that growing season transpiration in an average year to be much greater (and hence, lower streamflow) than stands dominated by ring-porous species such as oaks or hickories, and likewise be more responsive to climatic variation. In most cases, post-harvest or post-disturbance vegetation succession in the Appalachians is a complex mix of species in both space and time (Elliott and Vose 2011) which makes simple extrapolations difficult. For example, as eastern hemlock (*Tsuga canadensis*) declines and its basal area is reduced by attack from an invasive exotic insect, black birch, a diffuse-porous sapwood species, is dominating early successional trajectories of leaf and sapwood area response (Orwig et al. 2002). This shift in species composition has the potential to increase transpiration by 30% (and thus correspondingly decrease streamflow) (Daley et al. 2007).

To fully understand and predict how post-harvest shifts in the relative abundance of tree species regulate streamflow response (e.g., to explain the variation shown in

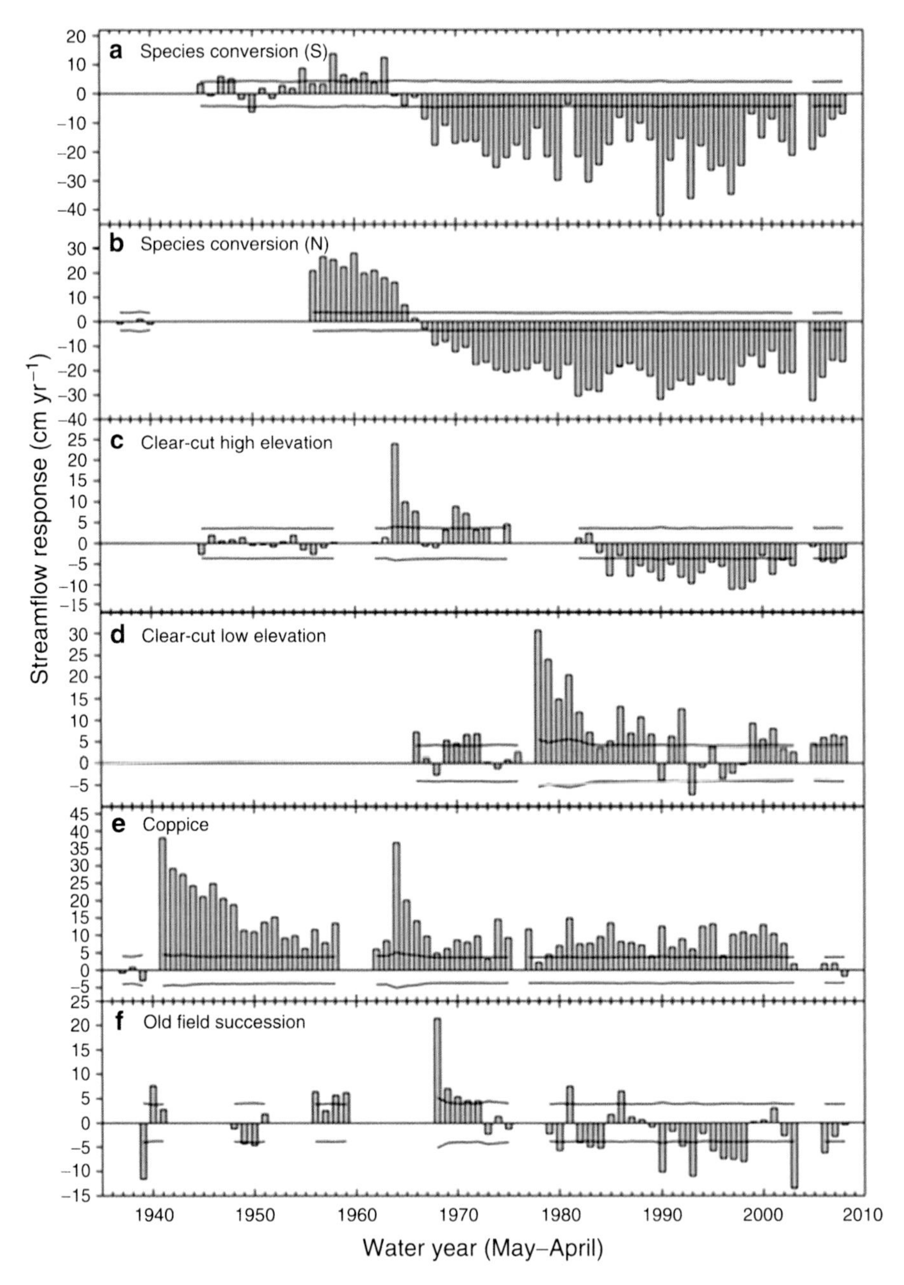

Fig. 14.1 Streamflow response (D, cm yr^{-1}) to forest cutting in the Southern Appalachians (see Swank and Crossley (1988) for site and treatment descriptions). Grey bars depict the calibration period and cyan bars depict streamflow response after treatments. Solid lines on either side of the zero line are 95% confidence intervals; data within the confidence intervals do not differ from zero. Species conversion treatments involved cutting hardwood species and planting *Pinus strobus* on north (N) and (S) facing watersheds (from Ford et al. 2011)

Table 14.2 Mean (standard error) growing season daily transpiration per unit leaf area (E_L, mm) for four hardwood species (Adapted from Ford et al. 2011). Within columns, species not sharing the same lowercase letters denote significant differences among species for that year. Within rows, years not sharing the same uppercase letters denote significant differences among years for that species

	Year		
Species	2004	2005	2006
Carya spp.	0.20 (0.03) b, A	0.19 (0.02) b, A	0.18 (0.02) c, A
Liriodendron tulipifera L.	0.45 (0.05) a, AB	0.39 (0.07) a, B	0.46 (0.03) a, A
Quercus prinus L.	0.21 (0.03) b, A	0.07 (0.01) b, B	0.10 (0.02) cd, AB
Quercus rubra L.	0.10 (0.02) b, A	0.07 (0.02) b, A	0.07 (0.01) c, A

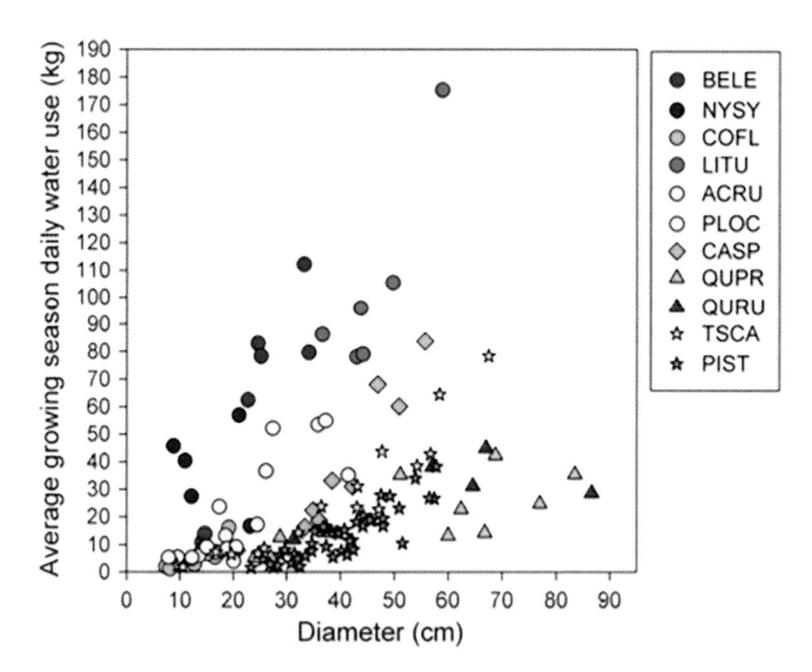

Fig. 14.2 Observed daily water use (DWU) estimated from sap flux density in trees of varying species (legend text denotes first two letters of Latin binomial: BELE *Betula lenta*, NYSY *Nyssa sylvatica*, COFL *Cornus florida*, LITU *Liriodendron tulipifera*, ACRU *Acer rubrum*, PLOC *Platanus occidentalis*, CASP *Carya* spp., QUPR *Quercus prinus*, QURU *Q. rubra*, TSCA *Tsuga canadensis*, PIST *Pinus strobus*) in reference watersheds at Coweeta (except PIST). Symbols represent the mean DWU of replicate trees in each species during the growing season for deciduous species, days of year 128–280 in 2006. Mean DWU during the entire annual period is shown for coniferous species (TSCA is during 2004, PIST is during 2006). LITU, QURU, QUPR, CASP, and PIST data are from (Ford et al. 2011). TSCA data are from Ford and Vose (2007). BELE, NYSY, COFL, ACRU, and PLOC are from (C. Ford and J. Vose, unpublished) but follow the methods in (Ford et al. 2011). Symbols: circles are species with diffuse porous xylem anatomy, diamonds are species with semi-ring-porous xylem anatomy, triangles are species with ring-porous xylem anatomy, stars are for species with tracheid xylem anatomy (from Vose et al. 2011)

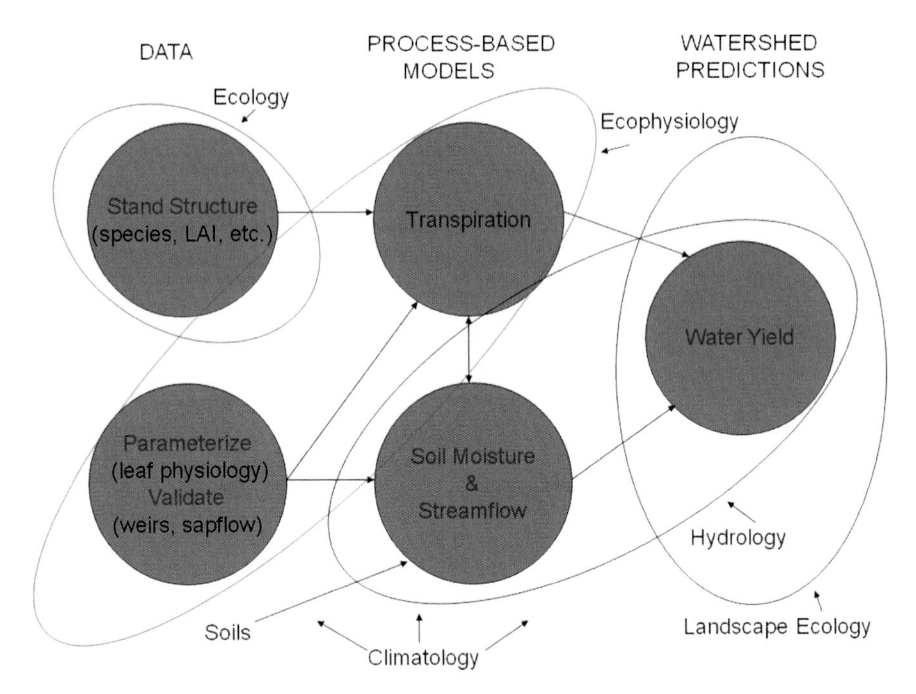

Fig. 14.3 Interdisciplinary approaches to understanding impacts of forest management and other disturbances on water yield requires linking species dynamics and physiology, soil moisture dynamics, and climate across scales ranging from leaves to landscapes (from Vose et al. 2011)

the empirical data shown in Fig. 14.1), we need to be able to link spatially explicit (i.e., cove, midslope, ridge, etc.) predictions of species composition and structure with: (1) species-specific physiology, (2) soil moisture and subsurface flow dynamics, and (3) microclimate. This is a significant departure from traditional hydrologic sciences and requires a multidisciplinary, multi-scale approach (Fig. 14.3).

14.4 Water Quality Responses

Considerable research has been conducted on the effects of forest harvesting on water quality in upland hardwood forests, as well as the development of BMPs to minimize impacts (Kochendorfer and Hornbeck 1999; Jackson et al. 2004; Sun et al. 2004). The most impacted water quality parameter is sediment load, although water temperature and dissolved nutrient concentrations can also be affected. The impact of all of these parameters can be reduced or eliminated with proper planning and BMP implementation. Thus, water quality from streams draining early successional forests can be as high from streams draining undisturbed forested catchments.

Sediment delivery to streams occurs primarily as a result of erosion from roads and skid trails associated with logging (Anderson et al. 1976; Swift 1988; Swank et al. 2001). For example, logging without BMPs resulted in annual sediment losses on the order of 3.1 MT ha^{-1} in the Central Appalachians compared to 0.04 MT ha^{-1} in uncut reference watersheds (Jackson et al. 2004). Careful layout and construction of roads and skid trails minimizes impacts (Swift 1988). However, roads and skid trails are particularly vulnerable to erosion during and shortly after construction, and stream crossings are the most likely locations for sediment delivery to streams. In a study in the Southern Appalachians examining the effectiveness of road construction BMPs, the majority of sediment was generated in two large storms that occurred shortly after new road construction and declined to pre-cut levels after road stabilization and reduced use after logging (Swank et al. 2001). Thus, it is critical to implement BMPs to ensure that newly constructed roads are quickly stabilized and that water and sediment moving from the forest roads and associated components such as ditches and cut banks is dispersed into areas that are disconnected from the streams to ensure infiltration and sediment trapping (Swift and Burns 1999). For example, in eastern Kentucky, BMPs such as streamside buffer strips and proper road construction and rehabilitation reduced suspended sediment considerably compared to a watershed clearcut without BMPs (Arthur et al. 1998). By contrast, other management activities that can be used to create early successional habitats without roads and skid trails (e.g., high intensity prescribed burning) are much less likely to cause a decline in water quality. For example, felling and burning low quality pine-hardwood stands in the Southern Appalachians resulted in no off-site movement of sediment (Swift et al. 1993).

Stream temperature, which affects dissolved oxygen concentration, may also be impacted by timber harvesting and the creation of early successional habitat. However, the magnitude and duration of the increase depends on the width of riparian buffers and the size of the harvested area. In the Central Hardwood Region, removal of forest canopy adjacent to forest streams increases maximum summer stream water temperatures by as much as 6°C (Swift and Messer 1971; Hornbeck and Federer 1975; Swift 1983; Clinton et al. 2010; Clinton 2011). However, maintaining a riparian forest buffer reduces or eliminates this effect (Hornbeck et al. 1986; Moore et al. 2005; Clinton 2011). For example, Clinton (2011) found that a buffer width as narrow as 10 m was adequate to prevent an increase in stream temperature after cutting. In addition, when only small areas of riparian forest canopy are removed, stream temperature responses are often dampened or eliminated within relatively short distances (e.g., 150 m) downstream (Clinton et al. 2010).

Disruption of terrestrial nutrient cycling processes through both alteration of soil abiotic conditions and reduced vegetation nutrient uptake can lead to nutrient transport into streams. Forest ecosystems are characterized by conservative nutrient cycling; most chemical constituents are limiting and tightly cycled by biogeochemical processes. Creating early successional habitats results in a considerable disruption to nutrient cycling processes and alters the environmental characteristics that regulate them. Opening the forest canopy increases soil temperature, and reduced transpiration rates increase soil moisture (Swank and Vose 1988). Both soil temperature and moisture influence nutrient cycling. For example, warmer and wetter soils result in increased nitrogen (N) mineralization and nitrification (Knoepp and Swank 2002; Knoepp and Vose 2007). Hence, these systems can transform N held tightly in organic matter to more mobile inorganic forms such as nitrate-N (NO_3^-).

In undisturbed forests, N typically limits productivity; most available N is used by the vegetation or immobilized by microbes. When nutrient uptake is disrupted by forest harvesting, combined with accelerated mineralization and nitrification, excess nutrients can be transported to streams. Studies examining changes in streamwater chemistry after timber harvesting have found that increases in nutrient concentrations can occur (especially for NO_3^-), losses are generally small relative to overall site nutrient pools and have little or no impact on water quality (Arthur et al. 1998; Martin et al. 2000; Swank et al. 2001). Nutrient responses tend to be greater in higher latitudes where nutrient cycling processes are more limited by temperature compared to responses at lower latitudes and elevations (Hornbeck et al. 1986). However rapid re-establishment of vegetation (both woody and herbaceous) plays a major in sequestering nutrients and re-establishing nutrient cycling processes. Indeed, major losses of nutrients (especially N, but also calcium and potassium) have been observed when vegetation regrowth is precluded by herbicides (Likens et al. 1970). Hence, one of the key BMPs to keep nutrients on site is to ensure rapid re-establishment of vegetation.

14.5 Potential Interactions with Climate Change

Because of the combination of biological and physical controls on hydrologic processes, climate change will both directly and indirectly impact the nation's water resources (Brian et al. 2004; Sun et al. 2008). The direct impacts of climate change on water resources will depend on how climate change alters the amount, type (e.g., snow vs. rain), and timing of precipitation; how this influences baseflow, stormflow, groundwater recharge, and flooding; and how these new hydrologic regimes interact with land use types (see Wear, Chap. 16). Long-term USGS streamflow data suggest that average annual streamflow has increased and this increase has been linked to greater precipitation in the eastern continental USA over the past 100 years (Lins and Slack 1999; Karl et al. 1995; IPCC 2007). However, fewer than 66% of all Global Circulation Models (GCMs) can agree on the predicted change in direction of future precipitation, e.g., wetter vs. drier (IPCC 2007). Inter- and intra-annual precipitation variability in the continental USA is a natural phenomenon related to large-scale global climate teleconnections (e.g., El Niño Southern Oscillation, Pacific Decadal Oscillation, North Atlantic Oscillation). Many regions of the USA have experienced an increased frequency of precipitation extremes over the last 50 years (Easterling et al. 2000a; Huntington 2006; IPCC 2007). As the climate warms in most GCMs, the frequency of extreme precipitation events increases across the globe (O'Gorman and Schneider 2009). However, the timing and spatial distribution of extreme events are among the most uncertain aspects of future climate scenarios (Karl and Knight 1998; Allen and Ingram 2002). Despite this uncertainty, recent experience with droughts and low flows in many areas of the USA indicate that even small changes in drought severity and frequency will have a major impact on society, including drinking water supplies (Easterling et al. 2000b; Luce and Holden 2009).

Most of the world's knowledge of the interactions among management, climate, vegetation, soils, and streamflow has been derived from long-term experiments on paired catchments. A key question is whether this knowledge, built primarily on empirical relationships under historical climate regimes, will allow robust predictions of responses under future climatic regimes. Creating early successional habitats has the potential to alter the hydrological responses to climate change again by influencing biological factors that determine evapotranspiration and physical factors that create soil disturbances or alter hydrologic flow paths. Management activities that favor or replace one species (or several species) over another can alter ET through direct and indirect changes in transpiration or interception (Ford et al. 2011, Stoy et al. 2006). For example, land management practices that favor high transpiration and interception may create conditions that mitigate the impacts of higher rainfall, but worsen the impacts of drought. As a result, streamflow responses (amount and timing) and recovery rates may be different under future climates. In general, hydrologic responses to climate change are larger in the humid Central Hardwood Region (McNab, Chap. 2). than in drier regions, and most climate models suggest the eastern USA will become more water-stressed (Sun et al. 2008). Thus, understanding the role of vegetation in hydrologic processes becomes increasingly important in the Central Hardwood Region as the climate gets warmer and more variable.

14.6 Summary

Because of the tight linkage between vegetation, soils, and water quantity and quality, creating early successional habitats has both direct and indirect impacts on water resources in the Central Hardwood Region. Decades of research using paired catchments in upland hardwood forests has shown:

1. Streamflow increases substantially in the first few years after cutting, but increases decline as sites revegetate and leaf area recovers. Streamflow increases are greater where precipitation is highest and where evapotranspiration represents a large portion of the overall site water budget.
2. The magnitude and rate of recovery to pre-disturbance streamflow depends on species composition and how species vary in transpiration and leaf and sapwood areas. Diffuse-porous species such as blackgum (*Nyssa sylvatica*), red maple, black birch, and yellow-poplar have the highest transpiration rates, while species with ring- or semi-ring porous sapwood, such as oaks and hickories, generally have the lowest transpiration rates for a given diameter. As such, watersheds dominated by the former would be expected to return to pre-cut streamflow levels faster than watersheds dominated by the latter; but depending on how the post-treatment vegetation differs from the pre-treatment vegetation, streamflow responses may be permanently higher or lower than reference conditions.

3. Stormflow increases by 10–20% following cutting and is directly proportional to the density and design of forest roads. However, these increases have not been shown to contribute to downstream flooding.
4. Sediment is the primary concern in terms of water quality responses to cutting and the primary sediment sources are roads and skid trails. BMPs have proven to be effective in reducing sediment.
5. Land managers will need to consider the potential interactions among future climate, changing vegetation structure and function, and physical impacts of forest operations on water resources.

As long as BMPs are properly implemented and maintained, creating early successional habitats in upland hardwood forests by harvesting trees is not likely to have a significant negative impact on either water quantity or water quality. However, it is also clear that forest operations associated with forest cutting (such as roads, stream crossings, culverts, etc.) can create permanent changes to hydrologic flow paths and serve as long-term sources of concern for water quantity and quality. In short, ensuring that BMPs are properly implesmented and functional requires a long-term commitment by land managers. Finally, much of what we know about the effects of disturbances on water resources (and the BMPs required to minimize those effects) has been developed from empirical data under historical climate regimes. Climatic conditions predicted for the eastern USA under climate change scenarios suggests the need for close monitoring of BMP effectiveness and the development of new BMPs applicable to more extreme climatic conditions in the future.

Literature Cited

Adams MB, Loughry L, Plaugher L (2008) Experimental forests and ranges of the USDA Forest Service. Gen Tech Rep NE-321, USDA Forest Service Northeastern Research Station, Newtown Square

Allen MR, Ingram WJ (2002) Constraints on future changes in climate and the hydrologic cycle. Nature 419:224–232. doi:10.1038/nature01092

Anderson H, Hoover M, Reinhart K (1976) Forest and water: effects of forest management on floods, sedimentation, and water supply. Gen Tech Rep PSW-18, USDA Forest Service Pacific Southwest Forest and Range Experiment Station, Berkeley

Andreassian V (2004) Waters and forests: from historical controversy to scientific debate. J Hydrol 291:1–27. doi:10.1016/j.jhydrol.2003.12.015

Arthur MA, Coltharp GB, Brown DL (1998) Effects of best management practices on forest streamwater quality in eastern Kentucky. J Am Water Resour Assoc 34:481–495. doi:10.1111/j.1752-1688.1998.tb00948.x

Bosch JM, Hewlett JD (1982) A review of catchment experiments to determine the effect of vegetation changes on water yield and evapotranspiration. J Hydrol 55:3–23. doi:10.1016/0022-1694(82)90117-2

Brian HH, Callaway M, Smith J, Karshen P (2004) Climate change and U.S. water resources: from modeled watershed impacts to national estimates. J Am Water Resour Assoc 40:129–148. doi:10.1111/j.1752-1688.2004.tb01015.x

Brown AE, Zhang L, McMahon TA, Western AW, Vertessy RA (2005) A review of paired catchment studies for determining changes in water yield resulting from alterations in vegetation. J Hydrol 310:28–61. doi:10.1016/j.jhydrol.2004.12.010

Calder IR (1993) Hydrologic effects of land use change. In: Maidment D (ed) Handbook of hydrology. McGraw-Hill, New York

Clinton BD (2011) Stream water responses to timber harvest: riparian buffer width effectiveness. For Ecol Manage 261:979–988. doi:10.1016/j.foreco.2010.12.012

Clinton BD, Vose JM, Fowler DL (2010) Flat Branch monitoring project: stream water temperature and sediment responses to forest cutting in the riparian zone. Res Pap SRS-51, USDA Forest Service Southern Research Station, Asheville

Daley MJ, Phillips NG, Pettijohn C, Hadley JL (2007) Water use by eastern hemlock (*Tsuga canadensis*) and black birch (*Betula lenta*): implications of effects of the hemlock woolly adelgid. Can J For Res 37:2031–2040. doi:10.1139/X07-045

Douglass JE (1983) The potential for water yield augmentation from forest management in the eastern United States. J Am Water Resour Assoc 19:351–358. doi:10.1111/j.1752-1688.1983.tb04592.x

Douglass JE, Swank WT (1975) Effects of management practices on water quality and quantity: Coweeta Hydrologic Laboratory, North Carolina. Municipal watershed management symposium proceedings. Gen Tech Rep NE-13, USDA Forest Service Northeastern Forest Experiment Station, Broomall

Douglass JE, Swank WT (1976) Multiple use in southern Appalachian hardwoods – a ten-year case history. In: Proceedings of the 16th international union of forestry research organization world congress, IUFRO, Oslo

Easterling DR, Evans JL, Groisman PY, Karl TR, Kunkel KE, Ambenje P (2000a) Observed variability and trends in extreme climate events: a brief review. Bull Am Meteorol Soc 81:417–425

Easterling DR, Meehl GA, Parmesan C, Changnon SA, Karl TR, Mearns LO (2000b) Climate extremes: observations, modeling, and impacts. Science 289:2068–2074. doi:10.1126/science.289.5487.2068

Eisenbies MH, Aust WM, Burger JA, Adams MB (2007) Forest operations, extreme flooding events, and considerations for hydrologic modeling in the Appalachians – a review. For Ecol Manage 242:77–98. doi:10.1016/j.foreco.2007.01.051

Elliott KJ, Vose JM (2011) The contribution of the Coweeta Hydrologic Laboratory to developing an understanding of long-term (1934–2008) changes in managed and unmanaged forests. For Ecol Manage 261:900–910. doi:10.1016/j.foreco.2010.03.010

Ellison AM, Bank MS, Clinton BD, Colburn EA, Elliott K, Ford CR, Foster DR, Kloeppel BD, Knoepp JD, Lovett GM, Mohan J, Orwig DA, Rodenhouse NL, Sobczak WV, Stinson KA, Stone JK, Swan CM, Thompson J, Holle BV, Webster JR (2005) Loss of foundation species: consequences for the structure and dynamics of forested ecosystems. Front Ecol Environ 9:479–486. doi:10.1890/1540-9295(2005) 003[0479:LOFSCF]2.0.CO;2

Enquist BJ, Brown JH, West GB (1998) Allometric scaling of plant energetics and population density. Nature 395:163–165. doi:10.1038/25977

Ford CR, Vose JM (2007) *Tsuga canadensis* (L.) Carr. mortality will impact hydrologic processes in southern Appalachian forest ecosystems. Ecol Appl 17:1156–1167. doi:10.1890/06-0027

Ford CR, Hubbard RM, Vose JM (2011) Quantifying structural and physiological controls on canopy transpiration of planted pine and hardwood stand species in the southern Appalachians. Ecohydrology 4:183–195. doi:10.1002/eco.136

Hewlett JD, Helvey JD (1970) Effects of forest clear-felling on the storm hydrograph. Water Resour Res 6:768–782

Hornbeck JW (1973) Storm flow from hardwood forested and cleared watersheds in New Hampshire. Water Resour Res 9:346–354

Hornbeck JW, Federer CA (1975) Effects of management practices on water quality and quantity: Hubbard brook experimental forest, New Hampshire. Municipal watershed management symposium proceedings. Gen Tech Rep NE-13, USDA Forest Service Northeastern Forest Experimental Station, Broomall

Hornbeck JW, Martin CW, Pierce RS, Bormann FH, Likens GE, Eaton JS (1986) Clearcutting northern hardwoods: effects on hydrologic and nutrient ion budgets. For Sci 32:667–686

Huntington TG (2006) Evidence for intensification of the global water cycle: review and synthesis. J Hydrol 319:83–95. doi:10.1016/j.jhydrol.2005.07.003

IPCC (2007) Contribution of working Groups I, II and III to the fourth assessment report of the intergovernmental panel on climate change. In: Core writing team, Pachauri RK, Reisinger A (eds) Climate change 2007: synthesis report. Geneva

Jackson CR, Sun G, Amatya DM, Swank WT, Riedel MS, Patric J, Williams T, Vose JM, Trettin CC, Aust WM, Beasley RS, Williston H, Ice GG (2004) Fifty years of forest hydrology in the southeast. In: Ice GG, Stednick JD (eds) A century of forest and wildland watershed lessons. Society of American Foresters, Bethesda

Jones JA, Post DA (2004) Seasonal and successional streamflow response to forest cutting and regrowth in the northwest and eastern United States. Water Resour Res 40:W05203. doi:10.1029/2003WR002952

Karl TR, Knight RW (1998) Secular trends of precipitation amount, frequency, and intensity in the USA. Bull Am Meteorol Soc 79:231–241

Karl TR, Knight RW, Plummer N (1995) Trends in high-frequency climate variability in the twentieth century. Nature 377:217–220. doi:10.1038/377217a0

Knoepp JD, Swank WT (2002) Using soil temperature and moisture to predict forest soil nitrogen mineralization. Biol Fertil Soils 36:177–182. doi:10.1007/s00374-002-0536-7

Knoepp JD, Vose JM (2007) Regulation of nitrogen mineralization and nitrification in Southern Appalachian ecosystems: separating the relative importance of biotic vs. abiotic controls. Pedobiologia 51:89–97. doi:10.1016/j.pedobi.2007.02.002

Kochendorfer JN, Hornbeck JW (1999) Contrasting timber harvesting practices illustrate the value of BMPs. In: Proceedings of the 12th central hardwood forest conference. Gen Tech Rep SRS-24, USDA Forest Service Southern Research Station, Asheville

Kochendorfer JN, Adams MB, Miller GW, Helvey JD (2007) Factors affecting large peakflows on Appalachian watersheds: lessons from the Fernow Experimental Forest. Res Pap NRS-3, USDA Forest Service Northern Research Station, Newtown Square

Likens GE, Bormann FH, Johnson NM, Fisher DW, Pierce RS (1970) Effects of forest cutting and herbicide treatment on nutrient budgets in the Hubbard Brook watershed-ecosystem. Ecol Monogr 40:23–47

Lins H, Slack JR (1999) Streamflow trends in the United States. Geophys Res Lett 26:227–230. doi:10.1029/1998GL900291

Luce CH, Holden ZA (2009) Declining annual streamflow distributions in the Pacific Northwest United States, 1948–2006. Geophys Res Lett 36:L16401. doi:10.1029/2009GL039407

Lull HW, Reinhart KG (1972) Forests and floods in the eastern United States. Res Pap NE-226, USDA Forest Service Northeastern Forest Experimental Station, Broomall

Martin CW, Hornbeck JW, Likens GE, Buso DC (2000) Impacts of intensive harvesting on hydrology and nutrient dynamics of northern hardwood forests. Can J Fish Aquat Sci 57:19–29. doi:10.1139/cjfas-57-S2-19

Meinzer FC, Bond BJ, Warren JM, Woodruff DR (2005) Does water transport scale universally with tree size? Funct Ecol 19:558–565. doi:10.1111/j.1365-2435.2005.01017.x

Moore RD, Spittlehouse DL, Story A (2005) Riparian microclimate and stream temperature response to forest harvesting: a review. J Am Water Resour Assoc 41:813–834. doi:10.1111/j.1752-1688.2005.tb03772.x

O'Gorman PA, Schneider T (2009) The physical basis for increases in precipitation extremes in simulations of 21st-century climate change. Proc Natl Acad Sci 106:14773–14777. doi:10.1073/pnas.0907610106

Orwig DA, Foster DR, Mausel DL (2002) Landscape patterns of hemlock decline in New England due to the introduced hemlock woolly adelgid. J Biogeogr 29:1475–1487. doi:10.1046/j.1365-2699.2002.00765.x

Patric JH, Reinhart KG (1971) Hydrologic effects of deforesting two mountain watersheds in West Virginia. Water Resour Res 7:1182–1188. doi:10.1029/WR007i005p01182

Perry CA, Combs LJ (1998) Summary of floods in the in the United States, January 1992 through September 1993. Water supply paper 2499, USGS, Desoto

Reinhart KG, Pierce RS (1964) Stream-gaging stations for research on small watersheds. Agri Handbook 268. USDA

Stednick JD (1996) Monitoring the effects of timber harvest on annual water yield. J Hydrol 176:79–95. doi:10.1016/0022-1694(95)02780-7

Stoy P, Katul G, Siqueira M, Juang J, Novick K, McCarthy HR, Oishi AC, Umbelherr J, Kim H, Oren R (2006) Separating the effects of climate and vegetation on evapotranspiration along a successional chronosequence in the southeastern US. Glob Change Biol 12:2115–2135. doi:10.1111/j.1365-2486.2006.01244.x

Sun G, Riedel M, Jackson R, Kolka R, Amatya D, Shepard J (2004) Influences of management of Southern forests on water quantity and quality. In: Rauscher HM, Johnsen K (eds) Southern forest sciences: past, current, and future. Gen Tech Rep SRS-75, USDA Forest Service Southern Research Station, Asheville

Sun G, McNulty SG, Myers JAM, Cohen EC (2008) Impacts of multiple stresses on water demand and supply across the southeastern United States. J Am Water Resour Assoc 44:1441–1457. doi:10.1111/j.1752-1688.2008.00250.x

Sun G, Noormets A, Gavazzi MJ, McNulty SG, Chen J, Domec JC, King JS, Amatya DM, Skaggs RW (2010) Energy and water balance of two contrasting loblolly pine plantations on the lower coastal plain of North Carolina, USA. For Ecol Manage 259:1299–1310. doi:10.1016/j.foreco.2009.09.016

Swank WT, Crossley DA Jr (1988) Introduction and site description. In: Swank WT, Crossley DA Jr (eds) Ecological studies, vol. 66: forest hydrology and ecology at Coweeta. Springer, New York

Swank WT, Douglass JE (1974) Streamflow greatly reduced by converting deciduous hardwood stands to pine. Science 185:857–859. doi:10.1126/science.185.4154.857

Swank WT, Johnson CE (1994) Small catchment research in the evaluation and development of forest management practices. In: Moldan B, Cerny J (eds) Biogeochemistry of small catchments: a tool for environmental research. Wiley, Chichester

Swank WT, Vose JM (1988) Effects of cutting practices on microenvironment in relation to hardwood regeneration. In: Guidelines for regenerating Appalachian hardwood stands: proceedings of a workshop. SAF Publication 88-034. West Virginia University Books, Morgantown

Swank WT, Vose JM, Elliott KJ (2001) Long-term hydrologic and water quality responses following commercial clearcutting of mixed hardwoods on a southern Appalachian catchment. For Ecol Manage 143:163–178. doi:10.1016/S0378-1127(00)00515-6

Swift LW Jr (1976) Algorithm for solar radiation on mountain slopes. Water Resour Res 12:108–112. doi:10.1029/WR012i001p00108

Swift LW Jr (1983) Duration of stream temperature increases following cutting in the southern Appalachian mountains. In: Proceedings of the international symposium on hydrometeorology. American Water Resource Association, Bethesda

Swift LW Jr (1988) Forest access roads: design, maintenance, and soil loss. In: Swank WT, Crossley DA Jr (eds) Ecological studies, vol. 66: forest hydrology and ecology at Coweeta. Springer, New York

Swift LW, Burns RG (1999) The three Rs of roads. J Forest 97:40–45

Swift LW Jr, Messer JB (1971) Forest cuttings raise temperatures in small streams in the southern Appalachians. J Soil Water Conserv 26:111–116

Swift LW Jr, Elliott KJ, Ottmar RD, Vihnanek RE (1993) Site preparation burning to improve southern Appalachian pine-hardwood stands: fire characteristics and soil erosion, moisture, and temperature. Can J For Res 23:2242–2254. doi:10.1139/x93-278

Vose JM, Ford CR, Laseter S, Sun G, Adams MB, Dymond S, Sebestyen S, Campbell J, Luce C (2010) Can forest watershed management mitigate climate change impacts on water resources? Unpubl Rep, USDA Forest Service, Southern Research Station, Asheville

Vose JM, Sun G, Ford CR, Bredemeier M, Otsuki K, Wei A, Zhang Z, Zhang L (2011) Forest ecohydrological research in the 21st century: what are the critical needs? Ecohydrology 4:146–158. doi:10.1002/eco.193

Wallace LL (1988) Comparative physiology of successional forest trees. In: Swank WT, Crossley DA Jr (eds) Ecological studies, vol. 66: forest hydrology and ecology at Coweeta. Springer, New York

Wullschleger SD, Hanson JJ, Todd DE (2001) Transpiration from a multi-species deciduous forest as estimated by xylem sap flow techniques. For Ecol Manage 143:205–213

Chapter 15
Carbon Dynamics Following the Creation of Early Successional Habitats in Forests of the Central Hardwood Region

Tara L. Keyser

Abstract Across a forested landscape, stand-level management actions or natural disturbances that create early successional habitats result in a short-term loss of carbon in any given stand, but are often offset by carbon gains in other, undisturbed stands. Standing carbon stocks and rates of sequestration vary with species, site productivity, stand age, and stand structure. The age distribution of forest stands has a particularly large effect on landscape-level carbon storage. Consequently, forest management activities, including creation of early successional habitats, have short-term implications for stand-level carbon storage, but their impact on forest- or landscape-level carbon storage ultimately depends upon the temporal distribution and spatial scale of young forest stands on the landscape.

15.1 Introduction

Anthropogenic activities, including burning fossil fuels and changes in land-use patterns, have increased atmospheric concentrations of greenhouse gases. Considered to be the most important anthropogenic greenhouse gas, atmospheric carbon dioxide (CO_2) concentration reached 387 ppm in 2009, significantly higher than the pre-industrial concentration estimated at 280 ppm (http://www.esrl.noaa.gov/gmd/ccggl/trends). Future emission scenarios suggest CO_2 concentrations will increase between 41 and 158% by 2100 (IPCC 2007).

The rise in CO_2 and other greenhouse gases have led to a 0.7 °C increase in global surface temperature during the twentieth century (IPCC 2007). Given current and

T.L. Keyser (✉)
USDA Forest Service, Southern Research Station, Bent Creek Experimental Forest, 1577 Brevard Road, Asheville, NC 28806, USA
e-mail: tkeyser@fs.fed.us

C.H. Greenberg et al. (eds.), *Sustaining Young Forest Communities*, Managing Forest Ecosystems 21, DOI 10.1007/978-94-007-1620-9_15,

predicted CO_2 levels over the next 100 years, future changes in global surface temperatures are expected to be even larger. Under all emission scenarios modeled by the Intergovernmental Panel on Climate Change (IPCC), fossil fuel use and CO_2 emissions will increase well into the 21st century. Mitigation efforts designed to capture and store carbon, however, can offset and regulate anthropogenic CO_2 emissions. The ability of forestland to sequester and offset CO_2 emissions has generated substantial interest in managing forests for increased CO_2 uptake and storage through activities such as afforestation, reforestation, and improved forest management.

Forests sequester atmospheric carbon in aboveground live and dead biomass, soil organic matter, roots, and surface detritus. They emit carbon back to the atmosphere through metabolic processes such as autotrophic and heterotrophic respiration, and physical processes such as herbivory, fire, insect and disease outbreaks, and timber harvesting. The forest carbon cycle is comprised of periods of carbon storage punctuated by periodic disturbance events (both natural and anthropogenic) that release carbon back to the atmosphere. At any given point in time, individual stands are either sources or sinks of carbon. This difference between gross carbon gained via photosynthesis, or gross primary productivity (GPP), and total carbon lost is net ecosystem production (NEP). When NEP is negative, a forest stand is a source of atmospheric carbon (Fig. 15.1). When NEP is positive, a forest stand is removing and storing carbon. Although individual stands are either sources or sinks of carbon at any specific point in time, carbon balance at the forest- or landscape-level is determined by summing the net carbon balance of individual stands.

At a global scale, forest ecosystems are a net carbon sink; over a defined timeframe, forests sequester more carbon than they emit. Forestland in the USA constitutes a particularly large carbon sink relative to the global carbon budget. Overall, forestland in the USA sequestered 216 Tg carbon in 2008, an 18% increase over that

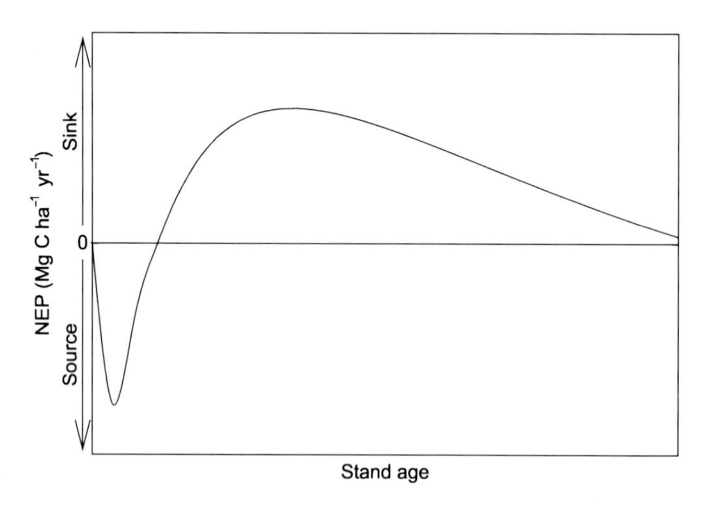

Fig. 15.1 Generalized pattern of net ecosystem production (NEP) as a function of stand age following stand replacing disturbance (e.g., regeneration harvest)

in 1990 (Environmental Protection Agency (EPA) 430-R-10-006). Much of this increase is the result of improved forest management practices, successful forest regeneration treatments, and afforestation and reforestation efforts, along with increases in aboveground live biomass in existing forest stands.

For purposes of inventory and accounting, carbon in forests is categorized into one of six pools: (1) aboveground biomass (all living biomass above the soil); (2) belowground biomass (all living biomass of roots >2 mm in diameter); (3) dead wood (non-living woody biomass); (4) litter [litter (O_i) and duff ($O_e + O_a$) layers in addition to woody material <7.5 cm]; (5) soil organic matter; and (6) harvested wood products (IPCC 2003, 2006). Because carbon is not directly measured as part of standard forest inventory procedures, methods exist to estimate carbon storage in the various carbon pools using forest inventory data. For example, aboveground live tree carbon is often estimated using species-specific allometric equations that relate tree size to biomass. Carbon is then calculated using a ratio that relates biomass (on a dry weight basis) to carbon [0.5 for aboveground live tree biomass (IPCC 2003)]. This biomass to carbon conversion factor varies based on the carbon pool analyzed [e.g., litter, dead wood, etc. (IPCC 2003)]. Excluding the harvested wood product carbon pool, which the size and longevity will vary with silvicultural prescription, harvesting system, species, and forest product type (Smith et al. 2006), aboveground biomass and mineral soil carbon pools constitute the greatest proportion of a forest stand's overall carbon stock; these are followed by belowground biomass, dead wood, and litter carbon pools.

Predominant forest types within the Central Hardwood Region are dominated by upland oak and hickory (*Quercus-Carya*) species (Johnson et al. 2002b). Associated canopy-tree species include mixed-mesophytic species such as yellow-poplar (*Liriodendron tulipifera*), ashes (*Fraxinus* spp.), maples (*Acer spp.*), and others (see Loftis et al., Chap. 5). As abandoned farmland returns to forestland and past timber harvests enter mid- to later stages of stand development, carbon continues to accumulate in these forests, making carbon capture and storage an added benefit of past management actions. Relative to other eastern forest types, second growth oak-hickory forests are a strong carbon sink; they are capable of sequestering 5.25 Mg carbon ha^{-1} $year^{-1}$ (Greco and Baldocchi 1996) compared to 3.7 Mg carbon ha^{-1} $year^{-1}$ in northern hardwood forests in the northeastern USA (Wofsy et al. 1993).

Upland hardwood forests of the Central Hardwood Region are managed for a multitude of ecosystem services. While meeting specific resource objectives, the creation of early successional habitats in mature upland hardwood forests can alter stand-level carbon storage and, depending on the temporal and spatial scale of management actions, landscape-level carbon dynamics (e.g., Campbell et al. 2004; Depro et al. 2008). Variations in climate, forest type, stand structure and species composition, stand age, and edaphic conditions among and within the physiographic regions of the central hardwoods interact to influence the rates of carbon sequestration and total carbon storage following both natural and silvicultural disturbance events. Although forest carbon storage is most meaningful when examined at the landscape-level (Harmon 2001), silvicultural prescriptions designed to achieve resource management objectives are implemented at the stand-level. Consequently, developing an understanding of the role forest management has on the various carbon pools and stand-level carbon storage is relevant to carbon management at the landscape-level. The following provides a synopsis of the literature associated with the potential effects of creating early successional habitat via silvicultural practices on stand-level biological carbon pools and dynamics in the Central Hardwood Region.

15.2 Carbon Dynamics Following the Creation of Early Successional Habitats

15.2.1 Aboveground Biomass

As described by Loftis et al. (Chap. 5), numerous silvicultural tools and methods can be used to create early successional habitats. Even-aged silviculture utilizing the clearcutting and shelterwood methods, a two-aged system or shelterwood with reserves, and the group selection method of uneven-aged management are all effective methods. In all cases, these regeneration harvests result in stands that lack closed or continuous canopy and are in the early stages of stand development (i.e., stand initiation).

Of the biological carbon pools after harvest (i.e., excluding the harvested carbon pool), carbon stored in aboveground biomass is the most dynamic as well as most easily quantified (Fahey et al. 2010). Following harvests and barring any managed or unmanaged disturbance events, the aboveground component of a newly regenerated forest stand accumulates carbon in a predictable sigmoid pattern (Hunt 1982) (Fig. 15.2). Young forests take up carbon rapidly and have high growth rates, but

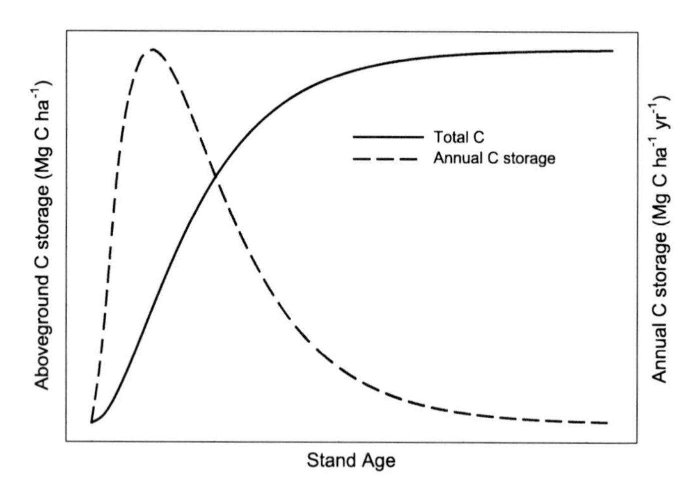

Fig. 15.2 Generic pattern of stand-level total carbon (C) storage and annual rate of C storage in the aboveground biomass pool following a regeneration harvest as a function of stand age

contain significantly less biomass and standing carbon stocks than mature forest stands. During canopy closure, net primary productivity and carbon uptake by aboveground vegetation are maximized. As stands age, rates of carbon sequestration and net primary productivity decline, but aboveground biomass and carbon storage continue to increase until biomass approaches a maximum defined, in part, by species composition and associated maximum size-density relations (Yoda et al. 1963) and environmental conditions (Gholz 1982; Johnson et al. 2000; Kranabetter 2009).

Carbon stored in the aboveground biomass pool is the largest (Li et al. 2007) or second largest (Bolstad and Vose 2005) carbon pool in upland forests of the Central Hardwood Region and is maximized when the frequency of anthropogenic and natural disturbances is low (Reinhardt and Holsinger 2010; Nunery and Keeton 2010; Harmon et al. 2009; Hudiburg et al. 2009; Harmon and Marks 2002; Janisch and Harmon 2002). The pattern of carbon accumulation observed throughout stand development (Fig. 15.2) is applicable to forest stands regardless of species composition (Spetich et al. 1998, Johnsen et al. 2001; Law et al. 2003; Taylor et al. 2007). However, the upper limit of forest stand-level carbon storage and the timeframe associated with maximizing carbon storage in aboveground biomass differ among forest stands due to variation in species composition (Grigal and Ohmann 1992; Caspersen and Pacala 2001; Bunker et al. 2005), disturbance patterns, and edapho-climatic conditions (Smithwick et al. 2002; Van Tuyl et al. 2005; Hudiburg et al. 2009). As a result of their history (see White et al., Chap. 3; Shifley and Thompson, Chap. 6), most regenerated upland forests in the Central Hardwood Region, are even-aged, second-growth forests (80–100 years old) and are at a point in stand development where they are still accumulating aboveground carbon (Brown et al. 1997; Brown and Schroeder 1999) albeit at a slower rate than occurred earlier in stand development (i.e., during stand initiation; Fig. 15.2).

Although second growth upland hardwood forests throughout the Central Hardwood Region are still accumulating carbon in aboveground biomass, aboveground carbon stocks are substantially less than in comparable old-growth forests. Throughout oak-hickory dominated regions, aboveground carbon stocks (calculated using a standard carbon to biomass ratio of 0.5) can range between 87.5 and 92.5 Mg carbon ha^{-1} for mature, sawtimber sized stands and 50–62.5 Mg carbon ha^{-1} for poletimber sized stands (Brown et al. 1997). Stands recently regenerated (i.e., seedling/sapling sized stands with early successional and young forest structure) generally store ≤25 Mg carbon ha^{-1} in aboveground live biomass (Brown et al. 1997). However, these carbon estimates reflect averages for oak-hickory forests across a broad range of physiographic regions and age classes. Aboveground carbon stocks vary with site productivity and species composition. In Indiana, for instance, aboveground carbon stocks in oak-hickory forests vary by as much as 50% between stands of low [oak site index (SI), base-age 50 between 16.8 and 18.3 m] and high (oak SI between 25.9 and 27.4 m) site quality (Kaczmarek et al. 1995).

If undisturbed, upland hardwood forests have the potential to sequester substantial quantities of carbon in aboveground vegetation. Old-growth forests across the Central Hardwood Region (Schmelz and Lindsey 1965; McClain and Ebinger 1968; Weaver and Ashby 1971; Muller 1982; Cho and Boerner 1991; Spetich and Parker

1998) possess aboveground carbon stocks that range from 98.5 Mg carbon ha^{-1} in oak-hickory forests in Illinois (McClain and Ebinger 1968) to 165 Mg carbon ha^{-1} in mixed-mesophytic forests of Kentucky (Muller 1982). As edaphic and climatic (i.e., edaphoclimatic) conditions improve, the capacity of a forest stand to accumulate and store aboveground biomass and carbon increases (Kranabetter 2009). A prime example of this positive productivity-carbon storage relationship is found in forests of Great Smoky Mountains National Park. Here, the edophoclimatic environment, species composition, and disturbance history have resulted in a substantial and prolonged accumulation of biomass with aboveground live tree carbon stocks estimated to range from 163 to197 Mg carbon ha^{-1} in some of the park's old-growth, cove hardwood forests (Busing et al. 1993). Old growth forests are a significant carbon sink (Luyssaert et al. 2008), but occupy only a small percentage of forested land relative to second-growth forests. Old growth forests represent the upper bounds of carbon storage that may not be attainable for the majority of second-growth forest stands, even if left unmanaged (Brown et al. 1997).

15.2.2 Belowground Biomass

The belowground biomass pool is composed of coarse and fine roots (IPCC 2003). Despite consuming a significant proportion of annual net primary productivity, fine roots contribute only a minor proportion of the total carbon storage in any given forest stand (Santantonio et al. 1977; Grier et al. 1981; Comeau and Kimmins 1989; John et al. 2001). Carbon is stored in coarse roots in two distinct pools: living vegetation and harvested or dead trees (hereafter referred to as residual coarse roots). Unlike fine roots, coarse roots represent a significant and long-term carbon pool (Resh et al. 2003; Miller et al. 2006; Yanai et al. 2006). Similar to aboveground biomass, carbon storage in coarse roots accumulates in a sigmoid pattern as stands progress through stages of stand development. In young, recently regenerated stands, small diameter trees require less structural support than large trees of older stands. Consequently, coarse root biomass and corresponding carbon stocks are relatively low early in stand development and are maximized during the later stages of stand development (Misra et al. 1998; Resh et al. 2003; Yanai et al. 2006).

Following a regeneration harvest, coarse roots of harvested trees remain on-site, and the carbon stored in them may have a relatively long residence time. The rate of decomposition of residual coarse root systems depends on tree species and wood quality (Fahey and Arthur 1994; Chen et al. 2001), size of residual coarse root material (Fahey et al. 1988; Janisch et al. 2005), and temperature and moisture availability (Fahey and Arthur 1994; Ximenes and Gardner 2006). Decomposition of residual coarse root systems is often modeled using a negative exponential function (Melin et al. 2009), suggesting that a portion of residual coarse root carbon is a long-term pool that must be accounted for when assessing the effects of management on stand-level carbon storage. In loblolly pine (*Pinus taeda*), for example, coarse- and tap-root biomass can persist upwards of 60 years after harvest (Ludovici et al. 2002) and between 10% and 50% of hardwood coarse root biomass can remain on-site 100 years following harvest in Australia (Ximenes and Gardner 2006).

Information regarding coarse root decomposition specific to upland hardwood tree species in the Central Hardwood Region is sparse. However, it is plausible that carbon stored in coarse roots systems following harvests in hardwood stands may be of greater importance to a stand's overall carbon budget than in conifer-dominated stands, given the propensity of hardwood species to allocate more carbon below-ground than managed conifer species (e.g., Miller et al. 2006). As a guideline, the IPCC suggests that carbon stored in the biomass of coarse root systems has a post-harvest residence time of only 10 years (IPCC 2003). From the limited data available on coarse root decomposition, (e.g., Ludovici et al. 2002; Ximenes and Gardner 2006) it would appear this 10-year residence time (IPCC 2003) underestimates carbon storage in residual root systems. In turn, this overestimation of residual root decay can lead to an overestimate of carbon flux to the atmosphere following forest management and underestimate stand-level carbon stocks over time.

15.2.3 Dead Wood Biomass

Dead and down coarse wood [defined as dead and down woody biomass greater than a specific diameter (often >7.6 cm in diameter or >10 cm in diameter)] is vital to proper maintenance of ecosystem structure and function. Dead and down wood fills an important niche in forested ecosystems, providing habitat for vertebrate species (e.g., Moorman et al., Chap. 11) as well as an energy and nutrient substrate for invertebrate, fungal, and microbial species (Harmon et al. 1986). Furthermore, the nutrients and energy released during the decomposition of wood contributes to the maintenance of long-term site productivity.

At the time of a regeneration harvest (i.e., stand initiation), down and dead coarse wood in a forest stand comes from one of two sources. Dead coarse wood mass present at the time of stand initiation is residual dead wood from the previous stand or is created by the regeneration harvest itself (Janisch and Harmon 2002). Immediately following creation of early successional habitats, or during stand initiation (Fig. 15.3, stand age = 0), carbon stored in dead coarse wood is typically abundant due to woody residues left on-site during harvest operations (Spetich et al. 1999, Duvall et al. 1999). For example, in Central Appalachian hardwood stands regenerated via clearcutting, dead coarse wood can approach 55 Mg ha^{-1} 2 years post-harvest (McCarthy and Bailey 1994), which exceeds dead coarse wood biomass in many old-growth forests.

As aboveground biomass begins to accumulate in recently regenerated stands, low mortality rates of the arborescent vegetation layer, coupled with decomposition of residual and post-harvest dead coarse wood, reduces carbon stored in dead coarse wood to a minimum (Fig. 15.3, I). Aggregation of coarse wood begins slowly during stem exclusion (Oliver and Larson 1996); however, inputs are relatively low, given mortality during this stage of stand development is limited to intermediate and suppressed (i.e., small diameter) trees (Fig. 15.3, II). During understory reinitiation

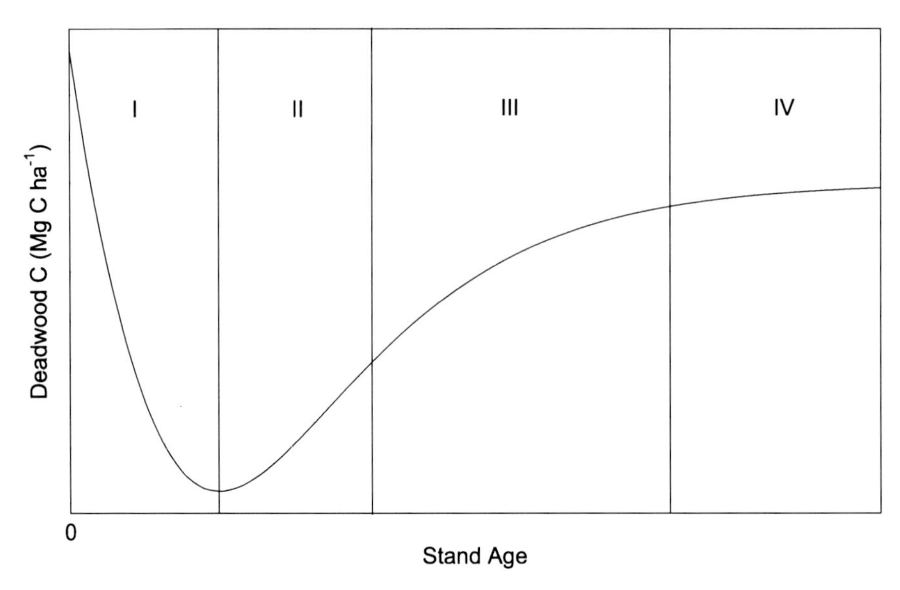

Fig. 15.3 Generic pattern of dead wood biomass and carbon (C) accumulation following a regeneration harvest as a function of stand age

(Oliver and Larson 1996), mortality of large-diameter trees beings to occur and the rate at which coarse wood carbon accumulates increases (Fig. 15.3, III) until an equilibrium between production and decomposition occurs during late (i.e., old-growth) stages of stand development (Fig. 15.3, IV). So, although regeneration harvests generally result in a short-term pulse of coarse wood to the system via logging residues, over time there is a reduction of both biomass and carbon that is sustained until stands enter the stand reinitiation or old-growth stages of stand development (Hardt and Swank 1997; Spetich et al. 1999; Duvall and Grigal 1999).

Although the U-shaped pattern (Fig. 15.2) of coarse wood accumulation following a regeneration harvest can be used to generalize coarse wood carbon dynamics across geographic regions and forest types (Harmon et al. 1986), the rate and amount of coarse wood carbon accumulation within any given forest stand varies as a function of time since disturbance or stand age (Sturtevant et al. 1997), type and severity/intensity of disturbance (Spies et al. 1988; McCarthy and Bailey 1994; Duvall and Grigal 1999), site quality (Harmon et al. 1986; Spetich et al. 1999), and decay rates, which are influenced by species, size of substrate, topographic position, climate, and site quality (MacMillan 1981; Harmon et al. 1986,1995; Janisch et al. 2005; Webster and Jenkins 2005; McCarthy et al. 2001; Spetich et al. 1999). The high degree of spatial variability in the coarse wood carbon pool is demonstrated by estimates of coarse wood volume (substitute for biomass) in old-growth forests that range from 32 m^3 ha^{-1} in Midwestern oak-hickory forests (Spetich et al. 1999) to 53 m^3 ha^{-1} in highly productive cove hardwood forests of Great Smoky Mountains National Park (Webster and Jenkins 2005).

15.2.4 Forest Floor

The forest floor comprises the upper portions of the soil profile defined by the litter (Oi) and duff (Oe+Oa) layers as well as small (<7.5 cm in diameter) dead woody biomass (IPCC 2006). Although vital to ecosystem function and nutrient cycling (Vitousek and Sanford 1986), the forest floor constitutes a minor component of a forest stand's overall carbon budget (Bolstad and Vose 2005; Li et al. 2007; Bradford et al. 2009). The distribution and abundance of the forest floor, as well as the response of this carbon pool to disturbance, is spatially and temporally variable (Wallace and Freedman 1986). Earlier models describing the response of the forest floor to disturbances suggested that as much as 50% of the forest floor mass is lost within the first 20 years, with recovery to pre-disturbance levels not occurring until approximately 50 years post-harvest (Covington 1981). Although this predicted pattern of forest floor mass and carbon loss and recovery has been observed elsewhere in eastern USA forests (Federer 1984; Mattson and Smith 1993; Griffiths and Swanson 2001), many studies in the Central Hardwood Region have documented long-term increases or no effect of disturbance on forest floor mass and, consequently, carbon storage, following regeneration harvests (Mattson and Swank 1989, Knoepp and Swank 1998, Johnson et al. 1991; Elliott and Knoepp 2005; Li et al. 2007). Increases in forest floor mass observed in forests across the Central Hardwood Region (Johnson et al. 1985; Mattson and Swank 1989; Johnson and Todd 1998, Knoepp and Swank 1998) following silvicultural treatments are attributed to logging debris deposited on the forest floor during harvest operations, increased fine-root mortality and production, and changes in species composition that may occur during stand development (Yin et al. 1989a, b; Knoepp and Swank 1997; Idol et al. 2000; Jandl et al. 2007; Nave et al. 2010).

Clearly, post-disturbance dynamics of the forest floor are variable. However, following a regeneration harvest, and provided successful regeneration occurs, a general pattern in which forest floor mass and, consequently carbon stored in the forest floor pool, increases until equilibrium between litter input and decomposition is achieved during the later stages of stand development. The recovery of forest floor mass following a stand initiating disturbance is a function of the quantity and quality of litter, climatic controls on decomposition (Prescott et al. 2000), and the relative speed at which aboveground biomass and leaf area of a regenerated stand develop (Bradford et al. 2009). Inputs of forest floor material can approach pre-harvest levels within a few years of a regeneration treatment (Covington and Aber 1980; Boring et al. 1988; Boring and Swank 1986), but this timeframe varies with site productivity (Vose and Allen 1988; Shi and Cao 1997; Frazer et al. 2000; Jokela and Martin 2000). Across the Central Hardwood Region, forest floor mass and the corresponding carbon pool are highly variable due to differences in decomposition rates and/or species composition and associated litter quality. In productive upland forests of the Southern Appalachians, forest floor mass has been found to range from 22.2 Mg ha^{-1} in 20 year old stands to 26.0 Mg ha^{-1} in an 85 year old stand to 26.7 Mg ha^{-1} in an old-growth stand (Vose and Bolstad 2007). In comparison, forest floor mass in less productive forest types is considerably less, ranging from only 13.1 Mg ha^{-1} in mature oak-hickory forests of Illinois (Luvall and Weaver 1986) to 20.0 Mg ha^{-1} in mixed-oak forests in southeastern Ohio (Graham and McCarthy 2006).

Forest floor dynamics following traditional approaches to creating early successional habitats, such as clearcutting, depend on the immediate effects of the treatment on the forest floor layer. Forest floor mass can decrease a result of decreased litter inputs or increased decomposition; alternatively, forest floor material can increase or be buried or mixed with mineral soil as a result of harvest operations. If treatment results in an immediate decrease in forest floor mass (Covington 1981), mass and carbon storage of the forest floor will decrease until, at some point in stand development, litter inputs exceed rates of decomposition. If, however, decomposition rates decrease or remain unchanged and the addition of logging debris offsets any short-term losses in litter inputs (until the point in stand development when litter input recovers to pre-disturbance levels), little to no change in forest floor mass or carbon storage capacity of the forest floor relative to pre-disturbance conditions can be expected over the short- and long-term (Knoepp and Swank 1997; Hall et al. 2006; Boerner et al. 2008).

15.2.5 Soil Organic Matter

Mineral soil carbon is commonly cited as the forest carbon pool that stores the greatest proportion of ecosystem carbon (e.g., Turner et al. 1995; Dixon et al. 1994). Although studies from mixed-hardwood forests in the Southern Appalachian region confirm this statement (e.g., Bolstad and Vose 2005), other forest types within the Central Hardwood Region store significantly more carbon in aboveground vegetation than in the mineral soil (Li et al. 2007). This suggests the capacity of the mineral soil to sequester carbon is spatially variable across physiographic regions. Accumulation of carbon in mineral soil depends on inputs from litterfall and fine root turnover, while release of carbon depends on mineral soil carbon quality (e.g., labile or stable condition) and edaphoclimatic conditions (Jandl et al. 2007). Despite the spatial (e.g., Sun et al. 2004) and temporal (e.g., Knoepp and Swank 1997) variability in size and significance of the mineral soil carbon pool across the Central Hardwood Region, independent studies (Mattson and Swank 1989; Knoepp and Swank 1997; Johnson et al. 2002a; Gilliam et al. 2004; Li et al. 2007; Boerner et al. 2008) and large-scale meta-analyses (Johnson and Curtis 2001) suggest that with successful vegetative recovery and accumulation of aboveground biomass, the mineral soil carbon pool is stable relative to pre-treatment conditions under a wide variety of silvicultural treatments, including regeneration treatments that create early successional habitats.

15.3 Conclusions

At the forest- or landscape-level, carbon storage is maximized when all stands comprising a forest are in an old-growth state. This, however, rarely occurs as natural or silvicultural disturbances create and maintain a heterogeneous mix of stands of various age classes and stages of stand development on the landscape. On a stand-by-stand basis, silvicultural disturbances can have a negative, albeit, short-term, impact on forest carbon storage relative to pre-disturbance levels, with the removal of aboveground biomass responsible for the greatest proportion of carbon lost. Forest management activities that may compensate for a proportion of the stand-level loss in carbon stocks include prescriptions that maintain partial canopy cover. Silvicultural methods that create early successional habitats while retaining structural heterogeneity in the post-harvest stand (e.g., Franklin 1989) may ameliorate some short-term losses in the aboveground carbon pool as well as alleviate some longer-term losses in the dead wood biomass pool that occur following clearcutting. Structural diversity can be increased by implementing a 2 aged silvicultural system or shelterwood with reserves (Smith et al. 1997), in which species of wildlife value (e.g., oaks in the Central Hardwood Region; see Greenberg et al., Chap. 8) are left in the overstory at densities low enough to regenerate a new cohort. As the newly regenerated cohort develops, the residual overstory or reserve trees provide structural diversity and wildlife habitat, as well as a source of large-diameter coarse woody debris, as these trees senesce and die throughout the rotation.

Although regeneration harvests used to create early successional habitats can convert stands from a carbon sink to a carbon source in the short-term, with successful vegetative recovery and adequate time (at least one full rotation) without disturbance(s), carbon storage will recover to pre-harvest levels. The timeframe for recovery of the various carbon pools will vary among forest types, edaphoclimatic conditions (Campbell et al. 2004), and the proportion of aboveground biomass removed (e.g., clearcutting versus single-tree selection). The timeframe during which an individual stand remains a carbon source will depend upon the amount of time required for carbon sequestration rates in the aboveground biomass pool to surpass carbon lost (Fig. 15.1) via decomposition of belowground, dead wood, forest floor, and/or mineral soil carbon pools.

In a landscape as diverse as the Central Hardwood Region, stand-level responses to disturbance, including responses that affect carbon storage, vary within and among physiographic regions. In highly productive forests, the amount of time during which a regenerated forest stand remains a net source of atmospheric carbon (i.e., negative net ecosystem productivity; Fig.15.1) may be minor due to the rapid accumulation of aboveground biomass and carbon (Davis et al. 2009). Less favorable edaphoclimatic conditions coupled with slower-growing species may result in a regenerated stand being a carbon source for a substantial period of time (Campbell et al. 2004). Regenerating high quality sites, however, has the potential to store more of the removed merchantable carbon in long-term products compared to less productive forest stands where timber quality and merchantability are lower (e.g., Gonzalez-Benecke et al. 2010). Because harvested wood products can constitute a large and long-term carbon sink (Skog 2008), failure to recognize and account for the carbon stored in durable, long-lived products would ultimately underestimate total carbon storage following forest management activities and overestimate carbon flux to the atmosphere (Nunery and Keeton 2010). When carbon storage in products is taken into consideration, the difference in carbon storage achieved and often maximized under a no management alternative versus that obtained through more active management is reduced (Johnsen et al. 2001; Seidl et al. 2007; Nunery and Keeton 2010).

When examining changes in carbon storage in relation to forest management, it is important to assess the carbon consequences of a single, stand-level management action at the landscape level (Harmon 2001). The effects of management on landscape-level carbon storage depend on the sum of stand-level changes in carbon stocks. Across a forested landscape, management actions that result in a short-term loss of carbon in any given stand are often offset by carbon gains in other, undisturbed stands (Harmon 2001; Ryan et al. 2010). The age-class distribution of forest stands has a particularly large effect on landscape-level carbon storage. Therefore, the impact of creating early successional habitats on forest- or landscape-level carbon storage will depend upon the temporal distribution and spatial scale of stands in young age classes on the landscape. A significant increase in the proportion of the forest or landscape in young age classes may negatively affect carbon storage in the short-term (e.g., Campbell et al. 2004; Depro et al. 2008). However, given stand-level reductions in carbon storage are largely attributed to reductions in aboveground biomass, upon successful regeneration and a period of re-growth equal to that of a full rotation length, stand-level carbon storage will approximate pre-disturbance levels. Although maximizing forest carbon is accomplished through eliminating disturbance or more passive management, the consequences of managing solely for carbon storage across a landscape may negatively impact other ecosystem services (e.g., Ryan et al. 2010), including creation and maintenance of wildlife habitat.

Literature Cited

Boerner REJ, Huang J, Hart SC (2008) Fire, thinning, and the carbon economy: effects of fire and fire surrogate treatments on estimated carbon storage and sequestration rate. For Ecol Manage 255:3081–3097

Bolstad PV, Vose JM (2005) Forest and pasture carbon pools and soil respiration in the Southern Appalachian Mountains. For Sci 51:372–383

Boring LR, Swank WT (1986) Woody biomass and net primary productivity following clearcutting in the Coweeta Basin. In: Proceedings of the 1986 Southern forest biomass workshop, Knoxville

Boring LR, Swank WT, Monk CD (1988) Dynamics of early forest successional structure and processes in the Coweeta Basin. In: Swank WT, Crossley DA Jr (eds) Forest hydrology and ecology at Coweeta. Springer, New York

Bradford J, Weishampel P, Smith ML, Kolka R, Birdsey RA, Ollinger SV, Ryan MG (2009) Detrital carbon pools in temperate forests: magnitude and potential for landscape-scale assessment. Can J For Res 39:802–813

Brown SA, Schroeder PE (1999) Spatial patterns of aboveground production and mortality of woody biomass for eastern US. For Ecol Appl 9:968–980

Brown SA, Schroeder P, Birdsey R (1997) Aboveground biomass distribution of US eastern hardwood forests and the use of large trees as an indicator of forest development. For Ecol Manage 96:37–47

Bunker DW, DeClerck F, Bradford JC, Colwell RK, Perfecto I, Phillips OL, Sankaran M, Naeem S (2005) Species loss and aboveground carbon storage in a tropical forest. Science 310:1029–1031

Busing RT, Clebsch EEC, White PS (1993) Biomass production of Southern Appalachian cove forests reexamined. Can J For Res 23:760–765
Campbell JL, Sun OJ, Law BE (2004) Disturbance and net ecosystem production across three climatically distinct forest landscapes. Glob Biogeochem Cycles 18:1–11
Caspersen JP, Pacala SW (2001) Successional diversity and forest ecosystem function. Ecol Res 16:895–903
Chen H, Harmon ME, Griffiths RP (2001) Decomposition and nitrogen release from decomposing woody roots in coniferous forests of the Pacific Northwest: a chronosequence approach. Can J For Res 31:246–260
Cho DS, Boerner REJ (1991) Canopy disturbance patterns and regeneration of *Quercus* spp. in two Ohio USA old-growth forests. Vegetatio 93:9–18
Comeau PG, Kimmins JP (1989) Above- and below-ground biomass and production of lodgepole pine on sites with differing soil moisture regimes. Can J For Res 19:447–454
Covington WW (1981) Changes in forest floor organic matter and nutrient content following clearcutting in northern hardwoods. Ecology 62:41–48
Covington WW, Aber JD (1980) Leaf production during secondary succession in northern hardwoods. Ecology 61:200–204
Davis SC, Hessl AE, Scott CJ, Adams MB, Thomas RB (2009) Forest carbon sequestration changes in response to timber harvest. For Ecol Manage 258:2101–2109
Depro BM, Murray BC, Alig RJ, Shanks A (2008) Public land, timber harvests, and climate mitigation: quantifying carbon sequestration potential on US public timberlands. For Ecol Manage 255:1122–1134
Dixon RK, Brown S, Houghton RA, Solomon AM, Trexier MC, Wisniewski J (1994) Carbon pools and the flux of global forest ecosystems. Science 264:185–190
Duvall MD, Grigal DF (1999) Effects of timber harvesting on coarse woody debris in red pine forests across the Great Lakes states, USA. Can J For Res 29:1926–1934
Elliott KJ, Knoepp JD (2005) The effects of three regeneration harvest methods on plant diversity and soil characteristics in the Southern Appalachian. For Ecol Manage 211:296–317
Environmental Protection Agency (EPA) (2010) Inventory of US greenhouse gas emissions and sinks: 1990–2008. US EPA report EPA-R-10-006. Available online at http://epa.gov/climatechange/emissions/usinventoryreport.html. Accessed 6 Jul 2010
Fahey TJ, Arthur MA (1994) Further studies of root decomposition following harvest of a northern hardwoods forest. For Sci 40:618–629
Fahey TJ, Hughes JW, Pu M, Arthur MA (1988) Root decomposition and nutrient flux following whole-tree harvest of northern hardwood forest. For Sci 34:744–768
Fahey TJ, Woodbury PB, Battles JJ, Goodale CL, Hamburg SP, Ollinger SV, Woodall CW (2010) Forest carbon storage: ecology, management, and policy. Front Ecol Environ 8:245–252
Federer CA (1984) Organic matter and nitrogen content of the forest floor in even-aged northern hardwoods. Can J For Res 14:763–767
Franklin JF (1989) Towards a new forestry. Am For (Nov-Dec):37–44
Frazer GW, Trofymow JA, Lertzman KP (2000) Canopy openness and leaf area in chronosequences of coastal temperate rainforests. Can J For Res 30:39–256
Gholz HL (1982) Environmental limits on above-ground net primary production, leaf area and biomass in vegetation zones of the Pacific Northwest. Ecology 63:469–481
Gilliam FS, Dick DA, Kerr ML, Adams MB (2004) Effects of silvicultural practices on soil carbon and nitrogen in a nitrogen saturated Central Appalachian (USA) hardwood forest ecosystem. Environ Manage 33:S108–S119
Gonzalez-Benecke CA, Martine TA, Cropper WP Jr, Bracho R (2010) Forest management effects on *in situ* and *ex situ* slash pine forest carbon balance. For Ecol Manage 260:795–805
Graham JB, McCarthy BC (2006) Forest floor dynamics in mixed-oak forests of south-eastern Ohio. Int J Wildland Fire 15:479–488
Greco S, Baldochhi DD (1996) Seasonal variations of CO_2 and water vapour exchange rates over a temperate deciduous forest. Glob Change Biol 2:183–197

Grier CC, Vogt KA, Keyes MR, Edmonds RL (1981) Biomass distribution and above- and below-ground production in young and mature *Abies amabilis* zone ecosystems of the Washington Cascades. Can J For Res 11:155–167

Griffiths RP, Swanson AK (2001) Forest soil characteristics in a chronosequence of harvested Douglas-fir forests. Can J For Res 31:1871–1879

Grigal DF, Ohmann LF (1992) Carbon storage in upland forests of the Lake States. Soil Sci Soc Am J 56:935–943

Hall SA, Burke IC, Hobbs NT (2006) Litter and dead wood dynamics in ponderosa pine forests along a 160-year chronosequence. Ecol Appl 16:2344–2355

Hardt RA, Swank WT (1997) A comparison of structural and compositional characteristics of Southern Appalachian young second-growth, maturing second-growth, and old-growth stands. Nat Areas J 17:42–52

Harmon ME (2001) Carbon sequestration in forests: addressing the scale question. J For 99:24–29

Harmon ME, Marks B (2002) Effects of silvicultural practices on carbon stores in Douglas-fir – western hemlock forests in the Pacific Northwest, USA: results from a simulation model. Can J For Res 32:863–877

Harmon ME, Franklin JF, Swanson FJ, Sollins P, Gregory SV, Lattin JD, Adnerson NH, Cline SP, Aumen NG, Sedell JR, Lienkaemper GW, Cromack K, Cummins KW (1986) Ecology of coarse woody debris in temperate ecosystems. Adv Ecol Res 15:133–202

Harmon MR, Wigham DF, Sexton J, Olmstead I (1995) Decomposition and mass of woody detritus in the dry tropical forests of the northeastern Yucatan Peninsula, Mexico. Biotropica 27:305–315

Harmon ME, Moreno A, Domingo JB (2009) Effects of partial harvest on the carbon stores in Douglas-fir/western hemlock forests: a simulation study. Ecosystems 12:777–779

Hudiburg T, Law B, Turner DP, Campbell J, Donato D, Duane M (2009) Carbon dynamics of Oregon and northern California forests and potential land-based carbon storage. Ecol Appl 19:163–180

Hunt R (1982) Plant growth curves. The functional approach to plant growth analysis. University Park Press, Baltimore

Idol TW, Pope PE, Ponder F Jr (2000) Fine root dynamics across a chronosequence of upland temperate deciduous forests. For Ecol Manage 127:153–167

IPCC (2003) Good practice guidance for land use, land-use change and forestry. IPCC National Greenhouse Gas Inventories Programme, Ch 3.2: forest land. Retrieved 8 May 2009 from http://www.ipcc-nggip-iges.or.jp/public/gpglulucf/gpglulucf.htm

IPCC (2006) Guidelines for national greenhouse gas inventories: agriculture, forestry and other land use. IPCC National Greenhouse Gas Inventories Programme, vol 4. Retrieved 15 Sept 2010 from http://www.ipcc-nggip.iges.or.jp/public/2006gl/vol4.html

IPCC (2007) Climate change 2007: the physical science basis. Contribution of working Group I to the fourth assessment report of the intergovernmental panel on climate change [Solomon S, Qin D, Manning M (eds)]

Jandl R, Lindner M, Vesterdal L, Bauwens B, Baritz R, Hagedorn F, Johnson DW, Minkkinen K, Byrne KA (2007) How strongly can forest management influence carbon sequestration? Geoderma 137:253–268

Janisch JE, Harmon ME (2002) Successional changes in live and dead wood carbon stores: implications for net ecosystem productivity. Tree Physiol 22:77–89

Janisch JE, Harmon ME, Chen H, Fasth B, Sexton J (2005) Decomposition of coarse woody debris originating by clearcutting of an old-growth conifer forest. Ecoscience 12:151–160

John B, Pandey HN, Tripathi RS (2001) Vertical distribution and seasonal changes of fine and coarse root mass in *Pinus kesiya* Royle Ex.Gordon forest of three different ages. Acta Oecologica 22:293–300

Johnsen KH, Wear DN, Oren R, Teskey RO, Sanchez F, Will RE, Butnor J, Markewitz D, Richter D, Rials T, Allen HL, Seiler J, Ellsworth D, Maier C, Katul G, Dougherty PM (2001) Carbon sequestration and southern pine forests. J For 99:14–21

Johnson DW, Curtis PS (2001) Effects of forest management on soil C and N storage: a meta analysis. For Ecol Manage 40:227–238

Johnson DW, Todd DE (1998) Effects of harvesting intensity on forest productivity and soil carbon storage in a mixed oak forest. In: Lal R, Kimble JM, Follett RF, Steward BA (eds) Management of carbon sequestration in soils. Advances in soil science. CRC, Boca Raton
Johnson JE, Smith DW, Burger JA (1985) Effects on the forest floor of whole-tree harvesting in an Appalachian oak forest. Am Midl Nat 114:51–61
Johnson CE, Johnson AH, Hungtinton TG, Siccama TG (1991) Whole-tree clear-cutting effects on soil horizons and organic-matter pools. Soil Sci Soc Am J 55:497–502
Johnson CM, Zarin DJ, Johnson AH (2000) Post-disturbance aboveground biomass accumulation in global secondary forests. Ecology 81:1395–1401
Johnson DW, Knoepp JD, Swank WT, Shan J, Morris LA, Van Lear DH, Kapeluck PR (2002a) Effects of forest management on soil carbon: results of some long-term resampling sites. Environ Pollut 116:201–208
Johnson PS, Shifley SR, Rogers R (2002b) The ecology and silviculture of oaks. CABI Publishing, New York
Jokela EJ, Martin TA (2000) Effects of ontogency and soil nutrient supply on production, allocation, and leaf area efficiency in loblolly and slash pine stands. Can J For Res 30:511–1524
Kaczmarek DJ, Rodkey KS, Reber RT, Pope PE, Ponder F Jr (1995) Carbon and nitrogen in oak-hickory forests of varying productivity. In: Gottschalk KW, Fosbroke SLC (eds) Proceedings of the 10th central hardwood forest conference. Gen Tech Rep NE-197, USDA Forest Service Northeastern Research Station, Newtown Square, pp 79–93
Knoepp JD, Swank WT (1997) Forest management effects on surface soil carbon and nitrogen. Soil Sci Soc Am J 61:928–935
Knoepp JD, Swank WT (1998) Forest management effects on surface soil carbon and nitrogen. Soil Sci Soc Am J 61:928–935
Kranabetter JM (2009) Site carbon storage along productivity gradients of a late-seral southern boreal forest. Can J For Res 39:1053–1060
Law BE, Sun OJ, Campbell J, Van Tuyl S, Thornton PE (2003) Changes in carbon storage and fluxes in a chronosequence of ponderosa pine. Glob Change Biol 9:510–524
Li Q, Chen J, Moorhead DL, DeForest JL, Jensen R, Henderson R (2007) Effects of timber harvest on carbon pools in Ozark forests. Can J For Res 37:2337–2348
Ludovici KH, Zarnock SJ, Richter DD (2002) Modeling in-situ pine root decomposition using data from a 60-year chronosequence. Can J For Res 32:1675–1684
Luvall JC, Weaver GT (1986) Organic matter and nutrient content of the forest floor of oak-hickory forests in southwestern Illinois. In: Dawson JO, Majerus KA (eds) Proceedings of the fifth central hardwood forest conference. Publ 85-05, Society of American Foresters
Luyssaert S, Shulze ED, Börner A, Knoh A, Hessenmöller D, Law BE, Ciais P, Grace J (2008) Old-growth forests as global carbon sinks. Nature 455:213–215
MacMillan PC (1981) Log decomposition in Donaldson's Woods, Spring Mill State Park, Indiana. Am Midl Nat 106:335–344
Mattson KG, Smith HC (1993) Detrital organic matter and soil CO_2 efflux in forests regenerating from clearcutting in West Virginia. Soil Biol Biochem 25:1241–1248
Mattson KG, Swank WT (1989) Soil and detrital carbon dynamics following forest cutting in the Southern Appalachians. Biol Fertil Soils 7:247–253
McCarthy BC, Bailey RR (1994) Distribution and abundance of coarse woody debris in a managed forest landscape of the Central Appalachians. Can J For Res 24:1317–1329
McCarthy BC, Small CJ, Rubino DL (2001) Composition, structure and dynamics of Dysart Woods, an old-growth mixed mesophytic forest of southeastern Ohio. For Ecol Manage 140:192–213
McClain WE, Ebinger JE (1968) Woody vegetation of Baber Woods, Edgar County, Illinois. Am Midl Nat 79:419–428
Melin Y, Petersson H, Nordfjell T (2009) Decomposition of stump and root systems of Norway spruce in Sweden – a modeling approach. For Ecol Manage 257:1445–1451
Miller AT, Allen HL, Maier CA (2006) Quantifying the coarse-root biomass of intensively managed loblolly pine plantations. Can J For Res 36:12–22

Misra RK, Turnbull CRA, Cromer RN, Gibbons AK, LaSala AV (1998) Below- and above-ground growth of *Eucalyptus nitens* in a young plantation I. Biomass. For Ecol Manage 106:283–293
Muller RN (1982) Vegetation patterns in the mixed mesophytic forest of eastern Kentucky. Ecology 63:1901–1917
Nave LE, Vance ED, Swanston CW, Curtis PS (2010) Harvest impacts on soil carbon storage in temperate forests. For Ecol Manage 259:857–866
Nunery JS, Keeton WS (2010) Forest carbon storage in the northeastern United States: net effects of harvesting frequency, post-harvest retention, and wood production. For Ecol Manage 259:1363–1375
Oliver CD, Larson BC (1996) Forest stand dynamics. Wiley, New York
Prescott CE, Maynard DG, Laiho R (2000) Humus in northern forests: friend or foe? For Ecol Manage 133:23–36
Reinhardt E, Holsinger L (2010) Effects of fuel treatments on carbon-disturbance relationships in forests of the northern Rocky Mountains. For Ecol Manage 259:1427–1435
Resh SC, Battaglia M, Worledge D, Ladiges S (2003) Coarse root biomass for eucalypt plantations in Tasmania, Australia: sources of variation and methods for assessment. Trees 17:389–399
Ryan MG, Harmon ME, Birdsey RA, Giardina CP, Heath LS, Houghton RA, Jackson RB, McKinley DC, Morrison JF, Murray BC, Pataki DE, Skog KE (2010) A synthesis of the science on forests and carbon for the US Forests. Issues Ecol 13:1–16
Santantonio D, Hermann RK, Overton WS (1977) Root biomass studies in forest ecosystems. Pedobiologia 17:1–31
Schmelz DA, Lindsey AA (1965) Size class structure of old-growth forests in Indiana. For Sci 11:258–264
Seidl R, Rammer W, Jäger D, Currie WS, Lexer MJ (2007) Assessing trade-offs between carbon sequestration and timber production within a framework of multi-purpose forestry in Austria. For Ecol Manage 248:64–79
Shi K, Cao QV (1997) Predicted leaf area growth and foliage efficiency of loblolly pine plantations. For Ecol Manage 95:109–115
Skog KE (2008) Sequestration of carbon in harvested wood products for the United States. For Prod J 58:56–72
Smith DM, Larson BC, Kelty MJ, Ashton PMS (1997) The practice of silviculture: applied forest ecology. Wiley, Hoboken
Smith JE, Heath LS, Skog KE, Birdsey RA (2006) Methods for calculating forest ecosystem and harvested carbon with standard estimates for forest types of the United States. Gen Tech Rep NE-343, USDA Forest Service Northeastern Research Station, Newtown Square
Smithwick EAH, Harmon ME, Remillard SM, Acker SA, Franklin JF (2002) Potential upper bounds of carbon stores in forests of the pacific northwest. Ecol Appl 12:1303–1317
Spetich MA, Parker GR (1998) Distribution of biomass in an Indiana old-growth forest from 1926 to 1992. Am Midl Nat 139:90–107
Spetich MA, Shifley SR, Parker GR (1999) Regional distribution and dynamics of coarse woody debris in Midwestern old-growth forests. For Sci 45:302–313
Spies TA, Franklin JF, Thomas TB (1988) Coarse woody debris in Douglas fir forests of western Oregon and Washington. Ecology 69:1689–1702
Sturtevant BR, Bissonette JA, Long JN, Roberts DW (1997) Coarse woody debris as a function of age, stand structure, and disturbance in boreal Newfoundland. Ecol Appl 7:702–712
Sun OJ, Campbell J, Law BE, Wolf V (2004) Dynamics of carbon stocks in soils and detritus across chronosequences of different forest types in the Pacific Northwest, USA. Glob Change Biol 10:1470–1481
Taylor AR, Wang JR, Chen HYH (2007) Carbon storage in a chronosequence of red spruce (*Picea rubens*) forests in central Nova Scotia, Canada. Can J For Res 37:2260–2269
Turner DP, Koerper GP, Harmon ME, Lee JJ (1995) A carbon budget for the forests of the conterminous United States. Ecol Appl 5:21–436
Van Tuyl S, Law BE, Turner DP, Gitelman AI (2005) Variability in net primary production and carbon storage in biomass across Oregon forests: an assessment integrating data from forest inventories, intensive sites, and remote sensing. For Ecol Manage 209:273–291

Vitousek PM, Sanford L (1986) Nutrient cycling in moist tropical forest. Annu Rev Ecol Syst 17:137–167
Vose JM, Allen HL (1988) Leaf area, stemwood growth, and nutrition relationships in loblolly pine. For Sci 34:547–563
Vose JM, Bolstad PV (2007) Biotic and abiotic factors regulating forest floor CO_2 flux across a range of forest age classes in the Southern Appalachians. Pedobiologia 50:577–587
Wallace ES, Freedman B (1986) Forest floor dynamics in a chronosequence of hardwood stands in central Nova Scotia. Can J For Res 16:293–302
Weaver GT, Ashby WC (1971) Composition and structure of an old-growth forest remnant in southwestern Illinois. Am Midl Nat 86:46–56
Webster CR, Jenkins MA (2005) Coarse woody debris dynamics in the Southern Appalachians as affected by topographic position and anthropogenic disturbance history. For Ecol Manage 217:319–330
Wofsy SC, Gouldin ML, Munger JW, Fan SW, Bakwin PS, Daube BC, Bassow SL, Bazzaz FA (1993) Net exchange of CO_2 in a mid-latitude forest. Science 260:1314–1317
Ximenes FA, Gardner WD (2006) The decay of coarse woody roots following harvest in a range of forest types. Prepared by the Cooperative Research Centre for Greenhouse Accounting for the Department of Environment and Heritage – Australian Greenhouse Office. Technical report no. 49
Yanai RD, Park BB, Hamburg SP (2006) The vertical and horizontal distribution of roots in northern hardwoods of varying age. Can J For Res 36:450–459
Yin X, Pery JA, Dixon RK (1989a) Fine-root dynamics and biomass distribution in a Quercus ecosystem after harvest. For Ecol Manage 27:159–177
Yin X, Perry JA, Dixon RK (1989b) Influence of canopy removal on oak forest floor decomposition. Can J For Res 19:204–214
Yoda K, Kira T, Ogawa H, Hozumi K (1963) Self-thinning in overcrowded pure stands under cultivated and natural conditions. J Biol Osaka City Univ 14:107–129

Chapter 16
Forecasting Forest Type and Age Classes in the Appalachian-Cumberland Subregion of the Central Hardwood Region

David N. Wear and Robert Huggett

Abstract This chapter describes how forest type and age distributions might be expected to change in the Appalachian-Cumberland portions of the Central Hardwood Region over the next 50 years. Forecasting forest conditions requires accounting for a number of biophysical and socioeconomic dynamics within an internally consistent modeling framework. We used the US Forest Assessment System (USFAS) to simulate the evolution of forest inventories in the subregion. The types and ages of forests in the Appalachian-Cumberland portions of the Central Hardwood Region are likely to shift over the next 50 years. Two scenarios bracket a range of forest projections and provide insights into how wood products markets as well as economic, demographic, and climate changes could affect these future forests. Shifts in the future age distributions of forests are dominated by projected harvest regimes that lead to qualitatively different forest conditions. The future area of young forests correlates with change in total forest area—as total forest area declines, so does the area of young forests. However, changes in the area of young forests and forest age class distributions are most directly altered by the extent of harvesting within the Appalachian-Cumberland subregion.

D.N. Wear (✉)
USDA Forest Service, Southern Research Station, Research Triangle Park, NC 27709, USA
e-mail: dwear@fs.fed.us

R. Huggett
Department of Forestry and Environmental Resources, North Carolina State University, Raleigh, NC 27695, USA
e-mail: rhuggett@fs.fed.us

C.H. Greenberg et al. (eds.), *Sustaining Young Forest Communities*, Managing Forest Ecosystems 21, DOI 10.1007/978-94-007-1620-9_16,

16.1 Introduction

Because forest conditions change over time, effective management guidelines address not only today's conditions but also the future trajectories of forest conditions. Future forests will be defined by a complex and interacting set of economic, biological and physical driving forces. We describe forecasts of how forest type and age distributions might change in the Appalachian-Cumberland subregion of the Central Hardwood Region over the next 50 years. We 'look forward' to provide a start for managing tomorrow's forests today. Our analysis is based on forest forecasts developed as part of the Southern Forest Futures Project (Wear et al. 2009), a regional assessment of the Southern USA addressing several questions regarding the future of forests and the benefits they provide.

Forecasting forest conditions requires accounting for a number of biophysical and socioeconomic dynamics within an internally consistent modeling framework. Biophysical factors include the influence of climate on species persistence and disturbance patterns along with the demographics of forest aging and mortality. Socioeconomic forces include the influence of population and economic growth on land use choices and associated loss (or gain) of forest area; the effects of timber harvesting patterns driven by demand for wood products; and the relative value of forest stands for providing wood products. Timber supply derives from evolving forest conditions and preferences of forest landowners regarding management of their lands.

Understanding the interrelated complex of change vectors requires a computer simulation framework; we use the US Forest Assessment System (USFAS) to simulate evolution of forest inventories. Forecasts of future forest conditions require a set of assumptions about the future course of climate and economic conditions, packaged as comprehensive scenarios. For our analysis, two scenarios bracket a range of forest age class and forest type projections, and provide insights into how wood products markets could influence availability and condition of early, mid and late successional forest habitats in the Appalachian-Cumberland subregion.

In the following sections of this chapter, we describe the structure of the USFAS and the information contained in its forecasts. We describe the structure of the two future scenarios and their derivation from international and national assessments. Forest forecasts are developed and discussed. We conclude with a discussion of implications as well as potential shortcomings and uncertainties inherent in our approach.

16.2 Methods

Forecasting forest conditions requires an integrated assessment approach that accounts for biological, physical, demographic, and economic changes. We utilized the US Forest Assessment System or USFAS (Wear 2010) developed by the US Forest Service for various national and regional resource assessments. The USFAS

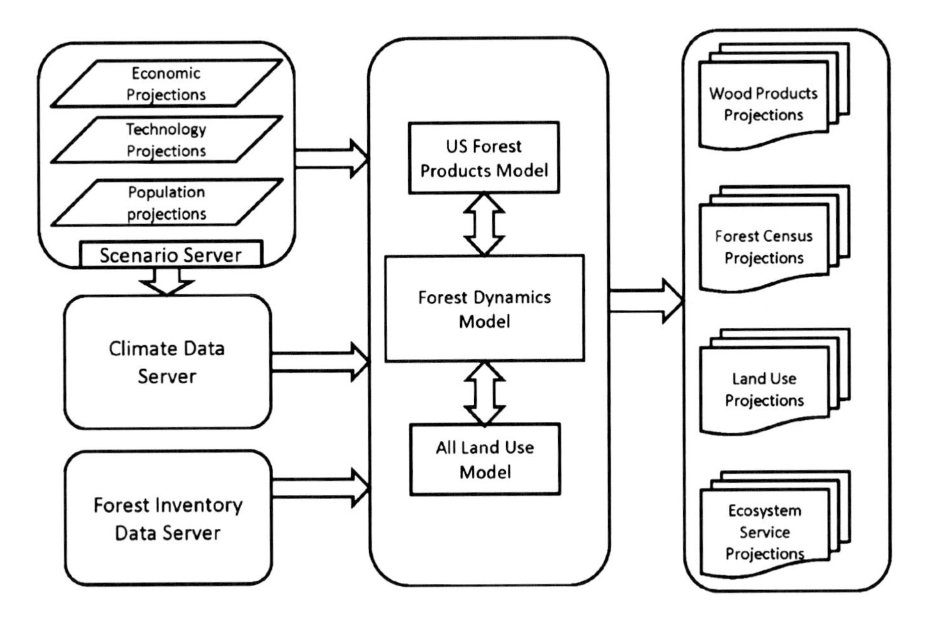

Fig. 16.1 Schematic of the US Forest Assessment System

addresses forest dynamics using scenario analysis to gauge uncertainty inherent in world views and model outputs. The basic inputs to the modeling system include various climate, economic, and demographic projections (Fig. 16.1, column 1) that are inputs to what is labeled a "scenario server". Most scenarios were constructed from a foundation defined by the International Panel on Climate Change (IPCC) storylines and scenarios. For the RPA national assessment[1] these projections of coupled economic, population, and climate changes were downscaled to a fine-scale for the continental USA. Within the USFAS we consider these data aggregated to the county level.

For the analysis described in this chapter we consider two future scenarios, both of which have the same economic and demographic growth. In particular, we use the population and economic growth projections from the IPCC's A1B storyline, which anticipates relatively high economic growth and a moderate level of population growth in the USA (IPCC 2007). In contrast, climate and timber market forecasts vary between the two scenarios. One scenario, labeled High Market, anticipates strong growth in the demand for wood products and applies a projection of climate

[1]The Forest and Rangeland Renewable Resources Planning Act (RPA) (P.L. 93–378, 88 Stat 475, as amended) was enacted in 1974. Section 3 of the Act requires the US Forest Service to provide a national renewable resource assessment to provide reliable information on the status and trends of the Nation's renewable resources on a 10-year cycle.

from the MIROC General Circulation Model (GCM)[2]. The other scenario, labeled Low Market, anticipates a steady decline in market demand and applies the CSIRO GCM. Change in demand is represented by exogenously imposed timber price trends.

The left column also contains the Forest Inventory and Analysis (FIA) Forest Inventory database (USDA Forest Service 2007), emphasizing that the USFAS simulates the development of detailed FIA inventories through time in response to these drivers. Current forest conditions are defined by the most recent panels of plot measurements in the FIA forest inventory and the condition of each plot is forecast forward based on the scenarios and a set of internal dynamics.

There are three simulation components of the USFAS (Fig. 16.1, middle column). At the top is the US Forest Products Model which links USA regional forest products markets to global market conditions and domestic timber supply conditions. Timber supply is provided by the Forest Dynamics Model which accounts for timber harvesting by public and private landowners and is described below. For the present simulation analysis we replaced the explicit market model with general price assumptions that imposed price trends that reflected expanding or contracting markets within the Appalachian-Cumberland subregion. This seems reasonable given the size of the market considered here and the nature of our analysis, which is to demonstrate a reasonable range of future trajectories. For the High Market scenario we increase timber prices at a compounded 1% per year rate throughout the 50 year forecast period. For the Low Market scenario, we decrease prices by a compounded 1% per year. This has the effect of altering harvest patterns as described below.

The Forest Dynamics Model simulated the evolution of all forested plots in the FIA inventory (Fig. 16.1, column 2). Modeled plot dynamics start with a harvest model that accounts for effects of price levels and forest conditions on propensity of forest owners to harvest their forests; separate models address harvesting by public and private forest ownerships (Polyakov et al. 2010). Harvesting allows for either partial or final harvesting, depending on economic and forest conditions, and is price responsive. That is, more timber is harvested when prices are higher. Plot age is incremented by the simulation time-step for unharvested plots, but is defined using historical patterns of age changes associated with forest plots that receive either a partial or final harvest in the FIA inventories. The broad forest management type of the site (upland hardwood, lowland hardwood, and three pine (*Pinus* spp.) types) is held constant for unharvested forest plots but allowed to change in response to harvesting. Forest planting after harvest is possible, but is tied to historical planting

[2]The emission outputs for various scenarios were used to initialize the atmospheric concentrations of GHG in numerous general circulation models (GCMs) to forecast the effects on climate variables. The RPA Assessment provides a downscaling of climate forecasts to the county level (Coulson et al. 2010) for three scenarios applied to three GCMs. In this analysis we apply the outputs from the MIROC GCM—the Model for Interdisciplinary Research on Climate from the Center for Climate System Research, University of Tokyo—and the CSIRO GCM—the Australian Commonwealth Scientific and Industrial Research Organization Mark 2 Global Climate Model.

frequencies for various forest types. Because planting is rare in the Appalachian-Cumberland subregion, post harvest planting has little influence on the forecasts.

Whole plot imputation, a resampling scheme, defines the future inventory (Wear 2010). Using this approach, a historical plot with comparable climate, forest management type, age, and harvest characteristics is selected to represent conditions for each future plot condition forecasted by the model. When forest management type is held constant, this resampling allows for shifts in the constituent forest types over time. For example, if a site becomes hotter and drier over time, the forest type might shift from yellow-poplar (*Liriodendron tulipifera*) to oak-hickory (*Quercus-Carya*). Forest type transitions are driven largely by changes in climate condition, and become more prevalent in later years of the simulations. Given this statistical approach to constructing future inventories, individual plot forecasts are less informative than summaries of changes in the inventories over large aggregates of plots. Several components of the model, including transitions and imputation schemes are probabilistic defining a stochastic modeling system. We run the models 25 times and then select one run with the greatest central tendencies for this set as a representative inventory for subsequent display and analysis.

A third component is the All Land Use Model (Fig. 16.1 middle column). This component simulates effects of population and economic growth on the distribution of land uses in each county in the Appalachian-Cumberland subregion. It starts by predicting effects of population and income growth on the amount of urban or developed land uses in the county. It then predicts changes to rural land uses, including forest, crops, range, and pasture land uses, in response to urbanization, timber prices, and average crop returns within the county. For this analysis, timber prices differ between the two scenarios (High Market and Low Market) by assumption, and crop returns are assumed to remain constant over time. Population and income changes are comparable across the two scenarios and reflect the growth modeled for the RPA/IPCC A1B storyline. Forecasts of land use changes at the county level are used to shrink or expand the "area expansion factor" proportionally for all plots within the county. These factors define the area represented by each plot in the inventory.

Outputs from the USFAS (Fig. 16.1, Column 3) include forecasts of complete inventories and detailed forest conditions that can be derived from inventory records, forecasts of land uses at the county level, and forecasts of forest removals determined by timber harvesting and land use changes. The modeling framework is stochastic and is used to generate multiple realizations of future inventories and examine uncertainty inherent in the modeled elements of the system. For this analysis, we focus on what we call the "representative inventory" which is the inventory simulation with the greatest central tendency defined as the minimum total percentage deviation from mean values for a vector of modeled variables (it is also possible to examine the variances associated with each forecasted variable by examining the full set of simulations). As described in Fig. 16.1, subsequent analysis of various ecosystem services and conditions can be supported by these forecasts; for example, our analysis of change in successional stage habitats.

For the present analysis of age class projections, we defined three age classes. The Early-Age class is defined as forests aged 0–20 years. Middle-Age class forests are between 21 and 70 years old. And the Old-Age class forests are greater than 70 years old. Other age breakdowns were possible, but we felt that the larger age bins (>15 years) provided the best accounting of broad trends in age class distributions. We summarized age class distributions of the various forest type groups using these age class breakdowns.

16.3 Study Area and Data

The simulated study area was the Appalachian-Cumberland subregion evaluated in the Southern Forest Futures Project (Wear et al. 2009). This subregion (Fig. 16.2) reflected a broad variety of geophysical and ecological conditions represented by Blue Ridge, Northern and Southern Ridge and Valley, Cumberland Plateau and Mountain and Interior Low Plateau ecological sections (Fig. 16.1; also see McNab, Chap. 2), but was limited Virginia, Kentucky, Tennessee, North Carolina, Alabama and Georgia. Although the boundaries generally followed standard ecophysiographic lines, their final definition was determined by a team of specialists working on various components of the project.

Our forecasts started with the most recent (2007–2009) forest inventories available for each state in the subregion. We link each to their immediate previous forest inventory to estimate the harvest /transition models for various ownership groups and forest types in each state (see Polyakov et al. 2010). Because of data limitations, the Tennessee harvest model was applied to Kentucky's plots.

Simulations generate forecasts of forest conditions across a number of variables. In this analysis, we focused on forecasts of change in area of various forest type groupings and their age classes. We started by examining the five forest management types used by FIA to aggregate forest types in the South: Naturally Regenerated Pine, Planted Pine, Mixed Pine-Hardwood, Upland Hardwood and Lowland Hardwood. The Appalachian-Cumberland subregion is dominated by the Upland Hardwood group (McNab, Chap. 2) and we focused most of our analysis on this broad group split out into four sub groups: Oak-Hickory, Yellow-Poplar, Maple-Beech-Birch, and Other Hardwoods. These groups are aggregates of forest types defined by FIA using dominant and associated tree species assemblages. We started with the Forest Type Groups defined by FIA but then modified our groupings to match the setting. We used FIA's definition of Oak-Hickory where generally all of these forest types have oak dominants (USDA Forest Service 2007). We separated types with Yellow-Poplar dominants from FIA's Oak-Hickory group to define a separate Yellow-Poplar group. These are generally found on moist soils (McNab, Chap. 2). We used FIA's definition of the Maple-Beech-Birch group which includes sugar maple (*Acer saccharum*), black cherry (*Prunus serotina*), hard maple, and red maple (*A. rubrum*) dominants. The Other Hardwoods group is dominated by FIA's mixed upland hardwood type which includes a variety of species without a clear species dominance to place the plot in one of the other types. FIA notes that these types are generally on upland sites.

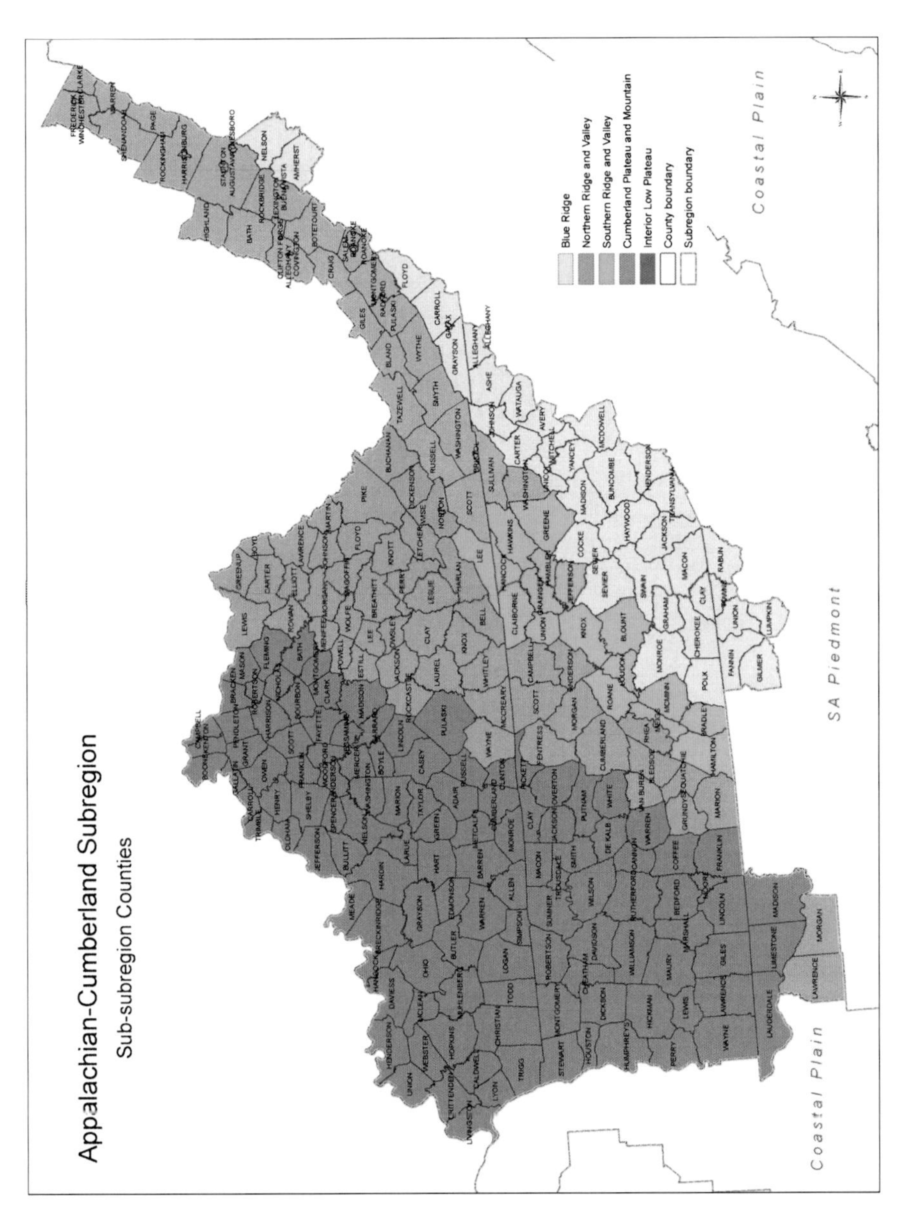

Fig. 16.2 Appalachian-Cumberland subregion, as defined by the Southern Forest Futures Project

16.4 Results

Anticipated population and income growth drive urbanization in the Appalachian-Cumberland subregion. In response, forest area in the subregion was projected to decline by 2.0 million acres under the High Market scenario and by 3.5 million acres under the Low Market scenario (Fig. 16.3). The loss was lower under the High Market scenario because higher prices shift rural land losses toward crops and pasture land rather than forests. Although not displayed here, the highest concentrations of forest losses were around Nashville Tennessee, the triangular area between Louisville, Lexington, and Cincinnati in Kentucky; and the area between Asheville, North Carolina and Knoxville, Tennessee. Little forest loss was projected for eastern Kentucky.

Among the five major forest management types in the South (Upland Hardwoods, Lowland Hardwoods, Natural Pine, Mixed Oak-Pine, and Planted Pine), Upland Hardwoods clearly dominated with 82% of total forest area (Fig. 16.4). Nearly all forest losses between 2010 and 2060 were also contained within this forest management type and we accordingly limited our subsequent analysis to understanding the dynamics of change in Upland Hardwoods.

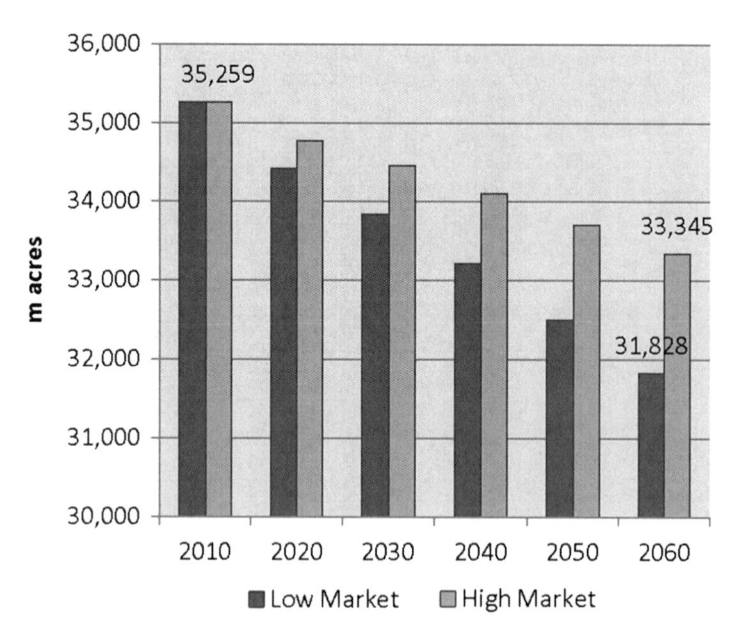

Fig. 16.3 Forecasts of forest area in the Appalachian-Cumberland subregion for High Market and Low Market scenarios, 2010–2060

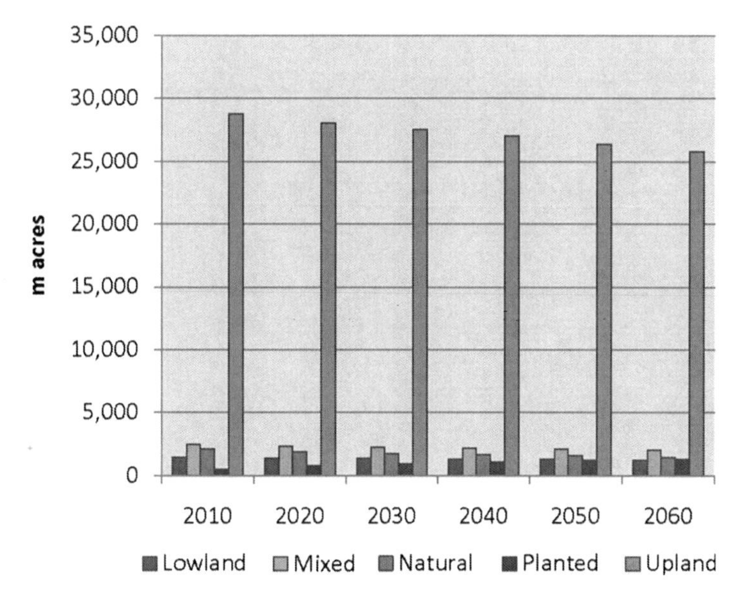

Fig. 16.4 Forecasts of forest area by broad management type for the Low Market scenario, 2010–2060

We defined four subgroups for the Upland Hardwood Forest Management group: Oak-Hickory, Yellow-Poplar, Maple-Beech-Birch, and Other Hardwoods (Fig. 16.5a). Under the Low Market scenario, the greatest change was in Upland Hardwood area (−10%); the Oak-Hickory type changed least, declining by 1% between 2010 and 2060 (Fig. 16.5b). Over the same period, the Maple-Beech-Birch group declined 9%, the Yellow-Poplar 24%, and Other Hardwoods 31%. The patterns of change were different for the High Market scenario where the area of Upland Hardwoods declined by 7% (Fig. 16.6) between 2010 and 2060. Under this scenario, the area of Oak-Hickory declined 5%, the area of Yellow-Poplar declined 12%, and the area of Other Hardwoods declined 17%. Under the High Market scenario, the area of Maple-Beech-Birch increased 8% between 2010 and 2060.

Combining forest area and forest transition dynamics with forest aging and disturbances yielded forecasts of age class distributions of these forests. For the Low Market scenario, area of both Early (0–20 years) and Middle (21–70 years) forest age classes declined. Early-Age class forests declined by 38% from 2.3 to 1.4 million acres and Middle-Age class forests declined by 51% from 18 to 9 million acres. As a result, the area of Old-Age class forests (>71 years) increased substantially between 2010 and 2060, from about 9 million acres to about 16 million acres (+76%) (Fig.16.7).

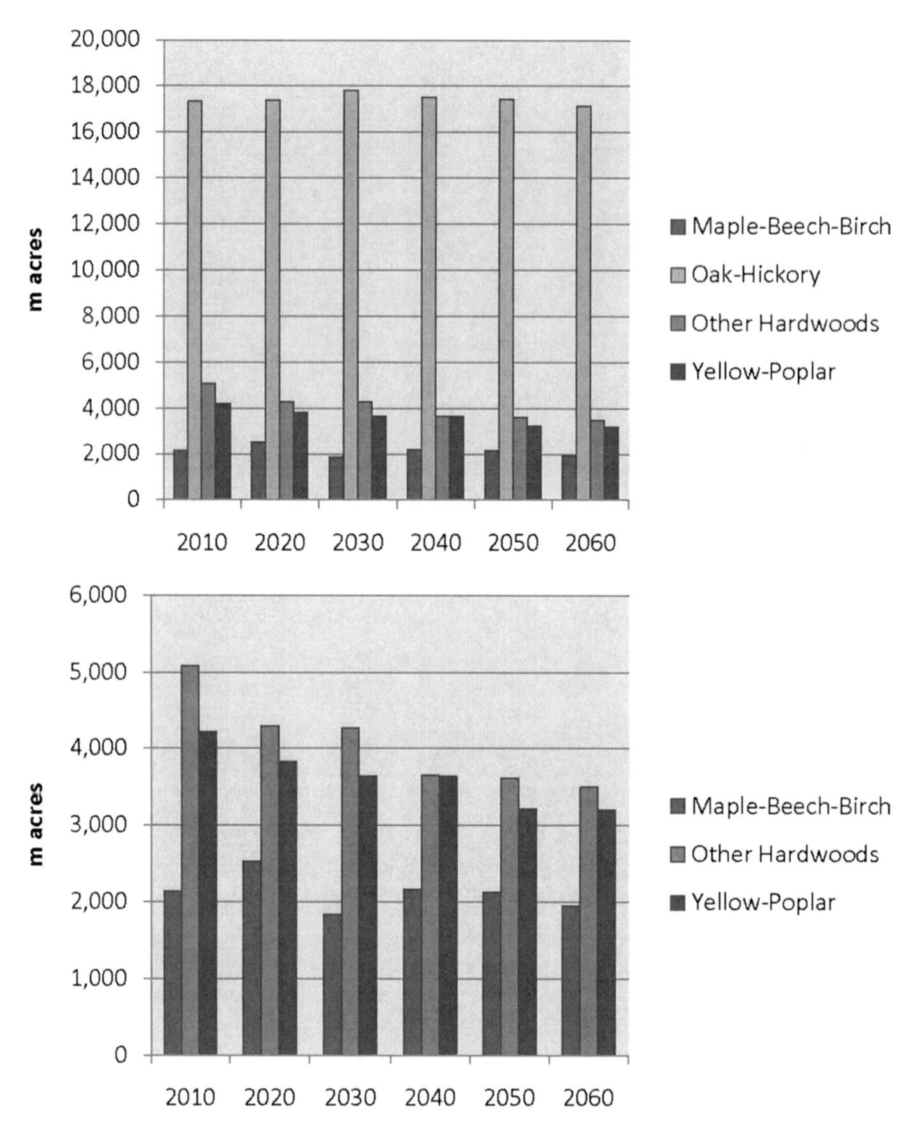

Fig. 16.5 Forecasts of forest area by forest type within the Upland Hardwood Forest Management group for the Low Market scenario, 2010–2060

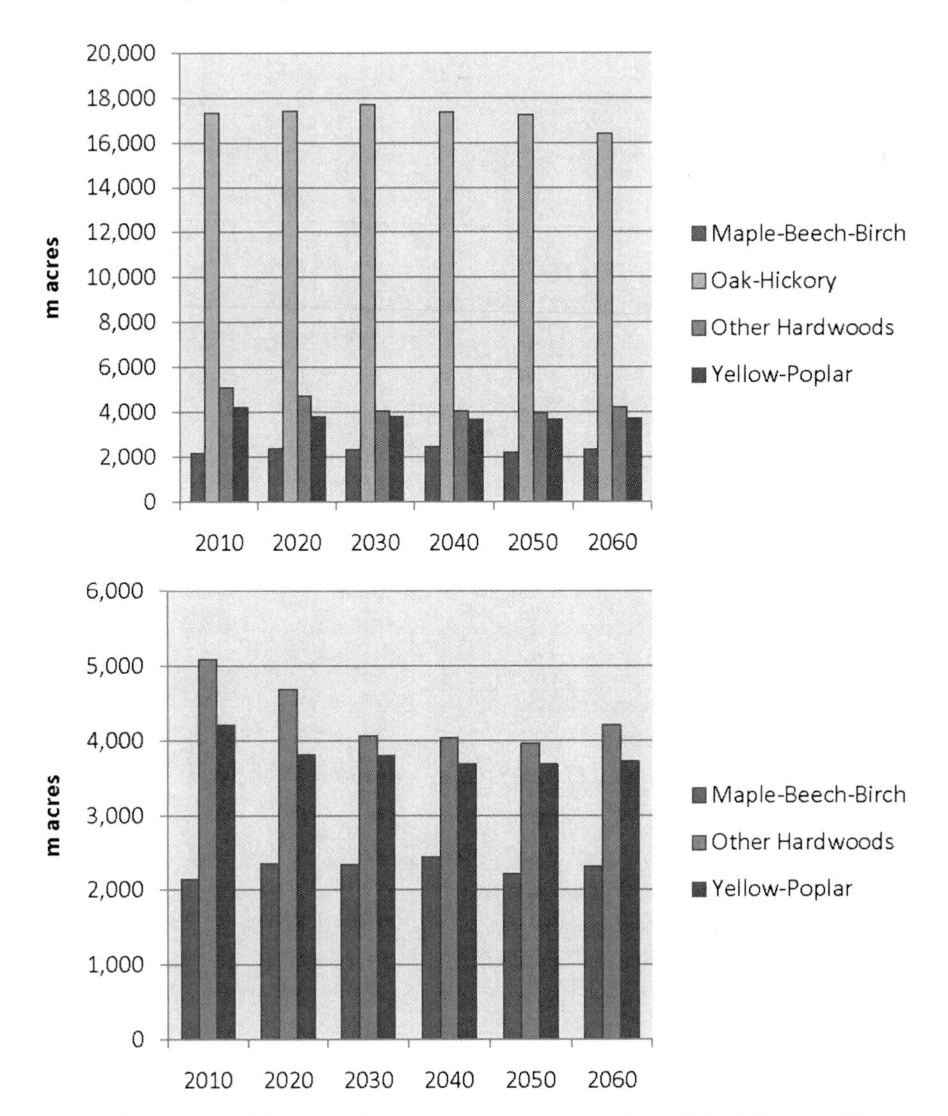

Fig. 16.6 Forecasts of forest area by forest type grouping within the Upland Hardwood Forest Management group for the High Market scenario, 2010–2060

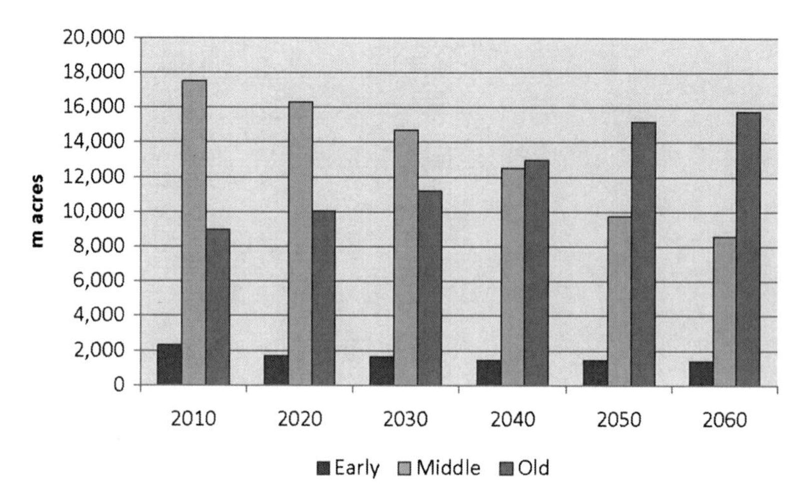

Fig. 16.7 Forecast of Early-, Mid-, and Old-Age class forests in the Appalachian-Cumberland subregion for the Low Market scenario, 2010–2060

Under the High Market scenario, area of Early- and Middle-Age class forests declined, but at a lower rate (Fig. 16.8). Loss of Early-Age class forests was less than one half of that forecasted for the Low Market scenario (−371,000 versus −878,000 acres). Loss of Middle-Age class forests and gain in Old-Age class forests was also dampened with this scenario: Middle-Age class forest area declines by 40% (versus 51% for the Low Market scenario) and Old-Age class forest area increases by 58% (versus 76%).

Forecasts of change in Early-Age class forests differed across forest types (Figs. 16.9 and 16.10). For the Low Market scenario, Other Hardwoods had the greatest loss in Early-Age forest area (−53%), decreasing substantially between 2010 and 2020 and gradually thereafter. The Other Hardwoods group was largely coincident with Early-Ages and was less frequent for Older-Ages because its species composition is largely indeterminate until later in stand development. Loss of this type would be expected with less forest harvesting over time. After Other Hardwoods, Yellow-Poplar forest types were forecasted to lose the most Early-Age forests (−33%), followed by Maple-Beech-Birch (−28%), and Oak-Hickory forests (−16%).

These patterns of change were different for the High Market scenario (Fig. 16.10). Here Other Hardwoods were also more likely to lose area of Early-Age forest but the difference between 2010 and 2060 was much smaller (−27%). Under this scenario, Oak-Hickory, Yellow-Poplar, and Maple-Beech-Birch forest types showed some oscillation in their Early-Age forest area but departed only slightly from their initial values between 2020 and 2060.

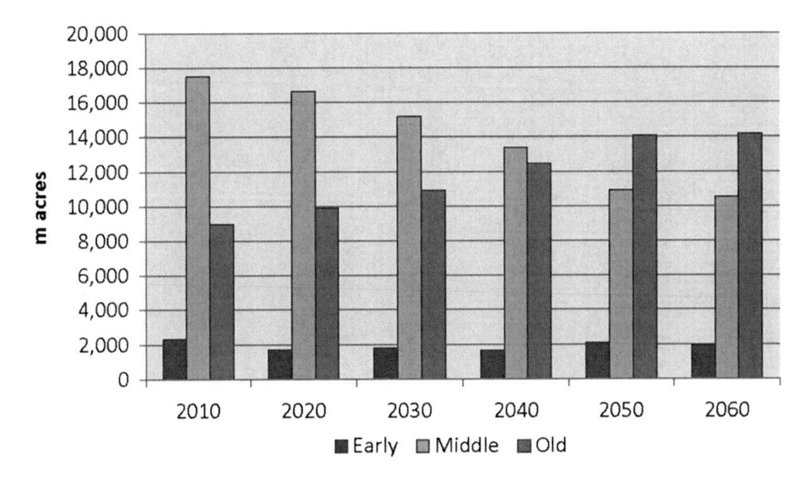

Fig. 16.8 Forecast of Early-, Mid-, and Old-Age class forests in the Appalachian-Cumberland subregion for the High Market scenario, 2010–2060

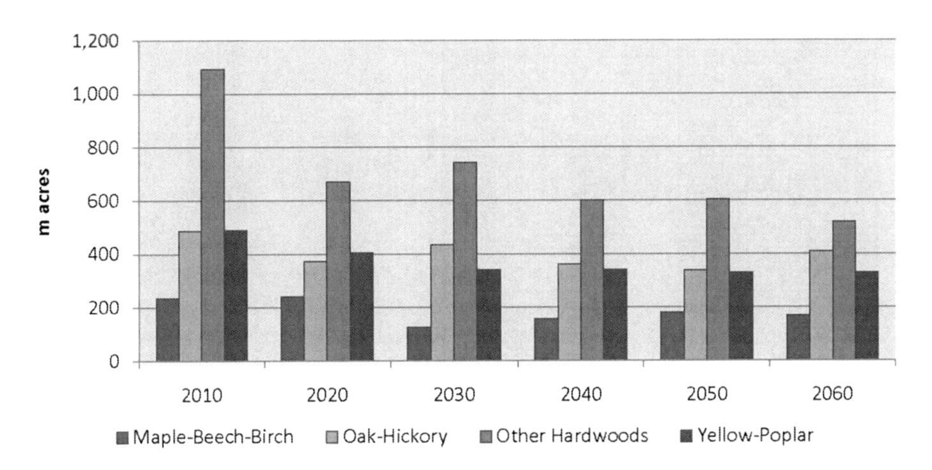

Fig. 16.9 Forecast of Early-Age class forests in the Appalachian-Cumberland subregion for various upland hardwood forest types, 2010–2060, under the Low Market scenario

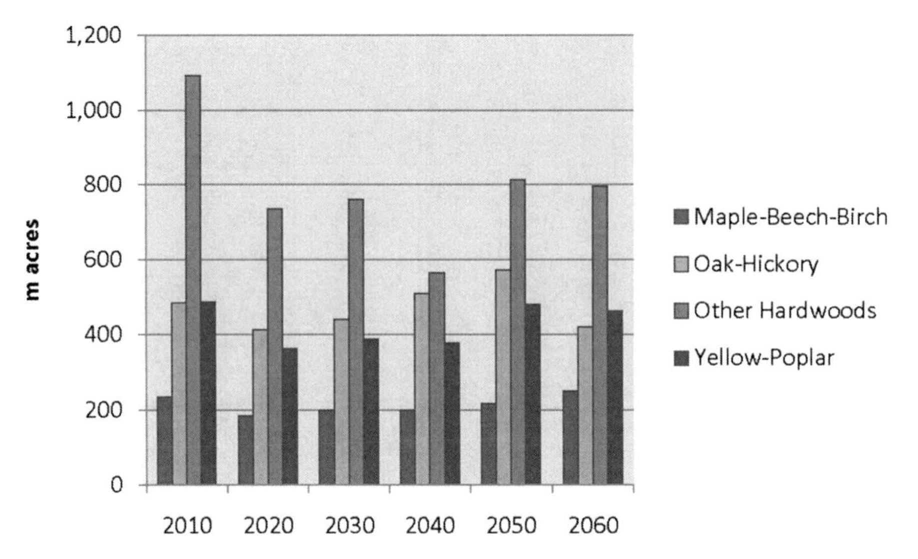

Fig. 16.10 Forecast of Early-Age class forests in the Appalachian-Cumberland subregion for various upland hardwood forest types, 2010–2060, under the High Market scenario

16.5 Discussion and Conclusions

Future forecasts highlighted how socioeconomic forces had a substantial influence on area and structure of forests in the Appalachian-Cumberland subregion of the Central Hardwood Region. These influences were expressed through forecasts of changes in land uses for the subregion, which were largely driven by forecasted increases in populations and incomes. Resulting urbanization drew down the area of rural land uses, primarily forests. For both High Market and Low Market scenarios, forest losses ranged between 2.1 and 3 million acres, or 7–10% of current forest area in the Appalachian-Cumberland subregion.

The area of individual forest types was differentially influenced by these changes in land uses coupled with changes in climate. In this subregion, temperatures increased throughout the projection period and the climate is somewhat drier, shifting growing conditions in important ways. In both scenarios, the Other Hardwoods and Yellow-Poplar forest types lost the highest percentage of their forests. Oak-Hickory, however, lost the lowest percentage of its area, even though it represents the largest share of forests within the subregion. This likely reflected both the location of forecasted urbanization, which could differentially affect the areas of different forest types, and the shift toward warmer and drier site conditions.

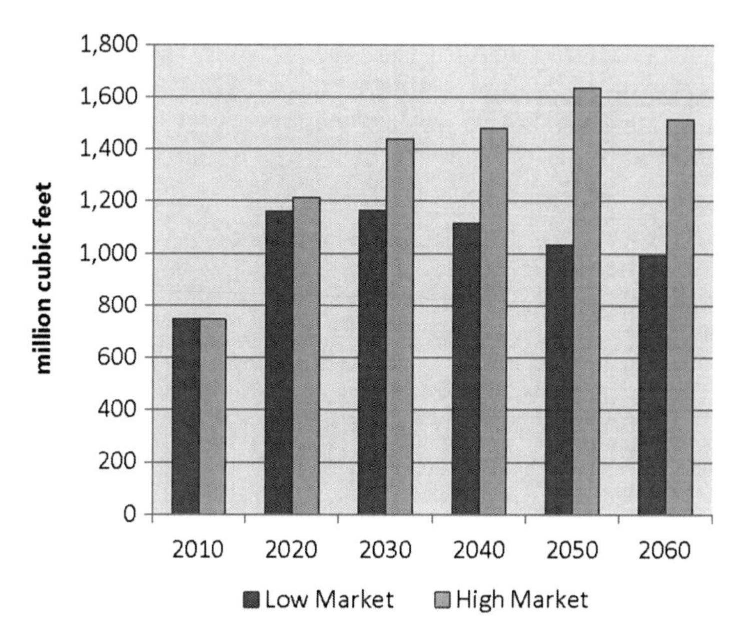

Fig. 16.11 Total hardwood removals forecasts in the Appalachian-Cumberland subregion for High Market and Low Market scenarios

Shifts in the age classes of forests were dominated by the projected harvest regimes. The High Market scenario produced a harvest rate (measured as annual hardwood removals) that was about 58% higher than the Low Market scenario by 2060 (Fig. 16.11). This leads to qualitatively different patterns of change for Early-Age forests. The Low Market scenario leads to a substantial (−38%) loss in Early-Age forests, which strongly favors Other Hardwoods and Yellow-Poplar forest types. The High Market scenario, in contrast, leads to a 16% loss in Early-Age forests over the projection period and favors Other Hardwoods. Changes in Early-Age forest area for all other forest types were forecasted to be minimal over time for the High Market scenario.

Clearly the area of Early-Age forests was scaled by the change in total forest area—as total forest area declined, so did the area of Early-Age forests. However, change in age class distributions of forest types was most directly altered by the extent of harvesting within the subregion. Although the forecasts were unambiguous in showing a decline in Early-Age forest area, comparison of the two scenarios indicated that the area of Early-Age forests was highly variable across what could be considered a moderate range of plausible forest market conditions.

Although the scenarios considered in this analysis are viewed as plausible, in that they reflect seemingly realistic projections of population and income along with the best knowledge regarding future climate and potential forest product prices, they should be viewed as uncertain. We argue that comparisons between the scenarios are the most informative aspect of the analysis, as they highlight the relative importance of vectors of change. Clearly, the future is unknowable, but forecasting models such as the one used here allow for a deliberate and informative consideration of the potential for critical changes in forest conditions.

Literature Cited

Coulson DP, Joyce LA, Price DT, McKenney DW, Siltanen RM, Papadopol P, Lawrence K (2010) Climate scenarios for the conterminous United States at the county spatial scale using SRES scenarios A1B and A2 and PRISM climatology. USDA Forest Service, Rocky Mountain Research Station, Fort Collins. Available online: http://www.fs.fed.us/rm/data_archive/dataaccess/US_ClimateScenarios_county_A1B_A2_PRISM.shtml. Accessed 10 Jan 2011

Intergovernmental Panel on Climate Change (2007) General guidelines on the use of scenario data for climate impact and adaptation assessment. Accessed online at: http://www.ipcc-data.org/guidelines/TGICA_guidance_sdciaa_v2_final.pdf. Last accessed 27 Jan 2010

Poylakov M, Wear DN, Huggett R (2010) Harvest choice and timber supply models for forest forecasting. For Sci 56:344–355

USDA Forest Service (2007) Forest inventory and analysis national core field guide: vol I. Field data collection procedures for phase 2 plots. (version 4.0)

Wear DN (2010) USFAS–The United States forest assessment system: analysis to support forest assessment and strategic analysis. Proposal and project plan (version 3). Study plan on file with David Wear at USDA Forest Service, Research Triangle Park Forestry Sciences Laboratory, RTP, 12 p

Wear DN, Greis JG, Walters N (2009) The southern forest futures project: what the public told us. Gen Tech Rep SRS-115, USDA Forest Service Southern Research Station, Asheville, 17 p

Index

C.H. Greenberg et al. (eds.), *Sustaining Young Forest Communities*, Managing Forest Ecosystems 21, DOI 10.1007/978-94-007-1620-9,

F

G

H

I